Springer Atmospheric Sciences

More information about this series at https://link.springer.com/bookseries/10176

Boris M. Smirnov

Global Energetics of the Atmosphere

Earth–Atmosphere Equilibrium, Greenhouse Effect, and Climate Change

Springer

Boris M. Smirnov
Joint Institute for High Temperatures
Russian Academy of Sciences
Moscow, Russia

ISSN 2194-5217 ISSN 2194-5225 (electronic)
Springer Atmospheric Sciences
ISBN 978-3-030-90007-6 ISBN 978-3-030-90008-3 (eBook)
https://doi.org/10.1007/978-3-030-90008-3

This Springer imprint is published by the registered company Springer Nature Switzerland AG
The registered company address is: Gewerbestrasse 11, 6330 Cham, Switzerland

The book is dedicated to my wife, E. A. Smirnova, in connection with the 60th anniversary of our life together

Preface

The analysis of atmospheric processes involving basic atmospheric components in the global atmosphere is combined with data of measurements of atmospheric parameters, now and in past, as well as with the contemporary understanding of global atmospheric problems. As a result, the physical picture of the global atmosphere is represented which combines the numerical parameters for various aspects of the atmosphere state and its evolution. In the first place, these aspects include the energetic balance of the Earth and atmosphere, as well as the global equilibrium between them for carbon and water. Following the principles of physics, the subsequent analysis is simple and transparent and leads to reliable conclusions in the form of numerical estimations, in contrast to climatological models with a larger number of parameters for the same problems. As a result, the basis of the Montreal protocol for ozone and Paris Agreement on climate is not correct. Using the contemporary physical theory and information, we obtain estimations for various parameters related to some aspects of global energetics of the atmosphere. In particular, an increase of the global temperature by approximately 10 °C leads to the greenhouse instability where each water molecule during its residence time in the atmosphere causes evaporation of more than one molecule from the Earth's surface.

Moscow, Russia

Boris M. Smirnov

Contents

List of Figures

Chapter 1
Introduction

In this book, we aim to give the physical picture of atmospheric processes which determine the heat state of the Earth and atmosphere, where the main role plays radiative processes including both infrared emission of the atmosphere and solar radiation. Basing on the principles of theoretical physics, we consider the atmosphere as a physical object. In this consideration, we take into account a general understanding of atmospheric physics which is contained in some books as [1–16]. In addition, we combine this understanding with the results of measurements within the framework of some scientific programs which include the monitoring of atmospheric carbon dioxide carried out since 1958 at the Mauna Loa Observatory (Hawaii, USA) [17–22], variations in the global temperature of the Earth [23, 24], the global carbon project [25] which results are given in [26–32], and also with the energetic balance of the Earth and its atmosphere [11, 33–39].

The physical picture of powerful atmospheric processes summarizes various aspects of this problem and includes some branches of physics as thermodynamics, quantum theory of radiation fields, physical kinetics, molecular spectroscopy and so on. Peculiarities of this analysis within the framework of theoretical physics are based on the simplicity, transparency and reliability. The simplicity consists in replacing the distribution function over any parameters by their average value, and an average is made over the globe and time. This decreases the accuracy of results and leads to the model of standard atmosphere [40] which operates with average parameters. In considering each property or process, we construct models with extracting the main factors in its origin that provides the transparency of the physical picture. The reliability of the conclusions and statements follows from numerical values of parameters which are estimated for problems under consideration.

This physical approach differs from climatological models which include computer codes with accounting for many factors simultaneously. These models call in question for used information and approaches, and we do not represent them below except the case of the greenhouse effect. In addition, due to the physical consideration, we refuse from description of methods for observations and measurements in the atmosphere, and we restrict by results of these measurements only. This allows

B. M. Smirnov, *Global Energetics of the Atmosphere*, Springer Atmospheric Sciences, https://doi.org/10.1007/978-3-030-90008-3_1

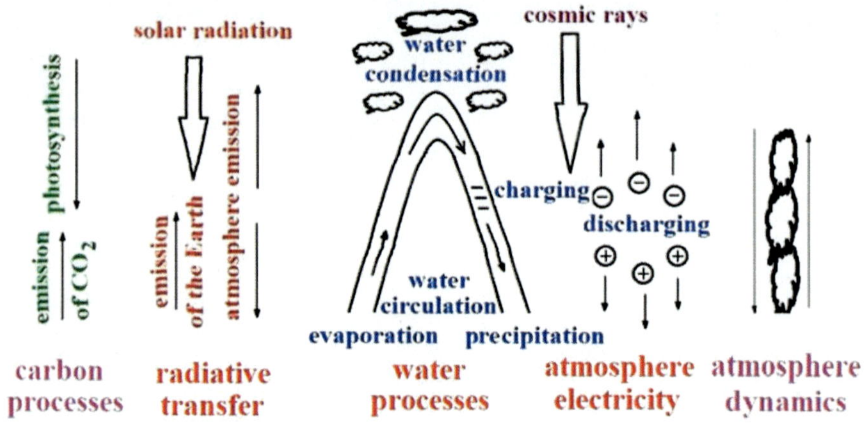

Fig. 1.1 Global powerful processes in the Earth's atmosphere [46, 47]

us to concentrate on physics of individual problems and to focus on the physical picture of powerful atmospheric processes.

The understanding of atmospheric problems depends on information which is used for their analysis. In the analysis of the greenhouse effect, the information is taken from the HITRAN data bank for parameters of radiative processes [41–45]. Constructing on the basis of physical principles, the theory of transport of infrared radiation in a weakly nonuniform plane gaseous layer which models the atmosphere, and using information from HITRAN data bank, one can analyze various aspects of atmospheric emission in the infrared spectrum range, as well as the transport of infrared radiation through the troposphere.

We present in Fig. 1.1 the groups of processes under consideration and some parameters of these processes which are taken from the author's books [46, 47]. Processes of Fig. 1.1 involving water molecules and water microdroplets or ice particles lead to circulation of water through the atmosphere due to atmospheric dynamics in the form of convection and air winds. The atmosphere may be divided into two parts, so that the small part of the Earth's surface is covered by cumulus clouds, while the basic part of the atmosphere is almost transparent and may be covered by rare stratus clouds. The condensation and electric processes proceed in cumulus clouds with a high density of atmospheric water. Formation and growth of water microdroplets through condensation of atmospheric water proceeds simultaneously with charging of microdroplets as a result of attachment to water microdroplets the atmospheric positive and negative ions which are formed under the action of cosmic rays.

Because all water microdroplets have the charge of the same sign, their falling leads to separation of the atmosphere charge that causes the creation of an atmospheric electric field. Note that it is accepted usually that collisions of water particles in different aggregate states are the only charging process in the atmosphere. But another charging process proceeds in a low-temperature plasma which results from the attachment of atmospheric ions to a particle. Then, a sign of the particle charge

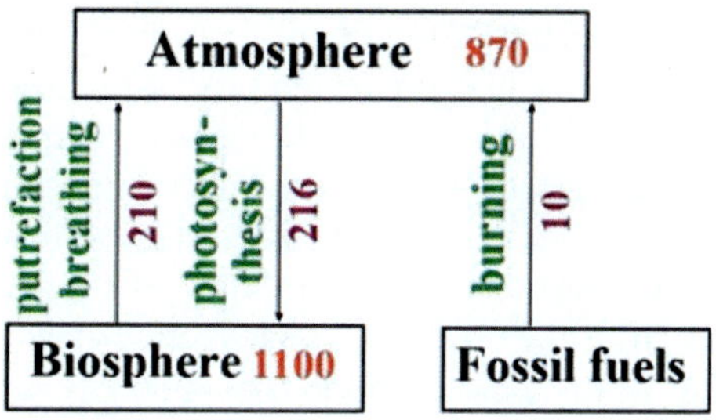

Fig. 1.2 Simplified scheme of carbon balance between the Earth and its atmosphere. The process rates for some transitions are shown near the arrows in units of GtC/year, and the carbon content is shown in the corresponding rectangles in units of GtC (GtC is billion tons of carbon)

corresponds to the sign of atmospheric ions of a larger mobility. Only this mechanism of particle charging can lead to separation of atmospheric charges and determine the creation of electric fields in the atmosphere. These processes proceed in cumulus clouds where the density of atmospheric water is large.

We focus in this book on global phenomena, mostly on the greenhouse effect of the atmosphere. Atmospheric electricity results from processes involving atmospheric water in air. The greenhouse phenomenon follows from radiative processes with participation of water molecules and water microdroplets of clouds, as well as carbon dioxide molecules. In this context, it is of importance that the concentration of atmospheric CO_2 molecules increases in time due to a human activity. Therefore, we consider the character of processes involving atmospheric carbon dioxide, and some results of the "global carbon project" [25] help us to analyze the carbon equilibrium between the Earth and its atmosphere by determining its parameters more precisely.

Figure 1.2 represents the character of the carbon balance between the atmosphere and the Earth's surface. This carbon equilibrium results mostly from processes of oxidation of organic compounds at the Earth's surface with formation of carbon dioxide molecules which are injected in the atmosphere, and the photosynthesis process resulted in the transformation of carbon dioxide molecules into the bound carbon of some plants under the action of solar radiation. Because the photosynthesis process is absent at the surface of neighboring planets, the atmospheres of these planets do not contain oxygen. Note that the contemporary composition of the Earth's atmosphere was formed approximately one billion years ago, when atmospheric oxygen resulted from the photosynthesis process. Processes involving oxygen cause the formation of ozone in the stratosphere that protects living organisms at the Earth surface from ultraviolet radiation.

The carbon equilibrium between carbon located in the atmosphere and at the Earth's surface is established during several years. In particular, the residence time of the CO_2 molecule in the atmosphere is approximately 4 years [48]. But the basic part of planet carbon is located inside the Earth. A typical time for the exchange of this carbon with the surface one is estimated in thousands years, so that one can ignore this exchange. The human activity violates partially this equilibrium as a result of extraction of fossil fuels from the Earth's interior. The flux of this carbon is small and equals approximately to 5% of that for natural processes due to the photosynthesis and oxidation ones as given in Fig. 1.2. But during a large time, this human activity becomes essential and changes the greenhouse effect of the atmosphere.

Atmospheric carbon dioxide is an important greenhouse component of the atmosphere. The greenhouse effect that resulted in infrared emission and absorption of the atmosphere is one of the elements of the energetic balance of the Earth and its atmosphere. The analysis of emission of the atmosphere is simplified because of a high pressure of atmospheric air as a buffer gas. This leads to thermodynamic equilibrium between atmospheric air and the radiation field. The latter means that the radiative temperature of emitted photons coincides with the atmospheric one at an altitude where they are created. Therefore, the radiative temperature of photons is expressed through the optical thickness of the atmosphere at a given frequency due to optically active atmospheric molecules.

According to the energetic balance of the Earth and its atmosphere, only 5% of the radiative flux created by the Earth's surface passes through the atmosphere. This allows one to separate the radiating atmosphere in two parts which are responsible for atmospheric emission to the Earth and outside. Therefore, in the first approximation, the radiative flux to the Earth's surface is independent on the outgoing radiation. This allows one to ignore the influence of external regions of the atmosphere in formation the radiative flux to the Earth's surface.

Thus, basing on the altitude distribution for the number density of optically active atmospheric molecules and using a certain temperature profile of the atmosphere within the framework of the model of standard atmosphere [40], one can determine the radiative flux from the atmosphere to the Earth with taking into account the data of the HITRAN bank [41–45] for parameters of radiative transitions of optically active atmospheric molecules. For the Earth's atmosphere, atmospheric molecules are H_2O and CO_2 optically active ones, as well as CH_4, N_2O and O_3, are trace molecules whose contribution to the radiative flux is small, but may be taken into account. As a result, one can determine the radiative flux from the atmosphere to the Earth at a given frequency, as well as the total one by integrating it over frequencies.

Along with optically active molecules, a certain contribution to the radiative flux toward the Earth follows from emission of water microdroplets which constitute atmospheric clouds. But clouds are formed and decay fast, and reliable information is absent about the density of clouds, their distribution over altitudes, as well as the average size distribution function for water microdroplets. Moreover, within the framework of the model of standard atmosphere [40], atmospheric condensed water is absent. Though we below analyze the character of condensation under the action of atmospheric winds in a wet atmosphere, we do not have reliable information for numerical global parameters of clouds. In order to overcome partially this trouble, we compared the total radiative flux to the Earth from the Earth's energetic balance and the total radiative flux due to atmospheric molecules. The difference of these fluxes gives the flux formed by clouds. This operation represents a reliable parameter which describes emission of water microdroplets of clouds.

Under such conditions, one can construct the following atmosphere model for evaluating the radiative flux from the atmosphere to the Earth. Clouds are located starting from a certain altitude, and above this altitude, the atmosphere is opaque for infrared radiation due to water microdroplets of clouds. This boundary altitude, as well as the cloud temperature, follows from the comparison of the total radiative flux from the atmosphere to the Earth, which results from the Earth's energetic balance and that created by optically active atmospheric molecules located in a gap between the Earth and clouds. This model allows one to determine some details of infrared radiation emitted and absorbed by the atmosphere. In particular, the contribution to the atmospheric radiative flux absorbed by the Earth equals to 62% from H_2O molecules, 19% from water microdroplets of clouds, 17% from CO_2 molecules, and 1% from CH_4, N_2O and O_3 molecules in total [49]. The accuracy of these values is several percent.

Along with radiative fluxes, this model allows one to evaluate the change in radiative fluxes as a result of variation of the atmosphere composition and this, in turn, the change of the global temperature. In particular, doubling of the concentration of atmospheric CO_2 molecules leads to the following change of the global temperature [46]

$$\Delta T = (0.6 \pm 0.3)\,^{\circ}\mathrm{C}, \tag{1.0.1}$$

and a large uncertainty in this value results from transition from the radiative flux change to the change of the global temperature. This differs by several times from this value from earlier evaluation and contemporary climatological models, and we exhibit the reason of this difference appealing to the history of this study.

This problem was set first by Arrenius [50] in the end of nineteenth century who noted that atmospheric carbon dioxide may influence on the climate. Practically, the Arrenius paper was devoted to another problem, namely to treatment of experimental data of Langley [51] for reflection of solar radiation with participation the resonant vibration transitions of H_2O and CO_2 molecules. This solar radiation is reflected from the Moon and is observed at the Earth. A reflected signal was observed at different relative positions of the Sun, the Moon and the Earth's observer. For CO_2 molecules, this study relates to the absorption band around 4.3 μm that is the strongest one for CO_2 molecules, while the thermal emission of CO_2 molecules for the real atmosphere proceeds in the range (12–18) μm.

Emission of the Earth's atmosphere for this absorption band (12–18) μm was considered in the subsequent analysis [52–54]. This analysis was fulfilled as new information was obtained about radiative parameters of CO_2 and H_2O molecules. At the first stage of the understanding of atmospheric emission created by CO_2 molecules, the assumption was used that the change of the radiative flux from the atmosphere as a result of the change of the concentration of CO_2 molecules coincides with the change of the flux created by CO_2 molecules. But according to the Kirchhoff law [55] radiators located in a buffer gas are simultaneously absorbers, and therefore, an increase of the radiative flux due to CO_2 molecules causes a decrease of the radiative flux created by H_2O molecules and clouds if spectra of these molecules are overlapped.

Therefore, the total change of the radiative flux created by CO_2 molecules differs from the change of the total radiative flux, and the ratio of these values depends on overlapping of spectra of these species. This was understood in [53, 54] in fifties where indicated the overlapping of spectra of CO_2 and H_2O molecules that leads to a decrease of the total change of the radiative flux from the atmosphere to the Earth in comparison with that created by CO_2 molecules. On the basis of information about spectra of these molecules in fifties, this effect was estimated as 20%. According to evaluations with using contemporary information on the basis of radiative parameters taking from the HITRAN database, this ratio is approximately five [56, 57].

Note that the indicated papers were based on the model of standard atmosphere [40] with average atmospheric parameters over time and globe. Practically, the model of standard atmosphere corresponds to replacing the distribution functions for some quantities by their average values. Hence, the error of this simplification may be estimated. In particular, the relative width of the distribution function over temperatures at each altitude is of the order of 10% of the average value, so that one can estimate the error from using the concept of the global temperature in 10%.

Climatological models can give, in principle, a more correct description of long-term atmospheric phenomena. In this approach, the globe is divided in separate sectors with an additional interaction between them. Such a consideration requires new information about transport processes between these sectors and a lot of additional information with the grounding of its reliability. In reality, this grounding is more difficult problem than preparing of the computer code. According to climatological models, as well as the conclusion of International Panel of Change Climate [58], the change of the global temperature as a result of doubling of the concentration for CO_2 molecules is

$$\Delta T = (3.0 \pm 1.5)\,^{\circ}\mathrm{C} \tag{1.0.2}$$

The uncertainty corresponds to averaging over different evaluations of this value.

Climatological models include complex computer codes and aim to account for various factors acted on the climate. Some versions of the climate change in future are analyzed in [59–64]. We use the physical principles which require to extract main factors which are of importance under certain conditions, and therefore, we deal with simple models. On the basis of this approach, we analyze the greenhouse effect of the real atmosphere [46] and atmospheric processes with water participation [47], in particular, atmospheric electricity. Here, we join these processes because atmospheric water is the key object both in atmospheric electricity and in the greenhouse effect. Because this consideration is combined with programs of atmosphere monitoring, this gives additional information for atmospheric problems under consideration. In particular, from this standpoint, we analyze a long-term evolution of the thermal Earth's state, i.e., the problem which is named in climatological models as a climate change.

Along with the analysis of a long-term change of the global temperature for different scenarios of evolution, we also consider the contemporary world energetics from the standpoint of the above understanding of atmospheric processes. Now, the most part of energy generated in the world results from combustion of fossil fuels (about 80%) and its portion decreases in time due to the creation of energy generators of other types. We show here that the power of emission of infrared radiation from the atmosphere to the Earth due to additional evaporation of water resulted from action of hydroelectric plants and wind turbines is comparable with the power generated by these systems. Our position is that all such factors must be taken into account in choosing a suitable type of the energy generator under certain conditions.

Thus, this book is devoted to the physical picture of powerful global atmospheric processes, mostly with participation CO_2 and H_2O atmospheric molecules. This physical picture is based on observational information and physical theories including the theory of transport of infrared radiation in gases. This physical picture is considered from the standpoint of physical principles. They include the simplicity, transparency and reliability of results. The simplicity means the extraction of main factors and ignoring other ones, and the connection between various aspects of the problem corresponds to its transparency, and the analysis leads to reliable conclusions which form the physical picture. This book intends for professionals and advance students who can consider this material critically.

References

1. J.T. Houghton, *The Physics of Atmospheres* (Cambridge University Press, Cambridge, 1977)
2. J.V. Iribarne, H.P. Cho, *Atmospheric Physics* (Reidel Publ, Dordrecht, 1980)
3. R.G. Fleagle, J.A. Businger, *Introduction to Atmospheric Physics* (Acadamic Press, San Diego, 1980)
4. R.M. Goody, Y.L. Yung, *Principles of Atmospheric Physics and Chemistry* (Oxford University Press, New York, 1995)
5. M.L. Salby, *Fundamentals of Atmospheric Physics* (Academic Press, San Diego, 1996)
6. J.H. Seinfeld, S.N. Pandis, *Atmospheric Chemistry and Physics* (Wiley, New York, 1998)
7. D.G. Andrews, *An Introduction to Atmospheric Physics* (Cambridge University Press, Cambridge, 2000)
8. J.H. Seinfeld, S.N. Pandis, *Atmospheric Chemistry and Physics* (Wiley, Hoboken, 2006)
9. J.M. Walace, R. Hobbs, in *Atmospheric Science. An Introductory Survey* (Amsterdam, Elsevier, 2006)
10. M.H.P. Ambaum, *Thermal Physics of the Atmosphere* (Wiley-Blackwell, Oxford, 2010)
11. M.L. Salby, *Physics of the Atmosphere and Climate* (Cambridge University Press, Cambridge, 2012)
12. I. Lagzi e.a. *Atmospheric Chemistry* (Budapest, Institute of Geography and Earth Science, 2013)
13. R. Caballero, *Physics of the Atmosphere* (IOP Publish, Bristol, 2014)
14. B.M. Smirnov, *Microphysics of Atmospheric Phenomena* (Switzerland, Springer Atmospheric Series, 2017)
15. G. Visconti, *Fundamentals of Physics and Chemistry of the Atmosphere* (Springer Nature, Switzerland, 2017)
16. B.M. Smirnov, in *Physics of Global Atmosphere* (Dolgoprudny, Intellect, 2017; in Russian)
17. Ch.D. Keeling, Tellus **12**, 200 (1960)
18. Ch.D. Keeling, R.B. Bacastow, A.E. Bainbridge et al., Tellus **28**, 538 (1976)
19. Ch.D. Keeling, J.F.S. Chin, T.P. Whorf, Nature **382**, 146 (1996)
20. https://www.wikiwand.com/en/Mauna-Loa-Observatory

21. https://www.esrl.noaa.gov/gmd/ccgg/trends/mlo.html
22. https://cdiac.ess-dive.lbl.gov/ftp/trends/co2/maunaloa-co2
23. J. Hansen, M. Sato, R. Ruedy. http://www.columbia.edu/~jeh1/mailing/2014/20140121-Temperature2013
24. J. Hansen, M. Sato, R. Ruedy et al., http://www.columbia.edu/~jeh1/mailing/2016/20160120-Temperature2015
25. http://www.globalcarbonproject.org
26. C. Le Quere, R. Moriarty, R.M. Andrew et al., Earth Syst. Sci. Data **7**, 47 (2015)
27. C. Le Quere, R. Moriarty, R.M. Andrew et al., Earth Syst. Sci. Data **7**, 349 (2015)
28. R.B. Jackson, J.G. Canadell, C. Le Quere et al., Nat. Clim. Change **6**, 7 (2016)
29. C. Quere, R.M. Andrew, J.G. Canadell et al., Earth Syst. Sci. Data **8**, 605 (2016)
30. G.P. Peters, C. Le Quere, R.M. Andrew et al., Nat. Clim. Change **7**, 848 (2017)
31. C. Le Quere, R.M. Andrew, P. Friedlingstein et al., Earth Syst. Sci. Data **10**, 405 (2018)
32. P. Friedlingstein, M. O'Sullivan1, M.W. Jones et al., Earth Syst. Sci. Data **11**, 1783 (2019)
33. *Understanding Climate Change* (Washington, Nat. Acad. Science, 1975)
34. J.T. Kiehl, K.E. Trenberth, Bull. Am. Meteorol. Soc. **78**, 197 (1997)
35. K.E. Trenberth, J.T. Fasullo, J.T. Kiehl, Bull. Am. Meteorol. Soc. **90**, 311 (2009)
36. K.E. Trenberth, J.T. Fasullo, Surf. Geophys. **33**, 413 (2012)
37. J.T. Fasullo, K.E. Trenberth, Science **338**, 792 (2012)
38. G.L. Stephens, J. Li, M. Wild et al., Nat. Geosci. **5**, 691 (2012)
39. M. Wild, D. Folini, Ch. Schär et al., Clim. Dyn. **40**, 3107 (2013)
40. *U.S. Standard Atmosphere* (Washington, U.S. Government Printing Office, 1976)
41. https://www.cfa.harvard.edu/
42. http://www.hitran.iao.ru/home
43. http://www.hitran.org/links/docs/definitions-and-units/
44. L.S. Rothman, I.E. Gordon, Y. Babikov et al., JQSRT **130**, 4 (2013)
45. I.E. Gordon, L.S. Rothman, C. Hill et al., JQSRT **203**, 3 (2017)
46. B.M. Smirnov, *Transport of Infrared Atmospheric Radiation* (de Gruyter, Berlin, 2020)
47. B.M. Smirnov, *Global Atmospheric Phenomena Involving Water* (Switzerland, Springer Atmospheric Series, 2020)
48. M. Grosjean, J. Goiot, Z. Yu, Global Planet Change **152**, 19 (2017)
49. D.A. Zhilyaev, B.M. Smirnov, JETP **133** (2021)
50. S. Arrhenius, Phil. Mag. **41**, 237 (1896)
51. S.P. Langley, Mount Whitney Expedition Rep. **303** (1884)
52. G.S. Calendar, Weather **4**, 310 (1949)
53. G.N. Plass, Tellus **8**, 141 (1956)
54. G.N. Plass, D.I. Fivel, Quant. J. Roy. Met. Soc. **81**, 48 (1956)
55. G. Kirchhoff, R. Bunsen, Annalen der Physik und. Chemie **109**, 275 (1860)
56. B.M. Smirnov, Int. Rev. At. Mol. Phys. **10**, 39 (2019)
57. B.M. Smirnov, J. Atmos. Sci. Res. **2**, N4, 21 (2019)
58. *Intergovernmental Panel on Climate Change.* Nature **501**, 297;298 (2013). http://www.ipcc.ch/pdf/assessment?report/ar5/wg1/WGIAR5-SPM-brochure-en.pdf
59. D. Martyn, *Climates of the World* (Elsevier, Amsterdam, 1992)
60. W.F. Ruddiman, *Earth's Climate: Past and Future* (Freeman, New York, 2000)
61. S.R. Weart, *The Discovery of Global Warming* (Harvard University Press, Harvard, 2003)
62. K.Y. Kondratyev, V.F. Krapivin, C.A. Varotsos, in *Global Carbon Cycle and Climate Change.* (Springer Praxis Publ., Chichester, 2003)
63. R.T. Pierrehumbert, *Principles of Planetary Climate* (Cambridge University Press, Cambridge, 2010)
64. D. Archer, *Global Warming: Understanding the Forecast* (Wiley, New York, 2012)

Chapter 2
Global Properties of Atmospheric Air

Abstract The model of standard atmosphere which is based on average atmospheric parameters is combined with transport phenomena in the atmosphere for the physical analysis of various properties for the Earth's atmosphere. Winds which are of importance for transpost parameters of the atmosphere, are measured in the Beaufort wind force scale. A stable atmospheric state is realized at low lapse rates, and the limiting adiabatic rate of dry air is 9.8 K/km. In contrast to other planets of the Sun system, the Earth's atmosphere contains water in the form of free molecules and the condensed phase, mostly as water microdroplets of clouds. Water is of importance for global properties of the Earth's atmosphere. Water microdroplets are formed mostly in cumulus clouds, i.e., in atmospheric regions with a heightened concentration of water. The equilibrium between water molecules and microdroplets takes place at the saturated pressure of a water vapor. Because of a large binding energy of water molecules in liquid, the process of water condensation is of importance for the energetic balance of the atmosphere.

2.1 Basic Components of Global Atmosphere

2.1.1 Model of Standard Atmosphere

In considering global processes in atmospheric air, we are based on the model of standard atmosphere [1] which deals with atmospheric parameters averaged over time and is guided by the USA atmosphere. This approach allows us to simplify the analyzed problems under consideration, but leads to their qualitative analysis. Dry atmospheric air contains nitrogen molecules (79%), oxygen molecules (20%) and argon atoms (0.9%). For simplicity, we use the concept of air molecules, assuming atmospheric air to be consisted of identical linear molecules with the molecular weight $m = 29$ a.m.u. These assumptions simplify the subsequent analysis, but lead to a decrease of its accuracy.

B. M. Smirnov, *Global Energetics of the Atmosphere*, Springer Atmospheric Sciences, https://doi.org/10.1007/978-3-030-90008-3_2

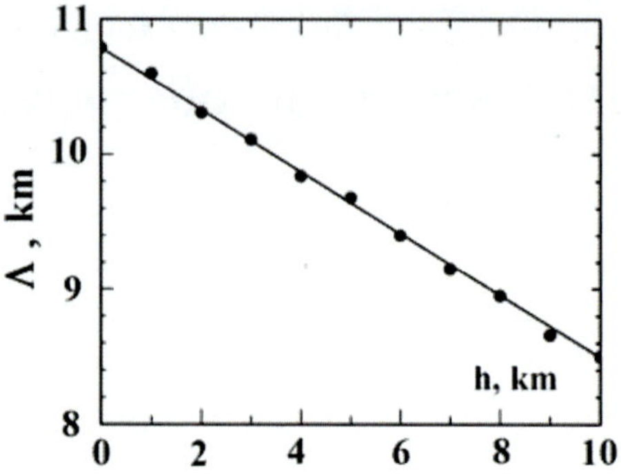

Fig. 2.1 Scale parameter Λ as a function of the atmosphere altitude. Filled circles correspond to the model of standard atmosphere [1], and the solid line is the approximation of the scale parameter $\Lambda = 10.4$ km according to formula (2.1.3)

Air molecules are located in the gravitation field of the Earth, and the distribution of their number density over altitudes is subjected to the Boltzmann distribution law [2–4]. The dependence of the number density of air molecules N on the altitude h according to the Boltzmann distribution law is given by

$$N(h) = N_o \exp\left(-\int \frac{dh}{\Lambda}\right) \approx N_o \exp\left(-\frac{h}{\Lambda}\right) \tag{2.1.1}$$

According to the model of standard atmosphere, the temperature of atmospheric air near the Earth's surface is $T_E = 288K$. Note that throughout this book, we consider Kelvin as the energetic units, i.e., represent the temperature as an energy without conversional coefficients between them. In motionless air, the parameter Λ in formula (2.1.1) is equal [5]

$$\Lambda = \frac{T}{mg}, \tag{2.1.2}$$

where m is the mass of an air molecule and g is the free fall acceleration. Figure 2.1 contains the altitude dependence for the scaling parameter constructed on the basis of the model of standard atmosphere [1]. From this, it follows for the average value of this parameter

$$\Lambda = (10.4 \pm 0.4)\ \text{km} \tag{2.1.3}$$

instead of the value $\Lambda = 8.4$ km according to formula (2.1.2). Note that according to the model of standard atmosphere, the number density of air molecules near the Earth's surface is $N_o = 2.55 \times 10^{19}\text{cm}^{-3}$. This corresponds to the total number of molecules in an atmospheric column as

$$n_a = \int N(h)dh = N_o\Lambda = 2.7 \times 10^{25}\text{cm}^{-2} \tag{2.1.4}$$

Let us assume atmospheric air to be consisted of linear molecules for which the energy of rotation excitation is small compared with a thermal energy, while the energy of vibration excitation is large compared with it. Under these conditions, the air molecule have five degrees of freedom (three translation and two rotation ones). Then considering the transition of molecules in regions of another altitude and number density of molecules N as an adiabatic process [5–8], one can obtain the connection between the number density N and temperature T of atmospheric air at various altitudes, as the altitude gradient temperature $\mathrm{d}T/\mathrm{d}h$ as

$$\ln\left(\frac{T}{N^{2/5}}\right) = \text{const}, \quad \frac{\mathrm{d}T}{\mathrm{d}h} = -\frac{0.4T}{\Lambda} = -14\,\mathrm{K/km} \tag{2.1.5}$$

Within the framework of the model of standard atmosphere [1] which is based on measured parameters, the altitude gradient temperature which is called the lapse rate [9] is equal

$$\frac{\mathrm{d}T}{\mathrm{d}h} = -6.5\,\mathrm{K/km} \tag{2.1.6}$$

The difference between values (2.1.5) and (2.1.6) is explained by water condensation in the atmosphere which has a nonregular character and depends on the moisture of the atmosphere. We below use the value (2.1.6) through the book as the temperature gradient in the troposphere.

Note that we deal through the book with the model of standard atmosphere where average parameters are used instead of the distribution functions over these parameters. There are other approaches to description of the atmosphere. Meteorology allows us to predict the weather for a close future on the basis of atmospheric parameters at a given time, and the main instrument of meteorology is transport equations for air mass, heat and moisture [10]. This approach is the basis of the weather forecasting [11, 12], and general principles of the meteorology are connected with solution of the system of transport equations for transport of air, heat and moisture under given initial conditions [13–15]). As a result, this allows one to predict the weather as a certain set of parameters of a low troposphere. It is clear that for this goal, the information is required about the atmosphere state for all the globe because atmospheric transport processes connect various points of the globe.

As is seen, the content of the meteorology includes complex computer problems and its goal is to describe the atmosphere state on the basis of its current parameters. This description is restricted in time due to Poincare instability [16, 17]. According to this instability, a very small shift in parameters of a dynamic system leads to their growth in an exponential way that leads to an enormous difference of two close states of this system through a large time. As for the Earth's atmosphere, a typical time of development of the Poincare instability, and, correspondingly, a time of the reliable weather forecast is estimated as two weeks [18]. One can see that the model of standard atmosphere and meteorological description of the atmosphere relate to different ways of the atmosphere description and have different aims. In contrast to the meteorological approach of the atmosphere which pretends on a correct short-term description of atmospheric processes, the model of standard atmosphere gives the transparent physical picture of some atmospheric properties as a physical object.

Climatological models have the same goal as the meteorological approach, but pretend on a long-term description of the atmosphere. In this case, additional factors can influence on the atmosphere state. For example, radiative processes including the greenhouse effect are of importance for the long-term evolution of the atmosphere. Other difference in climatological models based on complex computer codes from the meteorological and physical approach is the absence of enough information which is necessary for solution of computer problems. Because of the complexity of climatological models, the simplification is required. But the climatology is far from physics, whereas many atmospheric properties and its evolution are governed by physical laws. Therefore, ignoring physical laws in climatology may lead to mistakes. In particular, in the analysis of changes in the greenhouse effect due to an increase of the concentration of atmospheric carbon dioxide, the Kirchhoff law is neglected in climatological models. This leads to a large mistake in prediction of the global temperature change. Therefore, the caution is required with respect to climatological models.

Thus, the model of standard atmosphere is the basis of physical models for description various physical properties of the atmosphere on average in a simple and transparent form. According to the character of this model, such a description relates to global atmospheric properties only. Let us estimate the accuracy of such a description. Practically, the model of standard atmosphere corresponds to replacing the distribution function over atmospheric parameters by their average values. In particular, let us analyze this replacing in the case of the global temperature. Indeed, we use as the global temperature T_E of the Earth the temperature of the Earth's surface averaged over time and globe that is

$$T_E = 288\,K \tag{2.1.7}$$

In reality, the surface temperature at a given surface point differs from that given by formula (2.1.7) and oscillates due to daily and season variations. One can estimate this temperature difference as 10%. This means that one can consider the transition from a real atmosphere with a distribution over surface temperatures to an atmosphere with the average surface temperature (the global temperature) as an expansion over a small parameter that is of the order of 0.1 as the ratio of a typical temperature fluctuation to the global temperature. Therefore, below we estimate the accuracy of results for other atmospheric parameters to be better than 20%.

Note that because of the convective motion of atmospheric air, the number density of admixture atmospheric molecules has the same altitude dependence as that for air molecules given by formula (2.1.1). This distribution results from caption of molecules by vortices, where all molecules partake into convective motion of atmospheric air in the same manner, and therefore, the number densities of molecules of a different type are characterized by an identical altitude dependency. Hence, the altitude profile for the number density of molecules and aerosols of small sizes is

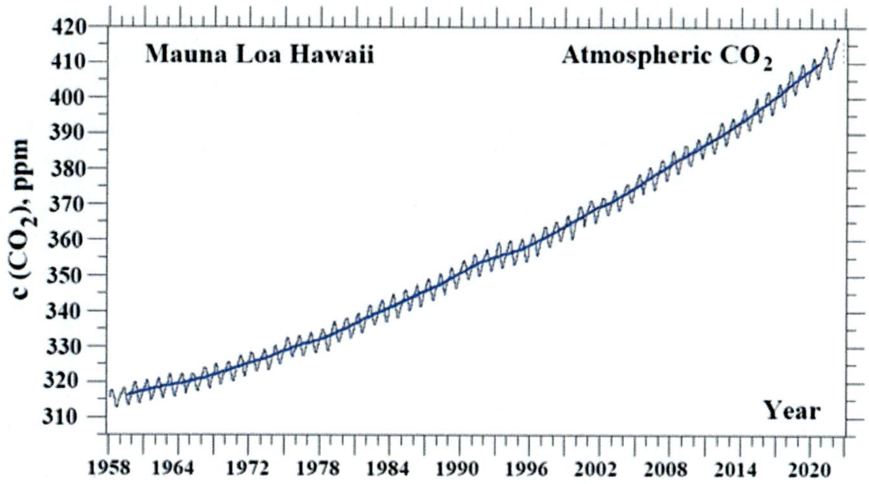

Fig. 2.2 Concentration of CO_2 molecules in atmospheric air according to measurements at the Mauna Loa Observatory [27–29]

identical if they do not partake in chemical processes. But the altitude profile of the number density of water molecules differs from that for air molecules because of water condensation in the atmosphere.

If molecules are inert in the atmosphere, their altitude profile repeats that of air molecules, and these molecules are characterized by a large residence time, i.e., the lifetime in the atmosphere. These properties relate to CO_2, CH_4 and N_2O molecules. These molecules penetrate in the atmosphere from the Earth's surface and are lost there in corresponding processes. In particular, the lifetime of atmospheric CO_2 molecules is 4 years [19]. Atmospheric CO_2 molecules are formed as a result of oxidation of carbon-containing materials at the Earth's surface and are lost by dissolving in oceans and in photosynthesis processes at land and oceans. Correspondingly, the residence time of CO_2 molecules in the atmosphere is an evidence about process at the Earth's surface involving these molecules.

Under these conditions, CO_2 molecules are mixed with atmospheric air completely, i.e., the concentration of CO_2 molecules is independent of the altitude. Because of a large residence time in the atmosphere, the concentration of carbon dioxide molecules is identical for any geographical point, if it is located far from sources and absorbers of carbon dioxide. These conditions are fulfilled for the Mauna Loa Observatory (Hawaii, USA) [20–25] which coordinates are 19°32′N, 155°35′W. This observatory is located at altitude 3400 m over the sea level. In addition, the measuring system includes four towers of 7 m height and one tower of 27 m height, and this measurement equipment is located far from sources or absorbers of carbon dioxide molecules [26].

Last time, the concentration of atmospheric carbon dioxide molecules varies in time due to the human activity. In 1750, the concentration of CO_2 molecules was equal

(277 ± 3) ppm, and in 1870, this value was (288 ± 3) ppm [30]. According to Mauna Loa Observatory measurements given in Fig. 2.2, the concentration of carbon dioxide molecules in atmospheric air increases from 316 ppm in 1959 up to 415 ppm in 2020. In addition, this figure shows season oscillations of the concentration of atmospheric CO_2 molecules that testifies about an intensification of the photosynthesis process in the Northern Hemisphere during crop ripening. It should be noted according to data of Fig. 2.2, that the rate of an annual increase of the carbon dioxide concentration grows in time from 0.7 ppm/yearr in 1959 up to approximately 2.4 ppm in 2020. In addition, from the contemporary increase of the concentration of atmospheric CO_2 molecules, the doubling of this concentration of atmospheric CO_2 molecules is expected through 110 year.

2.1.2 Atmospheric Winds

The Earth's atmosphere is nonuniform locally, and this causes atmospheric transport processes which consist in transport of air mass, temperature and moisture. The sum of these processes creates a local weather which varies in time and space. These processes are the object of study for meteorology [10]. Within the framework of global phenomena which are the object of this book, the transport of air proceeds through winds. In particular, within the framework of the model of standard atmosphere [1], the average moisture of atmospheric air is lower 100% at any altitudes, i.e., condensation in motionless air is absent. Therefore, formation of atmospheric clouds which consist of water microdroplets and microparticles is impossible in motionless air. Transport of the moisture in the atmosphere due to winds creates supersaturated air in some regions of the atmosphere, i.e., winds are responsible for formation of clouds which are of importance for electric and radiative properties of the atmosphere.

It is convenient to characterize wind speeds within the framework of the Beaufort wind force scale. Francis Beaufort was Britain's admiral who suggested the scale of winds which is presented in Table 2.1. We will use below for estimation a typical wind speed $v_w \sim 10$ m/s which corresponds to fresh breeze.

2.1.3 Atmospheric Water

Water is an important atmospheric component which is responsible for some atmospheric phenomena. Though the basic part of atmospheric water is found in the form of free water molecules, processes of water transitions into condensed phases influence on some properties of atmosphere. The total water mass into the atmosphere according to [32–35] is $M = 1.3 \times 10^{19}$ g compared to 5.1×10^{21} g for dry atmospheric air. An average concentration of water molecules in the near-surface atmosphere (the ratio of the number of water molecules to that of air molecules) is approximately 1.7% that exceeds the concentration of argon atoms (0.93%)). But the

Table 2.1 Name of winds and their velocity range according to the Beaufort wind force scale [31]

Beaufort number	Wind name	Wind speed, km/h	Wind speed, m/s
0	Calm	< 2	< 0.5
1	Light air	2–5	0.5–1.5
2	Light breeze	6–11	1.6–3.3
3	Gentle breeze	12–19	3.4–5.5
4	Moderate breeze	20–28	5.5–7.9
5	Fresh breeze	29–38	8–10.7
6	Strong breeze	39–49	10.8–13.8
7	High wind, moderate gale	50–61	13.9–17.1
8	Gale, fresh gale	62–74	17.2–20.7
9	Strong, severe gale	75–88	20.8–24.4
10	Storm, whole gale	89–102	24.5–28.4
11	Violent storm	103–117	28.5–32.6
12	Hurricane force	> 118	> 33

altitude profile for water molecules differs from that for air molecules because of condensation processes. The average concentration of water molecules in atmospheric air is approximately 0.4%.

It is convenient to express the amount of atmospheric water on the basis of the following operation. Let us convert all atmospheric water in the liquid state and distribute it uniformly over the globe. Then, the thickness of this liquid layer would be $h_w = 2.5$ cm [36]. But atmospheric water contains a small part of Earth's water which mass is $1.4 \times 10^{24} g$. If we apply the above operation to surface water, the thickness of the water layer will be $H_w = 2.7$ km, i.e., the main part of surface water is located underground. In addition, approximately 96% of Earth's water is salty.

Atmospheric water is found in equilibrium with the Earth's surface which consists in evaporation of water molecules from open water located on the Earth's surface and from a land, as well as precipitation to the Earth's surface in the form of rain, snow and water molecules. The total rate of falling of condensed water to the Earth's surface is 4.8×10^{20} g/year [34, 37–40] or $J = 1.5 \times 10^{13} g/s = 15 \times 10^6$ ton/s. Only 1.0×10^{18} g/year of the later value is found in the form of snow. Because of the equilibrium between atmospheric and surface water, the same rate refers to returning of evaporated water to the ground under the assumption that all evaporated water returns back only in the form of rain or snow. In addition, the average flux of water molecules j_w which falls on the Earth's surface in the form of rain or snow is equal

$$j = \frac{J}{Sm_o} = 1 \times 10^{17}\,\mathrm{cm}^{-2}\,\mathrm{s}^{-1}, \tag{2.1.8}$$

where $m_o = 3 \times 10^{-23}$ g is the mass of the water molecule and S is the area of the Earth's surface. One can add to this that the average density of water molecules n(H_2O) that correspond to the surface density 2.5 g/cm^2 is equal

$$n(H_2O) = 8 \times 10^{22}\,\mathrm{cm}^{-2} \tag{2.1.9}$$

It should be noted that the used value for the rate of water precipitation is based on the scheme, where evaporated water returns to the ground only in the form of the condensed water, i.e., as rain or snow. In reality, free water molecules give a small contribution to precipitated water. A typical time τ_w of water residence in the atmosphere is equal

$$\tau_w = \frac{n}{j} \approx 9\,\text{days}, \tag{2.1.10}$$

that is in accordance with [40].

The peculiarity of water behavior in the atmosphere connects with effective transitions between a water vapor and a condensed phase of water in the atmosphere. One can characterize the concentration of atmospheric water molecules in air by its moisture or humidity η [41] as

$$\eta = \frac{N_w}{N_{sat}(T)}, \tag{2.1.11}$$

where N_w is the number density of water molecules and $N_{sat}(T)$ is the number density of water molecules at the saturated vapor pressure $p_{sat}(T)$ for a given temperature, i.e., the number density of water molecules at the dew point [42]. At this temperature, the partial pressure of a water vapor is equal to the saturated vapor pressure for this temperature. Table 2.2 gives values of the saturated vapor pressure and the number density of water molecules in a temperature range that is of interest for the troposphere.

It is convenient to approximate the temperature dependence of the number density of water molecules at the saturated vapor pressure $N_{sat}(T)$ as

$$N_{sat}(T) = N_o \cdot \exp(-E_{sat}/T) \tag{2.1.12}$$

From data of Table 2.2, one can obtain $E_{sat} = (0.437 \pm 0.03)$ eV for the transition gas–liquid for temperatures below 273 K, and $E_{sat} = (0.48 \pm 0.8)$ eV for the transition gas–solid for temperatures above 273 K. We below take for parameters of formula (2.1.12)

$$E_{sat} = 0.437\,\mathrm{eV},\ \ N_o = 1.9 \times 10^{25}\mathrm{cm}^{-3},\ \ T > 273\,\mathrm{K} \tag{2.1.13}$$

and

Table 2.2 Parameters of a saturated water vapor in atmospheric air [43, 44]: p_{sat} is the water saturated vapor pressure at an indicated temperature, N_{sat} is the number density of water molecules in a saturated vapor, A_s is the ratio of a mass of a saturated water vapor to an air mass located in a given volume at atmospheric pressure

T, K	p_{sat}, Torr	N_{sat}, 10^{17}cm^{-3}	A_s, g/kg
223	0.030	0.013	0.024
228	0.054	0.023	0.044
233	0.096	0.040	0.079
238	0.168	0.068	0.137
243	0.285	0.113	0.233
248	0.475	0.185	0.388
253	0.775	0.296	0.633
258	1.24	0.463	1.01
263	1.95	0.716	1.59
268	3.02	1.09	2.46
273	4.58	1.62	3.74
278	6.55	2.27	5.35
283	9.22	3.14	7.52
288	12.8	4.29	10.4
293	17.6	5.78	14.3
298	23.8	7.70	19.4

$$E_{sat} = 0.48\,\mathrm{eV}, \; N_o = 1.9 \times 10^{26}\mathrm{cm}^{-3}, \; T < 273\,\mathrm{K} \tag{2.1.14}$$

On the basis of this, we take below the binding energy of a water molecule with a liquid surface as

$$\varepsilon_{ev} = 0.44\,\mathrm{eV} \tag{2.1.15}$$

As a matter, relations (2.1.12), (2.1.13), (2.1.14) are the Clasius–Clayperon equation [45, 46], where E_{sat} is the binding energy of a water molecule with a flat water surface, and this value has a jump at the temperature $T = 273$ K as a result of the liquid–solid transition. The accuracy of these approximations is approximately 20% for the liquid state of the surface $T > 273$ K (gas–liquid transition) and 30% for $T < 273$ K (gas–solid transition).

From this within the framework of the model of standard atmosphere [1], one can find the altitude dependence for the number density $N_{sat}(h)$. On the basis of the above approximations, we have

$$N_{sat}(h) \sim \exp\left(-\frac{h}{h_{sat}}\right), \; h_{sat} = \frac{T^2}{E_{sat} \cdot |\mathrm{d}T/\mathrm{d}h|} \approx 2.5\text{ km}, \tag{2.1.16}$$

where we take the air temperature near the Earth's surface $T = T_E = 288$ K, and the temperature gradient is $dT/dh = -6.5$ K/km for the model of standard atmosphere.

Evidently, the number density of atmospheric water molecules cannot exceed that at the saturated vapor pressure N_{sat}. From this, one can determine the maximum mass of a water vapor in the atmosphere in the form of free water molecules. Taking the maximum number density of water molecules $N_w = N_{sat}$ for the model of standard atmosphere, one can obtain the maximum water mass per unit area of the Earth's surface as

$$\rho_{max} = N_s h_{sat} m_o \approx 3.2 \text{ g/cm}^2, \quad (2.1.17)$$

where the number density of water molecules near the Earth's surface is $N_s = 4.3 \times 10^{17}\text{cm}^{-3}$ in accordance with Table 2.1 for standard atmosphere ($T = 288$ K), $m_o = 3.0 \times 10^{-23}$ g is the mass of a water molecule and h_{sat} is the parameter of formula (2.1.16). If we transfer this water in the liquid state and distribute uniformly over the Earth's surface, the average height of a formed water layer will be $h_{sat} = 3.2$ cm. Note that transformation of total atmospheric water of standard atmosphere into liquid and uniform distribution of atmospheric water over the globe leads to the height of $h_w = 2.5$ cm for a formed layer. One can see that the mass of atmospheric water in the form of free molecules is close to its limiting value, and the average atmospheric moisture (2.1.11) defined as h_w/h_{sat} is approximately 78%.

Note that condensation in the atmosphere with formation of water microdroplets and microparticles in the atmosphere is possible in a supersaturated vapor, where

$$\eta \geq 100\% \quad (2.1.18)$$

If atmospheric air contains water in the condensed phase, mostly in the form of microdroplets, and the equilibrium takes place between a water vapor and water microdroplets through processes of evaporation of microdroplets and attachment of water molecules to microdroplet surface, the moisture for this water vapor is 100%.

We give in Fig. 2.3 evolution in time for the atmospheric moisture averaged over the globe. Note that altitudes of Fig. 2.3 correspond to the air pressures of 1, 0.6 and 0.3 atm that allows one to compare the altitude dependence for the number densities of atmospheric water molecules. In addition, according to Fig. 2.3, season oscillations and random fluctuations accompany the average number density of water molecules. From this, we take the average atmosphere moisture for 1980–2020 to be correspondingly 78%, 44% and 40%. This information allows one to determine the atmosphere moisture at a given altitude averaged over the globe. Let us approximate the atmospheric number density N_w of water molecules as an altitude function by the dependence

$$N_w(h) = N(H_2O) \exp(-h/\lambda), \quad (2.1.19)$$

where $N(H_2O)$ is the number density of water molecules at the Earth's surface. We will find parameters of this formula from data of Fig. 2.3. Atmospheric parameters

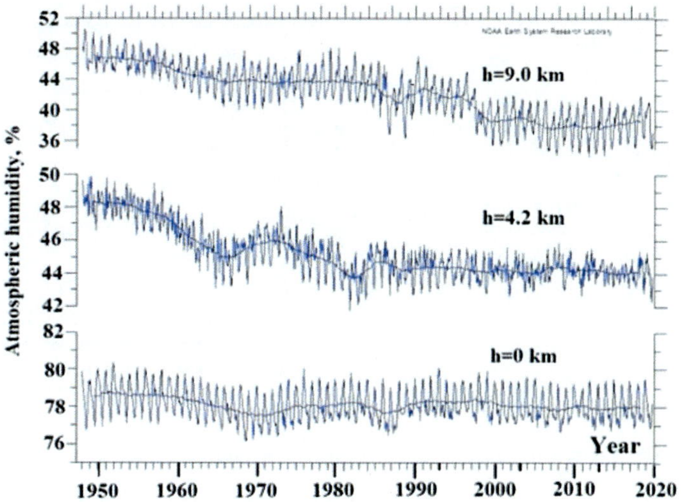

Fig. 2.3 Evolution in time for the global moisture of atmospheric air at indicated altitudes [41]

Table 2.3 Parameters of treatment of data of Fig. 2.3: T is the atmospheric temperature at an indicated altitude, N_a is the number density of air molecules at this altitude, η is the average atmospheric moisture for 1980–2020 according to data of Fig. 2.2, N_{sat} is the number density of water molecules at the saturated vapor pressure at this altitude, $N(H_2O)$ is the number density of water molecules for the model of standard atmosphere and data of Fig. 2.3

h, km	T, K	N_a, 10^{19}cm^{-3}	η, %	N_{sat}, 10^{16}cm^{-3}	$N(H_2O)$, 10^{16} cm^{-3}
0	288	2.55	78	43	34
4.2	261	1.7	44	6.0	2.6
9.0	230	0.97	40	0.68	0.27

at altitudes of Fig. 2.3 are presented in Table 2.3. In addition, according to formulas (2.1.18) and (2.1.19), the average moisture of atmospheric air varies as

$$\eta(h) = \eta_o \exp(-h/h_m), \ h_m = \left(\frac{1}{\lambda} - \frac{1}{h_{sat}}\right)^{-1}, \ h_m = (9 \pm 4)\,\text{km}, \qquad (2.1.20)$$

where $\eta_o = 0.78$ is the moisture at the Earth's surface.

On the basis of this figure, we have for the altitude profile of atmospheric water from the above data for the moisture

$$N(H_2O) = 3.4 \times 10^{17}\,\text{cm}^{-3}, \ \lambda = (1.9 \pm 0.2)\,\text{km} \qquad (2.1.21)$$

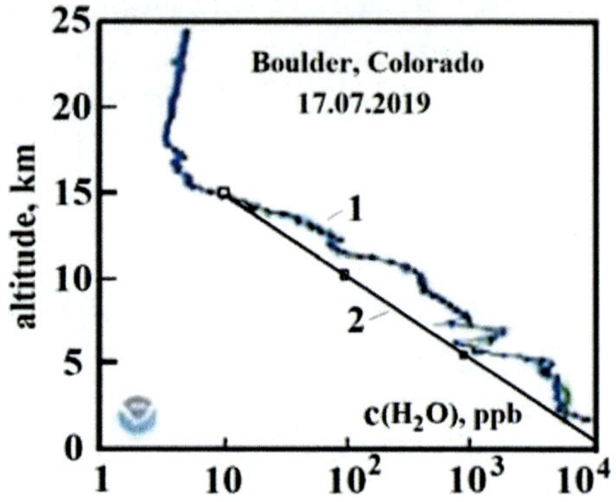

Fig. 2.4 Concentration of water molecules in the atmosphere as a result of balloon measurements at Boulder (Colorado, USA) [48] both during ascent and descent. 1—results of measurements, 2—formula (2.1.19)

Formula (2.1.19) allows one to determine the total mass of atmospheric water that is given by

$$M_w = N(H_2O)\lambda S m_w, \tag{2.1.22}$$

where $S = 5.1 \times 10^{18}\text{cm}^2$ is the area of the Earth's surface and $m_w = 3.0 \times 10^{-23}$ g is the mass of an individual water molecule. From this formula, it follows $M_w = 1.0 \times 10^{19}$ g, whereas from the analysis of atmospheric water [32–35], this value is equal 1.3×10^{19} g. The degree of coincidence of these value testifies about their accuracy.

From this, we have also for the number of water molecules per unit vertical column is equal

$$n(H_2O) = (6.5 \pm 1.0) \times 10^{23}\,\text{cm}^{-2} \tag{2.1.23}$$

The difference between values of formulas (2.1.20) and (2.1.23) characterizes the accuracy of these values. The variation of the number of water molecules per unit area of the vertical column n(H_2O) from the end of 19 century increases by $\Delta n(H_2O) = 3.0 \times 10^{21}\text{cm}^{-2}$, if we take in accordance with [47] that an increase of the amount of atmospheric water vapor is 4%.

There are various methods for measurement the local amount of atmospheric water both in the form of free water molecules and condensed water in clouds. Because the analysis has the qualitative character, we use only results of these measurements. In particular, Fig. 2.4) contains the result of a balloon measurement for the local concentration of an atmospheric water vapor. In these measurements [49, 50], a balloon rises up to the stratosphere, and then, it descents. In the course of balloon ascent and descent, the number density of water molecules in atmospheric air is measured on the basis of two-photon absorption of the resonant vibration transitions. This method allows one to realize the monitoring of the local altitude distribution of atmospheric water [51].

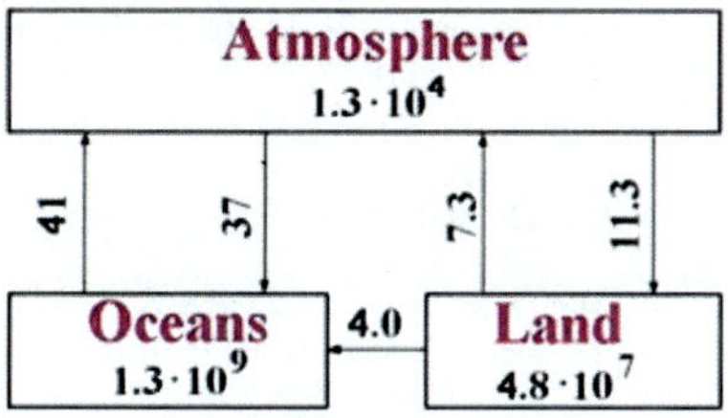

Fig. 2.5 Rates of exchange by water between land, oceans and atmosphere which are expressed in 10^{18} g/year and are given near arrows. The water amount is expressed in 10^{15} g (billion ton) and is indicated inside a corresponding rectangle [34, 37–40]

A result of a typical measurement of such type is given in Fig. 2.4. As it follows from this figure, a smooth altitude profile for the number density of atmospheric water molecules distorts due to condensation processes. Indeed, basing on the model of standard atmosphere and operating with average values, on the basis of formulas (2.1.1) and (2.1.19), we have for the altitude profile of the concentration $c(h)$ of atmospheric water molecules

$$c(h) = \frac{N_w}{N(h)} = c(0)\exp\left(-\frac{h}{\lambda^*}\right), \; \lambda^* = \left(\frac{1}{\lambda} - \frac{1}{\Lambda}\right)^{-1} = 2.5\ \text{km} \qquad (2.1.24)$$

As it follows from Fig. 2.4, fluctuations due to condensation processes distort an altitude profile for the local concentration of atmospheric water molecules.

We now consider water circulation through the atmosphere which in a simplified version [52] consists in water evaporation from the Earth's surface, their condensation in the atmosphere and subsequent falling to the Earth, mostly in the form of rain. On the basis of the distribution function of air molecules (2.1.1) and water molecules (2.1.19), we now determine the partial fluxes to the Earth in the form of free water molecules and through rain and snow. The evaporation flux is given by formula (2.1.8), and we assume this flux to be independent of condensation processes in the atmosphere. This is equivalent to that in the absence of condensation processes where one can represent it as

$$j_{ev} = -D_L\left(\frac{dN_w}{dh}\right) = \frac{D_L N(H_2O)}{\lambda} = 1.0 \times 10^{17}\,\text{cm}^{-2}\,\text{s}^{-1} \qquad (2.1.25)$$

As is seen, values of the fluxes of evaporated water according to formulas (2.1.8) and (2.1.25) coincide. We assume here that water molecules are captured by air vortices, and propagation for large distances has the diffusion character with an effective diffusion coefficient D_L. Next, N_w is the number density of free molecules at a given altitude, and $N(H_2O)$ is the number density of water molecules near the Earth's surface. We also have that the flux j_c of water molecules to the Earth's surface is given by

$$j_w = -D_L\left(\frac{dN_w}{dh}\right) = \frac{D_L N(H_2O)}{\lambda}, \qquad (2.1.26)$$

where the number density of water molecules N_w is given by formula (2.1.19). From this, we have for fluxes of water molecules j_w and water condensed phase j_c, mostly in the form of rain

$$j_w = \frac{\lambda}{\Lambda} j_{ev} = 2.3 \times 10^{16}\,\mathrm{cm}^{-2}\mathrm{s}^{-1}, \quad j_c = \frac{(\Lambda-\lambda)}{\Lambda} j_{ev} = 7.7 \times 10^{16}\,\mathrm{cm}^{-2}\mathrm{s}^{-1} \qquad (2.1.27)$$

Note that separation of water fluxes to the ground in the form of free molecules and condensed phase, we take into account that convection motion of water molecules in both cases is identical. In addition, this character of flux separation is valid for exponential altitude dependencies for the number densities of water molecules.

An equilibrium between atmospheric water and that located at the Earth's surface results from evaporation and precipitation processes. But the average specific rate of water evaporation from oceans is higher than that from the land, and this leads to a higher air moisture above oceans. This difference is compensated by a water flux from a land to oceans in the form of rivers, and the water balance at the Earth's surface is given in Fig. 2.5. Rivers which transfer water from the land to oceans compensate the difference of these fluxes and provide the water balance.

Along with season oscillations, the average air moisture increases in time slightly [53–55] in spite of a large average atmospheric moisture. Indeed, according to [47], the moisture increase during 20th century was approximately 4%. Some results of this type are shown in Fig. 2.6. According to data of this figure, the moisture change is expressed in Fig. 2.6 in units gram water per kilogram air, and this unit 1 g/kg corresponds to the partial water pressure 1.2 Torr. As it follows from Fig. 2.6, the rate of change of the amount of atmospheric water is equal near the Earth's surface correspondingly 0.06, 0.08 and 0.07 g/kg per decade according to [53–55]. From

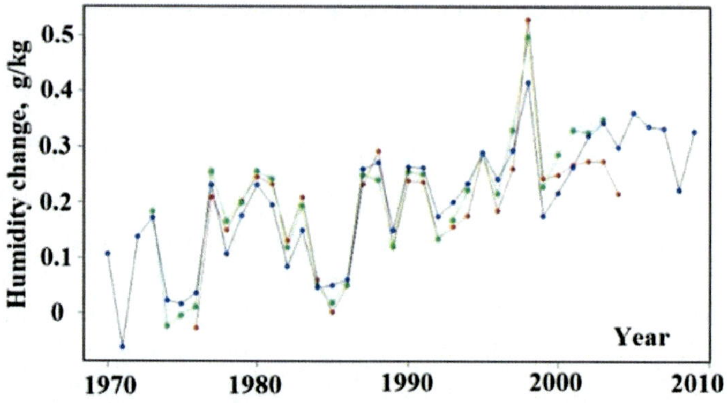

Fig. 2.6 Evolution of the average moisture of atmospheric air during 1970–2010 years. Red—[53], green—[54], blue—[55]

this, we have for the average change of the water amount in the atmosphere (0.09 ± 0.01) Torr/decade. For comparison, the average partial water pressure of standard atmosphere at the Earth's surface is equal approximately 12 Torr.

In conclusion of the analysis of data for the atmospheric water amount, we note that this information in a most degree is connected with laborious measurements and their treatment. Just this determines the reliability and accuracy of data which we use subsequently as a basis of the analysis of atmospheric processes.

2.1.4 Stability of Atmospheric Air

In this analysis, we are restricted mostly by the troposphere, i.e., by the lowest part of the Earth's atmosphere where global energetic and physical processes proceed excluding those involving ozone. The atmospheric temperature drops with an increasing altitude, and we below determine the temperature gradient $\mathrm{d}T/\mathrm{d}h$ in the troposphere for dry atmospheric air taking into account that the temperature change has an adiabatic character, i.e., atmospheric air do not change by energy with an environment. Take an air molecule and raise it by an altitude Δh. Then, its potential energy increases by $mg\Delta h$, where m_a is the mass of the air molecule. Simultaneously, the molecule thermal energy decreases by $c_\mathrm{p}\Delta T$, where $c_\mathrm{p} = 7/2$ is the heat capacity per one molecule. From the adiabatic character of this process, it follows [9, 56]

$$\left|\frac{\mathrm{d}T}{\mathrm{d}h}\right| = \frac{m_a g}{c_\mathrm{p}} \approx 9.8\,\mathrm{K/km}, \qquad (2.1.28)$$

where $m_\mathrm{a} = m/n$ is the mass of the air molecule. The value (2.1.28) is called the adiabatic lapse rate of a dry atmosphere.

Let us consider the stability with respect to a temperature gradient for the atmosphere as a gas located in the gravitation field of the Earth. For this goal, we extract an atmospheric column as it is shown in Fig. 2.7. This column is found under adiabatic

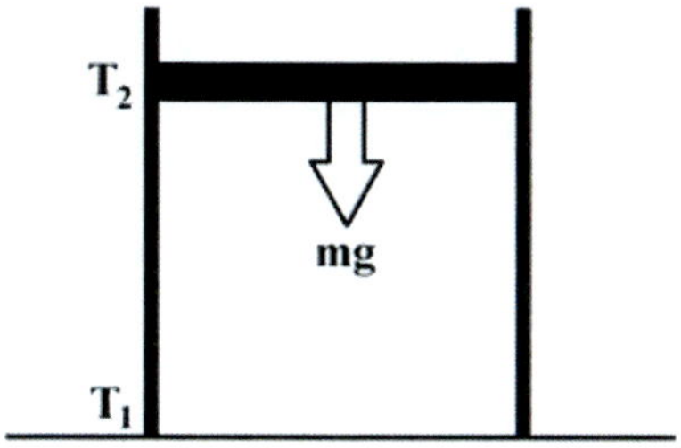

Fig. 2.7 Character of thermodynamic equilibrium for a column of atmospheric air in the gravitation field with a decreasing temperature as the altitude increases. The equilibrium violates if an increase of the gravitation potential as a result of displacement of an atmospheric layer is not compensated by the thermal energy change due to an increasing layer temperature

conditions, i.e., it is does not exchange by energy with an environment. Take the temperature of the column to be decreased with an altitude. The force mg acts on an extracted layer down, and this layer moves under action of this force, if its replace is thermodynamically profitable, i.e., the change of the potential energy $mgdh$ in the case of a shift by an altitude $\mathrm{d}h$ exceeds the thermal energy $nc_{\mathrm{p}}\mathrm{d}T$, where n is the number of molecules in an extracted layer and c_{p} is the heat capacity per one molecule. From this, the condition of the layer stability it follows $mg\mathrm{d}h < nc_{\mathrm{p}}\mathrm{d}T$, i.e., the stability condition takes the form

$$\left|\frac{\mathrm{d}T}{\mathrm{d}h}\right| \geq \frac{m_a g}{c_{\mathrm{p}}} = 9.8\,\mathrm{K/km} \tag{2.1.29}$$

If this condition is violated, atmospheric air is compressed, until the equality (2.1.28) will be fulfilled.

We now determine the temperature gradient for saturated atmospheric air by analogy with formula (2.1.28). It is necessary to include in this consideration the energy change of the saturated vapor due to the temperature variation for the number density of free water molecules due to their condensation or release. This value per unit volume is equal

$$\mathrm{d}E_{\mathrm{sat}} = E_{\mathrm{sat}}\frac{\mathrm{d}N_{\mathrm{sat}}}{\mathrm{d}T}\mathrm{d}T = \left(\frac{E_{\mathrm{sat}}}{T}\right)^2 N_{\mathrm{sat}}\mathrm{d}T,$$

where the number density N_{sat} of free water molecules at a given temperature is determined by formula (2.1.12). Repeating operations in derivation of formula (2.1.6) with accounting for free and bound water molecules, one can obtain the following stability criterion instead of (2.1.28)

$$\frac{\mathrm{d}T}{\mathrm{d}h} \leq \frac{m_a g}{c_{\mathrm{p}} + \left(\frac{E_{\mathrm{sat}}}{T}\right)^2 \frac{N_{\mathrm{sat}}}{N_{\mathrm{a}}}}, \tag{2.1.30}$$

where N_{sat} is the number density of free water molecules at this temperature in the saturated vapor and N_{a} is the number density of air molecules. In particular, at the Earth's surface within the framework of the model of standard temperature, this criterion gives

$$\left(\frac{\mathrm{d}T}{\mathrm{d}h}\right)_{\mathrm{sur}} \leq 3.9\,\mathrm{K/km} \tag{2.1.31}$$

According to formula (2.1.30), the lower boundary for the temperature gradient (lapse rate) corresponds to the saturated water vapor with parameters of standard atmosphere [1], and the lapse rate grows if the altitude increases. Within the framework of the model of standard atmosphere, the atmospheric temperature gradient is $\mathrm{d}T/\mathrm{d}h = 6.5\,\mathrm{K/km}$ independently on the altitude. Figure 2.8 contains the lapse rate of saturated atmospheric air. As is seen, the lapse rate of saturated atmospheric

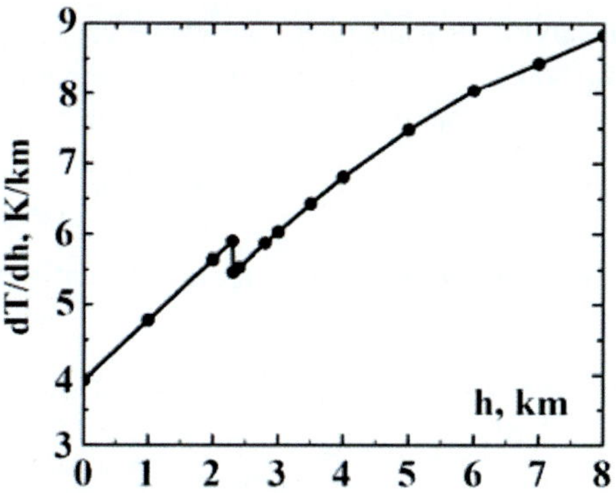

Fig. 2.8 Lapse rate for saturated atmospheric air as an altitude function according to formula (2.1.28) which parameters correspond to the model of standard atmosphere

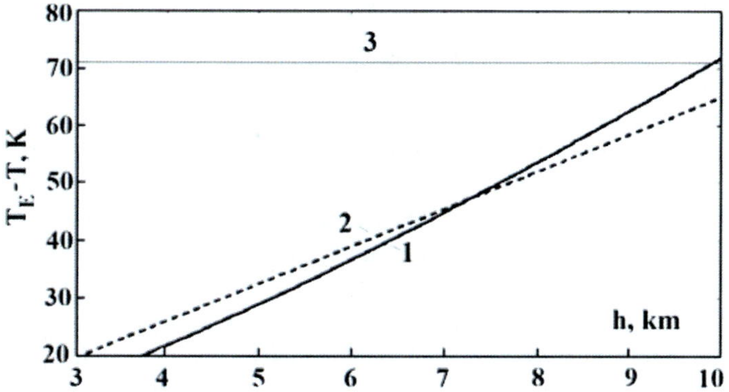

Fig. 2.9 Difference between the temperature of the Earth's surface ($T_E = 288$ K) and a current troposphere temperature T as an altitude function. 1—the boundary temperature in saturated air if heating of an upper troposphere proceeds due to ozone degradation, 2—the troposphere temperature for the model of standard atmosphere ($dT/dh = -6.5$ K/km), 3—the tropopause temperature T_p for this model ($T_p = 217$ K)

air increases with an increasing altitude, and a jump of this value takes place at the liquid–solid phase transition for a formed condensed phase of an atmospheric water vapor.

We thus obtain two forms of the thermal stability of the atmosphere. This stability is determined by transition of the gravitation (mechanical) energy of air molecules in the thermal one. In the case of dry or nonsaturated air, the thermal energy of air is connected with internal energy of air molecules only, whereas in the case of a saturated air, the energy of formation of condensed water is added to the internal thermal energy of air. Therefore, the stability boundary for dry air (2.1.29) is higher than that for saturated one (2.1.30). Figure 2.9 contains the temperature of the troposphere with saturated air and that for the model of standard atmosphere. The curve 1 corresponds to the instability in atmospheric air which causes a movement of atmospheric air. It is significant that motion inside atmospheric air in the form of winds is of importance for various properties of atmospheric air.

2.2 Water Microdroplets in Atmospheric Air

2.2.1 Water Microdroplet as System of Bound Molecules

We below consider properties of water microdroplets and processes with their participation. The total binding energy E_n of water molecules in a microdroplet contained $n \gg 1$ bound molecules is equal in the limit where one can ignore the thermal energy of bound molecules [57]

$$E_n = \varepsilon_o n - A n^{2/3} \tag{2.2.1}$$

This expression may be considered as an expansion over a small parameter $n^{-1/3}$. The first term is the energy of a bulk system, so that ε_o is the binding energy per molecule for their bulk bound system. This value is analogous to E_{sat} of formulas (2.1.21). Hence, if we account for thermal motion of bonded and free molecules in the course of its release from a water surface, we have approximately $\varepsilon_o = 0.44$ eV for the liquid state of the water surface and $\varepsilon_o = 0.48$ V for the its solid state for a range of temperatures and pressures under consideration.

The second term of formula (2.2.1) is the surface energy which decreases with an increasing droplet radius r. Indeed, the surface droplet energy is determined by surface molecules, and their number decreases with an increasing droplet radius. The parameter A may be connected with the surface tension of a bulk spherical droplet σ since the surface droplet energy E_{sur} may be given as

$$E_{sur} = A n^{2/3} = 4\pi r^2 \sigma,$$

that is

$$A = 4\pi r_W^2 \sigma \tag{2.2.2}$$

We use here the definition for the Wigner–Seits radius according to which the number of molecules n for a microdroplet of a radius r is equal [58, 59]

$$n = \left(\frac{r}{r_W}\right)^3 \tag{2.2.3}$$

In the case of liquid water, $r_W = 0.192$ nm [60]. In particular, taking the surface tension of water at the temperature $T = 273$ K as $\sigma = 0.076\,\mathrm{J/m^2}$ [61], one can find for a water droplet at temperatures in the vicinity of $0°C$

$$A \approx 0.22\,\mathrm{eV} \tag{2.2.4}$$

From formula (2.2.1), it follows for the binding energy of a large microdroplet consisting of n molecules

$$\varepsilon_\mathrm{n} = \frac{\mathrm{d}E_\mathrm{n}}{\mathrm{d}n} = \varepsilon_\mathrm{o} - \frac{2A}{3n^{1/3}} \tag{2.2.5}$$

For a microdroplet contained many molecules ($n \gg 1$), the second term of this expansion is relatively small. Indeed, if a microdroplet radius is $r = 1\,\mu\mathrm{m}$, the second term of formula (2.2.5) is 3×10^{-5} eV, whereas for a typical size for cumulus clouds [62] $r = 8\,\mu\mathrm{m}$, the contribution of the surface energy is 3.5×10^{-6} eV that is small compared to the binding energy for a water molecule in bulk liquid water $\varepsilon_\mathrm{o} = 0.44$ eV.

2.2.2 *Interaction of Water Microdroplet with Atmospheric Air*

We now consider the behavior of water microdroplets in atmospheric air being guided by microdroplet sizes $(1 - 10)\,\mu\mathrm{m}$. We deal with motion of a microdroplet under the action of the Earth's gravitation field that is characterized by velocities $v \sim 1$ cm/s, as well as with motion under the action of atmospheric winds which typical velocity is $v \sim 1$ m/s. In the last case, we have for the Reynolds number which describes the motion of atmospheric air near an individual microdroplet

$$Re = \frac{vr}{\nu} < 1, \tag{2.2.6}$$

where ν is the air viscosity coefficient. This criterion gives that a microdroplet does not change the state of surrounding atmospheric air and their motion has the laminar character.

We first consider relaxation of droplet motion in the atmosphere. Considering a microdroplet as a spherical particle of a radius r, one can obtain for its mass

$$M = 4\pi\rho r^3/3,$$

where $\rho = 1\,\mathrm{g/cm^3}$ is the mass density of water. Let us take this microdroplet to be located in an air flux of the velocity v_o, where at the beginning the microdroplet velocity differs from that of air, but the subsequent equilibrium corresponds to equality of these velocities. This is attained under the action of the Stokes force [63, 64]

$$F = 6\pi r\eta v_\mathrm{d} \tag{2.2.7}$$

and equation of motion of this microdroplet is given by

$$M\frac{d\mathbf{v}}{dt} = 6\pi r\eta(\mathbf{v}-\mathbf{v}_o) = -\frac{M(\mathbf{v}-\mathbf{v}_o)}{\tau_{rel}}, \quad \tau_{rel} = \frac{2\rho r^2}{9\eta} \tag{2.2.8}$$

The relaxation time τ_{rel} is responsible for establishment of the equilibrium microdroplet velocity. This value is proportional to the radius square r^2 of the microdroplet. Taking under normal conditions $\eta = 1.85 \times 10^{-4}\text{g}/(\text{cm s})$ for atmospheric air [65, 66], one can obtain

$$\tau_{rel}/r^2 = 1.2 \times 10^3\ \text{s}/\text{cm}^2$$

In particular, for a microdroplet radius $r \sim 10\,\mu\text{m}$, the relaxation time $\tau_{rel} \sim 10^{-3}$ s. Note that the velocity v_g of microdroplet falling in atmospheric air under the action of its weight is given by [64]

$$v_g = \frac{2g\rho r^2}{9\eta} \tag{2.2.9}$$

One can compare the relaxation time of a microdroplet τ_{rel} with a typical time τ_l of velocity variation in air vortices in the course of air convective motion. Taking parameters of air convective motion which are responsible for propagation of air to large distances as a typical vortex size to be $l \sim 100$ cm and a typical air velocity in this vortex $v_l \sim 10^3$cm/s, one can obtain a typical time of vortex variation $\tau_l \sim 0.1$ s. As is seen, a typical time of establishment of velocity of surrounding air for a microdroplet is small compared to a time of variation of vortex parameters. Thus, a microdroplet moves together with surrounding air in course of its convective motion.

Let us consider the character of absorption of water molecules by a microdroplet located in atmospheric air. Denote the number density of water molecules far from a microdroplet by N_w, and by $c(R)$ the relative concentration of water molecules at a distance R from the microdroplet. We have $c(\infty) = 1$ and $c(r) = 0$ if the sticking probability, i.e., the probability of molecule attachment to the microdroplet surface at their contact, is one. Then the flux of water molecules to the microdroplet surface through a spherical surface of a radius R is equal

$$j = -D_w N_w \frac{dc}{dR}$$

Correspondingly, the rate of molecule attachment to a microdroplet, i.e., the number of water molecules attached to the microdroplet surface per unit time, is equal

$$J = 4\pi R^2 j = -4\pi R^2 D_w N_w \frac{dc}{dR}$$

In derivation this formula, we assume that evaporation of molecules from the droplet surface is absent. In order to account for molecule evaporation, we replace the number density of water molecules far from the microdroplet N_w by $N_w - N_{sat}$, where N_{sat} is the number density at the saturated vapor pressure. Then, we account for an equilibrium for the fluxes of attaching and evaporating molecules if $N_w = N_{sat}$. This gives for the molecules flux through a spherical surface of a radius R

$$J = 4\pi R^2 j = -4\pi R^2 D_w (N_w - N_{sat}) \frac{dc}{dR}$$

Because water molecules are not formed in a space, as well as they are not absorbed there, one can consider this relation as an equation for $c(R)$. Solution of this equation

$$R^2 \frac{dc}{dR} = \text{const}$$

with using the above boundary conditions, one can obtain

$$c(R) = 1 - \frac{r}{R}$$

From this, it follows for the flux of attached molecule

$$J = 4\pi D_w r (N_w - N_{sat}) \tag{2.2.10}$$

In the limit where one can neglect by the evaporation process, this formula is transformed into the Smolukhowski formula (2.2.11) [67] for the diffusion flux of molecules to the surface of a particle surface

$$J = 4\pi D r N_w \tag{2.2.11}$$

2.2.3 *Water Microdroplets in Cumulus Clouds*

According to the above analysis of distribution of water molecules in the atmosphere, the average moisture of atmospheric air does not reach 100%. This means that condensation of atmospheric water vapor is absent in motionless air. From this, it follows that atmospheric water is found in the form of free molecules mostly, i.e., the water condensed phase gives a small contribution to the amount of atmospheric water.

But fluctuations in the number density of water molecules can provide formation of supersaturated water vapor in some regions in accordance with the criterion (2.1.18), i.e., condensation is possible in these atmospheric regions. Two ways of this can be realized. In the first case, the temperature of atmospheric air decreases, and an atmospheric air becomes supersaturated. This proceeds usually in autumn at night where the mist is formed consisting of water microdroplets. Tomorrow this mist descents to the Earth or disappears under the action of solar radiation.

Another way to form a supersaturated water vapor proceeds under the action of vertical winds which transfers warm wet air to larger altitudes and mixes it with cold air. Mixing of these layers-parcels leads to formation of clouds consisting of water microdroplets or water microparticles. Because clouds are responsible for atmospheric electricity and greenhouse atmospheric effect, we focus on this part of condensed atmospheric water. Moreover, we are concentrated on cumulus clouds where the most part of condensed atmospheric water is located. Moreover, because the basic part of condensed atmospheric water is found in the liquid aggregate state, we restrict ourselves by water microdroplets as the basic component of cumulus clouds. This and other simplifications make transparent the nature of processes under consideration.

Analyzing properties of clouds in the context of atmospheric electricity, we consider mostly electric and nucleation processes involving water microdroplets which are connected with this phenomenon. Note that processes with cloud participation are more rich and are described in a lot of atmospheric books [68–85]. Moreover, in this consideration, we are based on simplified models. In particular, below we represent cumulus clouds to be consisted of identical water microdroplets which radius r and the number density N_d are taken from measurements [86–89] and are equal

$$r = 8\mu\text{m}, \;\; N_d = 10^3\,\text{cm}^{-3}, \tag{2.2.12}$$

Let us use the connection between the number of bound water molecules of an individual microparticle n and its radius r through the Wigner formula (2.2.3). From this formula, it follows that a water microdroplet of a radius $r = 8\mu\text{m}$ contains $n = 7 \times 10^{13}$ water molecules, and the droplet mass is equal $m_d = 2.1 \times 10^{-9}$ g. In addition, the average number density of bound water molecules in microdroplets is $nN_d = 7 \times 10^{16}\,\text{cm}^{-3}$. The equilibrium of a vapor of free water molecules with water microdroplets requires that the moisture of atmospheric air contained this water vapor and its microdroplets to be

$$\eta = 100\%, \tag{2.2.13}$$

Comparing the above number density of bound water molecules with that of free ones according to data of Table 2.1, one can conclude that even inside cumulus clouds, free water molecules dominate if these clouds are located at not high altitudes.

We now consider motion of an individual droplet in an air flow. Denoting the droplet mass of a radius r by $m = 4\pi\rho r^3/3$, where ρ is the mass density of water, the droplet velocity by v, and the flow velocity by v_o, we have the following motion equation for the droplet in the flow direction

$$m\frac{dv}{dt} = 6\pi r(v_o - v) = \frac{(v_o - v)}{\tau_{rel}}, \tag{2.2.14}$$

where we use the Stokes force (2.2.7) as the resistance one. In this manner, we define the relaxation time for a droplet in an air flow which expression is given by formula

(2.2.8). For water microdroplets of a typical cumulus cloud with a droplet radius $r = 8\mu m$, this time is equal $\tau_{rel} = 0.8$ ms.

Let us consider falling of small particles to the ground being guided by water microdroplets. Along with falling under the action of the Earth's gravitation field, the particles drift toward the ground with the drift velocity v_L

$$v_L \approx D_L/\Lambda \approx 0.05\,\text{cm/s}, \tag{2.2.15}$$

where $D_L \approx 5 \times 10^4\,\text{cm}^2/\text{ s}$ is the diffusion coefficient described propagation for large distances, $\Lambda \approx 10$ km is the scale parameter given by formula (2.1.3). The drift valocities due to these mechanisms are equalized at the particle radius $r_L = 2.2\,\mu$m. At larger falling velocities of microdroplets, they are not captured by the convective motion of atmospheric air. Correspondingly, we have for the number densities of microdroplets instead formula (2.1.1), if these microdroplets are formed at an altitude h_o [90]

$$N(h) = N_o \cdot \exp\left(-\frac{h - h_o}{\lambda_a}\right), \tag{2.2.16}$$

where $N_o = N(h_o)$. Then, one can represent the parameter λ_a for a particle of radius r by [90]

$$\lambda_a = \frac{\Lambda}{1 + v_g/v_L} = \frac{\Lambda r^2}{r_L^2 + r^2}, \tag{2.2.17}$$

where $v_g(r_L) = v_L$, and for the model of standard atmosphere $r_L = 2.2\,\mu$m. According to formula (2.2.17), small particles are captured by air vortices of convective air, and their number densities as an altitude function are determined by formula (2.1.1). In another limiting case ($v_g \gg v_L$), the falling of large microparticles proceeds similar to that in motionless air.

2.2.4 Microdroplet in Heat Process

We now consider one more aspect of attachment of water molecules to a microdroplet which is connected with heat release. Indeed, attachment of one water molecule to the microdroplet surface results in energy release ε_n given by formula (2.2.5). This energy exceeds significantly a thermal energy, and hence, in spite of a small concentration of water molecules in atmospheric air, this process is of importance for the energy balance of atmospheric air. Therefore, the heat balance of a water microdroplet consists in energy release due to attachment of water molecules to a microdroplet, and heat removal proceeds through air thermal conductivity. As a result of these processes, the microdroplet temperature becomes above the temperature of air located far from the microdroplet.

For this heat balance of the behavior of a water microdroplet in atmospheric air, the heat flux q at a distance R from a microdroplet is given by

$$q = -\kappa \frac{dT}{dR},$$

where κ is the thermal conductivity coefficient of atmospheric air. Let us denote the air temperature far from the microdroplet by T_a, while the temperature at its surface is T_d, and we below determine this temperature depending on the power released at the microdroplet surface. We have for the power P which transfers through this sphere of a radius R

$$P = 4\pi R^2 \kappa \frac{dT}{dR}$$

Because of the absence of energy sources in a space, one can consider this relation as equation for the air temperature. Solution of this equation with using the above temperature values at the surface and far from the microdroplet gives the following relation under the assumption that the thermal conductivity coefficient κ varies slowly in the temperature range of surrounding air

$$\Delta T \equiv T_d - T_a = \frac{P}{4\pi\kappa r} \quad (2.2.18)$$

This formula describes the heat equilibrium for a microdroplet, so that it is heated due to some internal or surface energy source with the power P of heat release of a microdroplet, and this causes an increase of the microdroplet temperature by ΔT. We now estimate a typical time of establishment of this thermal equilibrium based on the heat balance equation

$$C_p m_o \frac{d\Delta T}{dt} = -P = -4\pi r \kappa \Delta T,$$

where $C_p m_o$ is the heat capacity of the microdroplet, $C_p = 1\text{cal}/(\text{gK})$ is the heat capacity of water per unit mass, $m_o = 4\pi r^3 \rho/3$ is the microdroplet mass, so that r is its radius, $\rho = 1\text{g}/\text{cm}^3$ is the water mass density. One can reduce this equation to the form

$$\frac{d\Delta T}{dt} = -\frac{\Delta T}{\tau_r}, \quad \tau_r = \frac{r^2 \rho C_p}{3\kappa}, \quad (2.2.19)$$

Formula (2.2.19) is the definition of heat relaxation of a microdroplet. The reduced relaxation time for water microdroplets in atmospheric air is equal $\tau_r/r^2 = 5.5 \times 10^3\ \text{s}/\text{cm}^2$ that gives for a typical radius $r = 8\,\mu\text{m}$ of water microdroplets of cumulus clouds $\tau_r = 0.9$ ms.

We now consider the case where the process of heat release or heat absorption is determined by attachment of water molecules to the microdroplet surface or their evaporation from it. Each act of attachment or evaporation leads to the change of the microdroplet energy by the binding energy ε_o (we neglect the contribution to the binding energy from surface tension). Then, on the basis of formula (2.2.10), we have for the power of heat release related to a test microdroplet

$$P = 4\pi\varepsilon_o D_w r(N_w - N_{sat}) \tag{2.2.20}$$

We above take into account that the molecule binding energy ε_o in a microdroplet exceeds remarkably a typical thermal energy T of molecules ($\varepsilon_o \gg T$), i.e., the thermal effect results in formation of new bonds of water molecules with the microdroplet or their breaking. Correspondingly, formula (2.2.18) for the temperature change of a microdroplet compared to surrounding air is given by

$$\Delta T = \frac{\varepsilon_o D_w(N_w - N_{sat})}{\kappa} \tag{2.2.21}$$

Let us consider the feedback effect in the heat balance of a microparticle. Indeed, the evaporation rate for water molecules from the microdroplet surface is determined by the microdroplet temperature and relates to the second term in brackets of formula (2.2.21), whereas other parameters of this formula relate to the temperature of surrounding atmospheric air. Using the relation

$$N_{sat}(T_d) = N_{sat}(T_a)\exp\left(\frac{\varepsilon_o \Delta T}{T^2}\right),$$

one can reduce equation (2.2.21) to the form

$$\Delta T = \frac{\varepsilon_o D_w\left(N_w - N_{sat}\exp\left(\frac{\varepsilon_o \Delta T}{T^2}\right)\right)}{\kappa} \tag{2.2.22}$$

Thus, one can represent equation of the heat balance for a microdroplet in the form (2.2.21)

$$\Delta T = \frac{\varepsilon_o D_w(N_w - N_{sat})}{\Phi_o(T_a)\kappa}, \quad \Phi_o(T) = 1 + \left(\frac{\varepsilon_o}{T}\right)^2 \frac{D_w N_{sat}}{\kappa}, \tag{2.2.23}$$

Parameters of this formula relate to the air temperature, and $\Delta T \ll T^2/\varepsilon_o$. The function $\Phi_o(T)$ takes into account the heat feedback that provides a decreasing temperature difference. This function is presented in Fig. 2.10 for parameters of atmospheric air within the framework of the model of standard atmosphere [1] at atmospheric pressure. As is seen, the heat feedback under consideration is essential in the heat balance of a water microdroplet in atmospheric air.

We now estimate the role of radiative processes in the behavior of a microdroplet in atmospheric air. For simplicity, we assume the blackbody character of interaction between the radiation field of the atmosphere and a test water microdroplet.

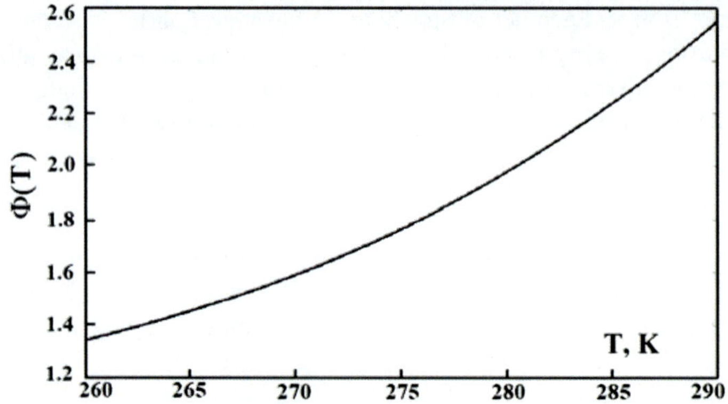

Fig. 2.10 Thermal factor $\Phi_o(T)$ that accounts for heating of the water droplet in atmospheric air as a result the growth process [91, 92]

Then, introducing ΔT as the temperature difference between the microdroplet and surrounding atmospheric air, we assume the air temperature to be coincided with that of atmospheric radiation. Then taking the radiative flux $J_{\rm rad}$ of the blackbody emission for the microdroplet surface which temperature is T, we have according to the Stephan–Boltzmann law

$$P_{\rm rad} = \sigma T^4, \tag{2.2.24}$$

where $\sigma = 5.67 \times 10^{-8}\mathrm{W}/(\mathrm{m}^2\,\mathrm{K}^4)$ is the Stephan–Boltzmann constant.

Note that the Stephan–Boltzmann law holds true if the wave length of radiation is small compare do the particle size. Assuming this criterion to be valid, we take heat release for the microdroplet to be resulted from its thermal emission. Under these conditions, we have for the radiative power $P_{\rm rad}$ for emission and absorption by the microdroplet

$$P_{\rm rad} = -4\sigma T^3 \Delta T \cdot 4\pi r^2,$$

where $4\pi r^2$ is the area of the microdroplet surface. The heat release due to thermal conductivity is equal in this case

$$P = 4\pi r \kappa \Delta T$$

Let us introduce the part of thermal energy ξ which releases through radiation

$$\xi = \frac{P_{\rm rad}}{P} = \frac{4\sigma T^3 r}{\kappa} \tag{2.2.25}$$

One can see that the contribution of radiation into the heat balance of a microdroplet grows with an increasing microdroplet radius. Formula (2.2.25) gives for a typical microdroplet radius $r = 8\mu m$ in clouds and room temperature $T = 288$ K, where the thermal conductivity coefficient of air is equal $\kappa = 2.5 \times 10^{-4}\,W/(cm\,K)$ [65] at room temperature. We have for this parameter $\xi = 1.7 \times 10^{-3}$. As it follows from this estimation, radiation of typical microparticles located in atmospheric air under atmospheric conditions is negligible.

References

1. *U.S. Standard Atmosphere.* (Washington, U.S. Government Printing Office, 1976)
2. L. Boltzmann, Wiener Berichte **58**, 517 (1868)
3. https://en.wikipedia.org/wiki/Boltzmann-distribution
4. L.D. Landau, E.M. Lifshitz, *Statistical Physics*, vol. 1 (Pergamon Press, Oxford, 1980)
5. C. Kittel, H. Kroemer, *Thermal Physics* (Wiley, New York, 1980)
6. P.M. Morse, *Thermal Physics* (Benjamin Inc., New York, 1964)
7. D. Haar, H. Wergeland, *Elements of Thermodynamics* (Addison-Wesley, Reading, 1967)
8. R. Kubo, *Thermodynamics* (North Holland, Amsterdam, 1968)
9. https://www.sciencedirect.com/topics/earth-and-planetary-sciences/adiabatic-lapse-rate
10. https://en.wikipedia.org/wiki/Weather-and-climate
11. https://en.wikipedia.org/wiki/Weather-forecasting
12. https://www.encyclopedie-environnement.org/en/air-en/weather-forecasting-models/
13. J.M. Wallace, P.V. Hobbs. *Atmospheric Science. An Introductory Survey.* (New York, Academic Press, 1977)
14. J. Mcilveen, *Fundamentals of Weather and Climate* (Chapman and Hall, London, 1992)
15. J.R. Holton, *An Introduction to Dynamic Meteorology* (Elsevier, Amsterdam, 2004)
16. H.J. Poincare, Acta Mathematica **13**, 1 (1890)
17. H.J. Poincare, in *Les methodes nouvelles de la mecanique celeste.* (Vols 1-3. Paris, Gauthiers-Villars, 1892, 1893, 1899)
18. F. Zhang, Y.Q. Sun, L. Magnusson et al., J. Atm. Sci. **76**, 1077 (2019)
19. M. Grosjean, J. Goiot, Z. Yu, Global Planet Change **152**, 19 (2017)
20. Ch.D. Keeling, Tellus **12**, 200 (1960)
21. Ch. D. Keeling, R.B. Bacastow, A.E. Bainbridge e.a. Tellus **28**, 538 (1976)
22. Ch.D. Keeling, J.F.S. Chin, T.P. Whorf, Nature **382**, 146 (1996)
23. R.B. Bacastow, Ch.D. Keeling, T.P. Whorf, J. Geophys. Res. **90**, 10529 (1985)
24. Ch.D. Keeling, T.P. Whorf, M. Wahlen, J. van der Plicht. Nature **375**, 666 (1995)
25. Ch. D. Keeling, Ann. Rev. Energy Environ. **23**, 25 (1998)
26. https://www.esrl.noaa.gov/gmd/ccgg/trends/mlo.html
27. https://www.wikiwand.com/en/Mauna-Loa-Observatory
28. https://www.co2.earth/monthly-CO2
29. https://cdiac.ess-dive.lbl.gov/ftp/trends/co2/maunaloa-co2
30. F. Joos, R. Spahni, Proc. Nat. Acad. Sci. USA **105**, 1425 (2008)
31. https://en.wikipedia.org/wiki/Beaufort-scale
32. I.A. Shiklomanov, in *Water in Crisis : A Guide to the World's Fresh Water Resources,* ed. by P.H. Gleick (Oxford, Oxford University Press, 1993), pp. 13–24
33. I.A. Shiklomanov, J.C. Rodda (eds.), *World Water Resources at the Beginning of the Twenty-First Century* (Cambridge University Press, Cambridge, 2003)
34. R.W. Healy, T.C. Winter, J.W. Labaugh, O.L. Franke, in *Water budgets : Foundations for Effective Water-Resources and Environmental Management.* (Reston, Virginia, U.S. Geological Survey Circular 1308, 2007)
35. https://en.wikipedia.org/wiki/Atmosphere-of-Earth
36. https://water.usgs.gov./edu/watercyrcleatmosphere.html

37. A. Baumgartner, E. Reichel, *The World Water Balance* (Elsevier, Amsterdam, 1975)
38. J.P. Peixoto, A.H. Oort, *Physics of Climate* (Am er. Inst. Phys, Washington, 1992)
39. K.E. Trenberth, L. Smith, T. Qian et al., J. Hydrometeorol. **8**, 758 (2007)
40. https://en.wikipedia.org/wiki/water-circle
41. https://en.wikipedia.org/wiki/Moisture
42. https://en.wikipedia.org/wiki/Dew-point
43. *Handbook of Chemistry and Physics*, edn. 86, ed. by D.R.Lide (London, CRC Press, 2003-2004)
44. A. Wexler, J. Res. Nat. Bur. Stand. **80A**, 775 (1976)
45. R. Clausius, Annalen der Physik **155**, 500 (1850)
46. W. Kenneth, *Thermodynamics* (McGraw-Hill, New York, 1988)
47. https://tamino.wordpress.com/2011/05/17/hot-and-wet/
48. https://www.esrl.noaa.gov/gmd/dv/iadv/graph.php?code=HIH.program=wvap.type=vp
49. https://en.wikipedia.org/wiki/Weather-balloon
50. https://public.wmo.int/en/resources/bulletin/observing-water-vapour
51. *Monitoring Atmospheric Water Vapours,* ed. by N. Kãmpfer (Bern, Springer Nature, 2013)
52. R. Braham, J. Meteorol. **9**, 227 (1952)
53. A. Dai, J. Clim. **19**, 3589 (2006)
54. K.M. Willett, P.D. Jones, N.P. Gillett, P.W. !orne, J. Clim. **21**, 5364 (2008)
55. D.I. Berry, E.C. Kent, Bull. Amer. Meteor. Soc. **90**, 645 (2009)
56. https://en.wikipedia.org/wiki/Lapse-rate
57. S. Ino, J. Phys. Soc. Japan. **27**, 941 (1969)
58. E.P. Wigner, W.F. Seits, Phys. Rev. **46**, 509 (1934)
59. E.P. Wigner, Phys. Rev. **46**, 1002 (1934)
60. B.M. Smirnov, *Clusters and Small Particles in Gases and Plasmas* (Springer NY, New York, 1999)
61. H.R. Pruppacher, S.D. Klett, *Microphysics of Clouds and Precipitation* (Kluwer, New York, 1997)
62. https://en.wikipedia.org/wiki/Cumulus-cloud
63. G.G. Stokes, Trans. Cambridge Phil. Soc. **9**, Part II, 8 (1851)
64. L.D. Landau, E.M. Lifshits, *Fluid Mechanics* (Pergamon Press, London, 1959)
65. N.B. Vargaftic, *Tables of Thermophysical Properties of Liquids and Gases* (Halsted Press, New York, 1975)
66. B.M. Smirnov, *Reference Data on Atomic Physics and Atomic Processes* (Springer, Heidelberg, 2008)
67. M.V. Smolukhowski, Zs. Phys. **17**, 585 (1916)
68. H.L. Green, W.R. Lane, *Particulate Clouds: Dust, Smokes and Mists* (Princeton, Van Nostrand, 1964)
69. H.R. Byers, *Elements of Cloud Physics* (The University of Chicago, Chicago, 1965)
70. N.H. Fletcher, *The Physics of Rainclouds* (Cambridge University Press, London, 1969)
71. B.J. Mason, *Clouds, Rain and Rainmaking* (Cambridge University Press, Cambridge, 1975)
72. S. Twomey, *Atmospheric Aerosols* (Elsevier, Amsterdam, 1977)
73. P.C. Reist, *Introduction to Aerosol Science* (Macmillan Publ. Comp, New York, 1984)
74. R.R. Rogers, M.K. Yau, *A Short Course in Cloud Physics* (Pergamon Press, Oxford, 1989)
75. F.H. Ludlam, *Clouds and Storms: The Behavior and Effect of Water in the Atmosphere* (University Park, Penn State University Press, 1990)
76. K. Young, *Microphysical Processes in Clouds* (Oxford University Press, New York, 1993)
77. K. Friedlander, *Smoke, Dust, and Haze. Fundamentals of Aerosol Dynamics* (Oxford, Oxford Univ.Press, 2000)
78. M. Satoh, *Atmospheric Circulation Dynamics and General Circulation Models* (Springer-Praxis, Chichester, 2004)
79. H.R. Pruppacher, J.D. Klett, *Microphysics of Clouds and Precipitation* (Kluwer, New York, 2004)
80. J. Straka, *Clouds and Precipitation Physics* (Cambridge University Press, Cambridge, 2009)

81. B.J. Mason, *The Physics of Clouds* (Oxford University Press, Oxford, 2010)
82. H. Pruppacher, J. Klett, *Microphysics of Clouds and Precipitation* (Springer, Dordrecht, 2010)
83. D. Lamb, J. Verlinde, *Physics and Chemistry of Clouds* (Cambridge University Press, Cambridge, 2011)
84. P.K. Wang, *Physics and Dynamics of Clouds and Precipitation* (Cambridge University Press, Cambridge, 2013)
85. A.P. Khain, M. Pinsky, *Physical Processes in Clouds and Cloud Modeling* (Cambridge University Press, Cambridge, 2018)
86. B.J. Mason, *The Physics of Clouds* (Claredon Press, Oxford, 1971)
87. J. Warner, Tellus **7**, 450 (1955)
88. W.R. Leaitch, G.A. Isaak, Atmosp. Environ. **25**, 601 (1991)
89. http://en.wikipedia.org/wiki/Liquid-water-content
90. B.M. Smirnov, *Microphysics of Atmospheric Phenomena* (Switzerland, Springer Atmospheric Series, 2017)
91. B.M. Smirnov, EPL **99**, 13001 (2012)
92. B.M. Smirnov, Phys. Usp. **57**, 1041 (2014)

Chapter 3
Transport Processes in Atmospheric Air

Abstract Transport phenomena in motionless atmospheric air are analyzed and include mostly diffusion, thermal conductivity and viscosity of air. Kinetic coefficients for these transport processes due to collisions of molecules are expressed through the mean free path of molecules in air. Criteria of the Rayleigh–Taylor instability because of a temperature gradient are analyzed for a gas located in an external field. These results are applied to the convection motion of atmospheric air including propagation for large distances and transport of a water vapor. This allows one to analyze water circulation in the atmosphere which consists of water evaporation from the Earth's surface in the form water molecules, its partial condensation in air and returning back both in the form of free water molecules and condensed water phase as rain or snow. Propagation of air for large distances has the diffusion character, and the corresponding diffusion coefficient follows from the altitude distribution of atmospheric water molecules and is approximately $5 \times 10^4\,\mathrm{cm}^2/\mathrm{s}$ near the Earth's surface.

3.1 Transport Phenomena in Gases

3.1.1 Diffusion of Atomic Particles in Gases

We consider two types of transport processes in gases being guided by atmospheric air. In the first case, this transport is realized by individual atoms or molecules, and a distance resulted from an elementary event is of the order of mean free path λ of atoms or molecules in a gas. In the second case, the transport process is determined by fluxes, streams or vortices which include macroscopic volumes of a transported gas. Then transport of a gas takes place for larger distances than that in the first case, but generation of macroscopic motion of a gas is possible if certain requirements are fulfilled.

We start from transport processes in a motionless gas resulted from transfer of individual atoms or molecules. Transport of atoms or molecules in a buffer gas is determined by collisions between these atomic particles and gas molecules. In this consideration, the mean free path of atomic particles transported in a gas is relatively small.

B. M. Smirnov, *Global Energetics of the Atmosphere*, Springer Atmospheric Sciences, https://doi.org/10.1007/978-3-030-90008-3_3

$$\lambda \ll L, \tag{3.1.1}$$

where L is a typical size of the region under consideration. We below derive expressions for kinetic coefficients for transport of atomic particles in gases and then apply the results to transport properties of atmospheric air.

Let us introduce the diffusion coefficient D as the proportionality coefficient between the flux $\mathbf{j}$ of atomic particles and the gradient of the concentration c of atomic particles according to the relation

$$\mathbf{j} = -\mathrm{DN}\nabla c, \tag{3.1.2}$$

where N is the total number density of atoms or molecules of a buffer gas.

We now estimate the diffusion coefficient of atomic particles of a given type in a buffer gas taking the diffusive flux j as the difference of fluxes in opposite directions which reach a given point from a distance $\sim \lambda$ from it. At this distance the concentration difference between available points is $\Delta c \sim \lambda \nabla c$. Hence, the diffusion flux is estimated as $j \sim \upsilon_T \mathrm{N}\Delta c \sim \upsilon_T \mathrm{N}\lambda \nabla c$. Thus, the diffusion coefficient of atomic particles can be estimated as

$$D \sim \upsilon_T \lambda \sim \lambda \sqrt{\frac{T}{m}}, \tag{3.1.3}$$

where T is the gas temperature expressed in energetic units, m is the mass of transported atomic particles under the assumption that this mass is of the same order of magnitude of the mass of the elementary atomic particle for a buffer gas.

We now define the diffusion coefficient D in a gas more precisely, considering the diffusion process as a result of subsequent collisions of a test atomic particle with gaseous molecules. We then introduce the differential cross section $d\sigma$, so that scattering of colliding particles on a gas particle proceeds in the range of scattering angles from ϑ up to $\vartheta + d\vartheta$ in the center-of-mass frame of reference. This quantity depends on the relative collision velocity $\mathbf{g} = \mathbf{v} - \mathbf{v}_\mathrm{m}$, where $\mathbf{v}$ is the velocity of a test atomic particle, $\mathbf{v}_\mathrm{m}$ is the velocity of a gas molecule which partakes in the collision act.

In the case of the behavior of atomic particles in gases at room temperature, the behavior of an atomic particle in a gas is determined by its elastic scattering. The diffusion coefficient D for an atomic particle may be determined on the basis of the Chapman-Enskog approximation which is expansion over a small numerical parameter and is given by [1–3]

$$D = \frac{3T^2}{\mu^2 \langle g^3 \sigma^*(g) \rangle} = \frac{3\sqrt{\pi T}}{8N\sqrt{2\mu}\overline{\sigma}}, \quad \overline{\sigma} = \frac{1}{2}\int_0^\infty \exp(-t) t^2 \sigma^*(t) dt, \quad t = \frac{\mu g^2}{2T} \tag{3.1.4}$$

Here $\mu = m_1 m_2/(m_1 + m_2)$ is the reduced mass of colliding particles with masses m_1 and m_2, $\sigma^*(g)$ is the diffusion cross section of particle scattering according to the following integration over angles

$$\sigma^* = \int (1 - \cos\vartheta) d\sigma$$

Colliding atomic particles are modeled by rigid spheres in their thermal collisions. This means that the diffusion cross section of atomic particles $\sigma^* = \pi(r_1 + r_2)^2$, where r_1 and r_2 are radii of colliding particles, and the cross section σ^* is independent of the collision velocity. In this case of the model of rigid spheres we have from formula (3.1.4)

$$D = \frac{3\pi}{32} \cdot v_T \lambda \approx 0.3 v_T \lambda, \ v_T = \sqrt{\frac{8T}{\pi\mu}}, \ \lambda = \frac{1}{N\sigma^*} \tag{3.1.5}$$

In this case of rigid spheres, the temperature dependence has the form $D \sim T^{-1.5}$. The diffusion cross section $\sigma_g = \sigma^*$ which follows from formula (3.1.5) is named as the gas-kinetic cross section.

Let us represent this dependence as

$$D = D_o \left(\frac{T}{273}\right)^{\gamma} \tag{3.1.6}$$

We below consider diffusion of nitrogen and oxygen molecules in atmospheric air, as well as some admixture molecules [4, 5]. Parameters of collisions involving nitrogen and oxygen molecules are nearby and coincide within the accuracy of determination of these values. Therefore, below we join nitrogen and oxygen molecules in air molecules. Parameters of formula (3.1.6) for diffusion coefficient of some molecules in atmospheric air are given in Table 3.1. In addition, this Table contains the gas-kinetic cross sections in diffusion processes at the temperature $T = 273$ K. This information will be used in various estimations for air processes.

Let us consider the character of diffusion motion of a particle in a gas. The continuity equation for the number density $N_i = Nc_i$ of atomic particles of the sort under conditions that these particle do not partake in chemical processes, including processes of absorption of particles in a space and their generation, has the form

Table 3.1 Parameters of formula (3.1.6) for diffusion of indicated molecules in atmospheric air at the atmospheric pressure

Parameter/molecule	D_o, cm^2/s	γ	σ_g, 10^{-15} cm^2
N_2, O_2	0.19	1.8	3.8
H_2O	0.22	1.8	3.8
CO_2	0.14	1.85	4.8

$$\frac{\partial N_{\rm i}}{\partial t} + \mathbf{j}_i = 0,$$

and using formula (3.1.2) for the diffusion flux, one can obtain the equation describing the diffusive motion of particles

$$\frac{\partial N}{\partial t} + D\Delta N = 0 \qquad (3.1.7)$$

To study the diffusion motion of a test particle in a gas, it is convenient to introduce the probability $W(\mathbf{r}, t)$ that a test particle will be located at point $\mathbf{r}$ at moment t. The normalization condition for this probability has the form

$$\int W(\mathbf{r}, t)d\mathbf{r} = 1,$$

where the integration is fulfilled over all the space. In the case of a uniform gas, the probability W satisfies to equation (3.1.7), that takes the following form in the spherically symmetrical case

$$\frac{\partial W}{\partial t} = \frac{D}{r}\frac{\partial^2}{\partial r^2}(rW)$$

Let us find the average particle parameters by multiplication of this equation by $4\pi r^4 \mathrm{d}r$ and integration the result over r [6]. One can obtain for the left-hand side of this equation

$$\int_0^\infty 4\pi r^4 \mathrm{d}r \frac{\partial W}{\partial t} = \frac{d}{\mathrm{d}t}\int_0^\infty r^2 W 4\pi r^2 \mathrm{d}r = \frac{d}{\mathrm{d}t}\overline{r^2},$$

where $\overline{r^2}$ is the mean square of the distance from the origin where a particle is located at the beginning. Integrating twice by parts and using the above normalization condition, one can transform the right-hand side of the equation into

$$D\int_0^\infty 4\pi r^4 \mathrm{d}r \frac{1}{r}\frac{\partial^2}{\partial r^2}(rW) = -3D\int_0^\infty 4\pi r^2 \mathrm{d}r \frac{\partial}{\partial r}(rW) = 6D\int_0^\infty W 4\pi r^2 \mathrm{d}r = 6D$$

The resulting equation takes the form

$$\mathrm{d}\overline{r^2} = 6D\mathrm{d}t$$

Since at the beginning the particle under consideration is located at the origin, one can obtain as the solution of this equation

$$\overline{r^2} = 6Dt \tag{3.1.8}$$

Because the diffusive motion of a particle in different directions of a uniform gas is random, from this it follows

$$\overline{x^2} = \overline{y^2} = \overline{z^2} = 2Dt$$

This relation will be used below.

One can introduce the probabilities for each coordinate in this spherically symmetric case

$$W(\mathbf{r}, t) = w(x, t)w(y, t)w(z, t)$$

with the normalization condition

$$\int W(\mathbf{r}, t)\mathrm{d}x\mathrm{d}y\mathrm{d}z = 1,$$

or for each coordinate

$$\int w(x, t)\mathrm{d}x = 1$$

Introducing $\Delta^2 = \langle x^2 \rangle = 2Dt$ and accounting for the random character of motion, one can express through this parameter the distribution function $w(x, t)$ in the form of the normal distribution for the probabilities

$$w(x, t) = (2\pi\Delta^2)^{-1/2} \exp\left(-\frac{x^2}{2\Delta^2}\right), \tag{3.1.9}$$

and it is valid for each coordinate. From this we have

$$W(r, t) = (4\pi Dt)^{-3/2} \exp\left(-\frac{r^2}{4Dt}\right) \tag{3.1.10}$$

We now use the above results for the problem where admixture molecules are located in a space and attach to a hard spherical particle in the case of their diffusion motion in a buffer gas where this particle and admixture molecules are located. The particle radius r is large compared with the mean free path λ of a test admixture molecule in a buffer gas. We have for the flux of admixture molecules j to the particle surface at a distance R from the particle

$$j = -D\frac{\mathrm{d}N}{\mathrm{d}R}$$

where N is the number density of particles. From this we have for the number of particles passed through a sphere of a radius R per unit time

$$J = 4\pi R^2 D \frac{dN}{dR}$$

Because this value is independent of the sphere radius R, one can consider this relation as an equation $J = \text{const}$. Taking the number density of admixture molecules far from the particle to be $N(\infty) = N_w$ and the sticking probability, i.e., the probability to attach of admixture molecules as a result of a contact with the particle surface, to be one. This leads to another boundary condition for admixture molecules $N(r) = 0$. The solution of the above equation with given boundary conditions has the form

$$N(R) = N_w \left(1 - \frac{r}{R}\right)$$

Correspondingly, the flux of admixture molecules to the surface of a spherical particle is given by the Smolukhowski formula (2.2.11) [7]

$$J = 4\pi D r N_w$$

We now apply this formula to water molecules which interact with liquid droplets. Along with attachment of free water molecules to a liquid droplet, evaporation of a droplet partakes in the balance of a water vapor located in air along with its droplets. The equilibrium between evaporation and attachment processes takes place, if the number density of free water molecules N_w is equal to that $N_{sat}(T)$ at the saturated vapor pressure. Taking this into account, one can obtain the balance equation for the number density of free water molecules due to a droplet of a radius r

$$\frac{dN_w}{dt} = 4\pi r D \left(N_{sat} - N_w\right) \tag{3.1.11}$$

Let us introduce the mobility b of molecules in a gas which is defined as the ratio of the molecule drift velocity $\mathbf{v}$ to the force $\mathbf{F}$ acted on it as

$$\mathbf{v} = b\mathbf{F} \tag{3.1.12}$$

Let a gas be subjected to an action from a field of the potential $U(\mathbf{R})$ ($\mathbf{R}$ is a space coordinate). We assume that the diffusion character of motion for molecules of a given sort in a space is not violated by the external field. Hence along with the diffusion motion, admixture molecules are drifted under the field. Under the equilibrium, the drift motion of admixture molecules is compensated by the diffusion one. In this case, the drift flux of molecules $b\mathbf{F}N$ is equalized by the diffusion flux $-D\nabla N$, where $N(\mathbf{R})$ is the number density of admixture molecules in a space. Since the Boltzmann distribution of admixture molecules in a space has the form $N \sim \exp(-U/T)$, the force acted on a molecule is $F = -\nabla U$. From this one can obtain, equalizing the diffusion and drift fluxes of molecules, the following relation between the mobility b and diffusion coefficient D of molecules in a buffer gas

$$D = bT \tag{3.1.13}$$

This relation is named as the Einstein relation [8–10]. In particular, from the Einstein relation and formula (3.1.3) it follows

$$b \sim \frac{1}{N\sigma\sqrt{mT}}$$

We also determine the diffusion coefficient of large microparticles in a gas if their radius r is large compared with the mean free path of gas molecules λ which is equal for atmospheric air $\lambda \approx 0.1\ \mu$m. Using the Stokes formula (2.2.7) for the resistance force, one can obtain for the mobility of a macroscopic particle in the diffusion regime

$$b = \frac{v}{F} = \frac{1}{6\pi r\eta}$$

Correspondingly, the diffusion coefficient of a macroscopic particle in a gas is equal [11]

$$D = bT = \frac{T}{6\pi r\eta}, \quad r \gg \lambda \tag{3.1.14}$$

In particular, the diffusion coefficient for a microdroplet of a radius $r = 1\mu m$ in atmospheric air is equal $D = 7.5 \times 10^{-8}\text{cm}^2/\text{s}$, and the diffusion coefficient is inversely proportional to the microdroplet radius under the used condition $r \gg \lambda$.

3.1.2 Heat Transport in Atmospheric Air

Heat transport in gases is realized in two ways. In the first case, a hot molecule is transported in a cold region and carries heat in the form of its kinetic energy. In the second case, an excitation is created in a hot region and is transferred in a cold region where it is quenched, so that the excitation energy transfers in the kinetic energy of molecules, i.e., in the heat. Let us introduce the thermal conductivity coefficient κ as the proportionality factor between the heat flux $\mathbf{q}$ and the temperature gradient ∇T

$$\mathbf{q} = -\kappa \nabla T \tag{3.1.15}$$

Below we estimate the thermal conductivity coefficient for the first case, where heat is transported by hot molecules. We have that like to the case of diffusion transport of molecules in a gas, a point attained without collisions is located at a distance of the order the mean free path λ from an initial point of molecule location. The temperature difference is of the order of $\Delta T = \lambda \nabla T$ between points of the molecule pass from which molecules go to this point without collisions. The heat flux is $q \sim j\Delta T$, where the molecule flux between these points is $j \sim N v_T$ (N is the number density of molecules, $v_T = \sqrt{8T/(\pi m)}$ is the thermal velocity of molecules). From this, one can obtain the following estimation for the thermal conductivity coefficient

$$\kappa \sim \frac{v_T}{\sigma}, \tag{3.1.16}$$

As is seen, the thermal conductivity coefficient is independent of the number density of molecules. Indeed, an increase in the number density of molecules leads to an increase in the number of molecules which partake in transfer of heat, but this simultaneously causes a decrease in a typical distance of this transport. These two effects are mutually canceled.

The thermal conductivity coefficient in the first Chapman–Enskog approximation is given by analogy with formula (3.1.4) [1–3]

$$\kappa = \frac{25\sqrt{\pi T}}{32\overline{\sigma_2}\sqrt{m}}, \tag{3.1.17}$$

where the average cross section of elastic collision of molecules is determined by the following formula

$$\overline{\sigma_2} = \int\limits_0^\infty t^2 \exp(-t)\sigma^{(2)}(t)\mathrm{d}t, \ t = \frac{\mu g^2}{2T}, \ \sigma^{(2)}(t) = \int (1 - \cos^2\vartheta)\mathrm{d}\sigma \tag{3.1.18}$$

It is convenient to represent the temperature dependence for the thermal conductivity coefficient of atmospheric air near room temperature in the form

$$\kappa = \kappa_o \left(\frac{T}{273}\right)^\beta, \ \kappa_o = 2.4 \times 10^{-4}\frac{\mathrm{W}}{\mathrm{cm\,K}}, \ \beta = 0.88 \tag{3.1.19}$$

where the temperature T is expressed in Kelvin.

One can consider as an example of the second mechanism of heat transport, where an atomic particle transfer an excitation, the case of supersaturated water vapor located in atmospheric air. Then an excess of the saturated water vapor is transformed in the liquid or solid phase in the form of liquid droplets or solid particles, and new water molecules penetrated in a cold region attach to these droplets or particles and transfer the extracted binding energy to air. We assume that an equilibrium between free and bonded water molecules establishes fast, and hence the number density N_w of free water molecules is equal to $N_{sat}(T)$ that is the number density of water molecules at the saturated vapor pressure and given temperature.

We now determine the thermal conductivity coefficient κ_{sat} due to the effect under consideration. The flux of water molecules to the region of a lower temperature is given by

$$j = -D_w \nabla N_w = -D_w \nabla N_{sat}(T)$$

Correspondingly, the heat flux is equal

$$q = \varepsilon_o j,$$

where ε_o is the binding energy of a water molecule in a water droplet (for definiteness, we restrict ourselves by the liquid aggregate state of water). Evidently, the temperature dependence for the saturated number density of bonded water molecules has the form

$$N_{sat}(T) \sim \exp(-\varepsilon_o/T),$$

and from this it follows

$$\mathbf{q} = -D_w \left(\frac{\varepsilon_o}{T}\right)^2 N_{sat}(T)\nabla T$$

Thus, the thermal conductivity coefficient in this case is given by the expression

$$\kappa_{sat} = D_w \left(\frac{\varepsilon_o}{T}\right)^2 N_{sat}(T) \tag{3.1.20}$$

This corresponds to regions contained the condensed water phase.

3.1.3 Momentum Transport in Atmospheric Gas

If a gas flux moves along a solid surface, the average velocity of molecules near the surface is zero, while far from it the gas velocity is nonzero. Hence, the average velocity of different layers is different, and transition of molecules between these layers leads to braking of the flux in layers with a heightened velocity and to acceleration of layers with a low velocity, i.e., the friction force occurs. The viscosity coefficient η is the proportionality factor between the frictional force acting on a unit area of a moving gas, and the gradient of the mean gas velocity in the direction perpendicular to the surface of moving layers. Let us take the average gas velocity $\mathbf{v}$ to be parallel to the x-axis and to be vary in the z-direction. The friction force is proportional to $\partial v_x/\partial z$ and acts on an xy surface of gaseous fluxes. In this definition, we have for the force p_{xy} which acts per unit area in the direction x and is similar to the pressure

$$p_{xy} = -\eta \frac{\partial v_x}{\partial z} \tag{3.1.21}$$

Let us estimate the viscosity coefficient η on the basis of the above method to estimate the coefficients of diffusion and thermal heat conductivity in gases. The force acting per unit area as a result of the momentum transport due to motion of molecules is $F \sim N v_T m \Delta v_x$, where $j = N v_T$ is the flux of molecules, and $m\Delta v_x$ is the difference of the average momenta carried by molecules which are moving in opposite directions with respect to a given point. Since molecules reached this point without collisions and located from it at distances of the order of the mean free path λ, we have for a transferred momentum $m\Delta v_x \sim m\lambda \partial v_x/\partial z$. Hence, the force acting per unit area of a moving gas is $F \sim N v_T m\lambda \partial v_x/\partial z$. From this, one can estimate the gas viscosity coefficient as

$$\eta \sim \frac{\sqrt{mT}}{\sigma} \tag{3.1.22}$$

As is seen, the viscosity coefficient is independent of the number density of molecules. As was proved for the thermal conductivity coefficient, this independence comes from the compensation of opposite effects occurring with the momentum transport. The number of momentum carriers is proportional to the number density of molecules, while a typical transport distance is inversely proportional to it. The effects offset each other.

The viscosity coefficient in the first Chapman–Enskog approximation by analogy with formulas (3.1.4) and (3.1.17) is given by [1–3]

$$\eta = \frac{5\pi m v_T}{48\sqrt{2\overline{\sigma_2}}}, \quad \overline{\sigma_2} = \int\limits_0^\infty t^2 \exp(-t)\sigma^{(2)}(t)dt, \quad t = \frac{\mu g^2}{2T}, \quad \sigma^{(2)}(t) = \int (1 - \cos^2\vartheta)d\sigma \tag{3.1.23}$$

As is seen, the average cross section of elastic collisions of molecules averaged over the Maxwell distribution of molecules over velocities is the same as in the case of the thermal conductivity coefficient according to formula (3.1.18).

One can represent the temperature dependence for the viscosity coefficient of atmospheric air in vicinity of room temperature in the same manner as for other kinetic coefficients, namely,

$$\eta = \eta_o \left(\frac{T}{273}\right)^n, \quad \eta_o = 1.7 \times 10^{-4}\mathrm{g/cm\,s}, \quad n = 0.83 \tag{3.1.24}$$

where the temperature T is expressed in Kelvin. We note nearby temperature dependencies (3.1.19) and (3.1.24) for the thermal conductivity and viscosity coefficients. In addition, the effective cross section of molecule scattering in air that follows from the value of the viscosity coefficient according to formula (3.1.23) is equal $\overline{\sigma_2} = 4.6 \times 10^{-15}\,\mathrm{cm}^2$.

The gas viscosity is responsible for a friction force which occurs in the course of gas motion near a solid surface. Indeed, the average velocity of molecules is zero near the surface, and the braking force result from friction between neighboring gas layers moved with different velocities. We below evaluate the force acted on a solid spherical particle by a radius r under the criterion $r \gg \lambda$ where many molecules collide simultaneously with the particle surface.

This force is proportional to the particle velocity v_d with respect to a gas that depends also from the particle radius r. From this and the dimensionality consideration one can find

$$F \sim r\eta v_d$$

More precisely solution of this problem with accounting for the velocity distribution near the particle allows one to determine the numerical coefficient in this formula that leads to the Stokes formula (2.2.7)

$$F = 6\pi r\eta v_d \tag{3.1.25}$$

One can determine also the diffusion coefficient of a spherical solid particle D in a gas on the basis of the Stokes formula and the Einstein relation $D = bT$, that is the definition of the mobility. As a result, we obtain for the diffusion coefficient of a microparticle [11]

$$D = T \cdot \frac{v}{F} = \frac{T}{6\pi r\eta} \tag{3.1.26}$$

In addition, one can obtain the velocity of particle falling under the action of its weight $P = mg = 4\pi\rho g r^3/3$, where ρ is the density of a particle material. One can find from this the velocity of a falling droplet (2.2.9)

$$v_g = \frac{P}{6\pi r\eta} = \frac{2r^2\rho}{9\eta}$$

In particular, in the case of falling of water microdroplets in atmospheric air this formula gives $v_g/r^2 = 1.2 \times 10^6 \mathrm{s}^{-1}\,\mathrm{cm}^{-1}$.

3.2 Convection in Atmospheric Air

3.2.1 Equations of Macroscopic Motion of a Gas

We above consider transport processes in gases due to transfer of individual molecules. Such a transport takes place in laboratory systems where gases are located in a restricted space. Transport for large distances in large gaseous volumes occurs as a result of propagation of fluxes or streams which include macroscopic masses of a gas. Just these transport processes take place in the Earth's atmosphere where transfer of a gas results from the convection. In considering this problem, we first consider macroscopic equations [12–16] of a gas which is located in an external field and its elements move with some velocity $\mathbf{v}(\mathbf{R})$, where $\mathbf{R}$ is a space coordinate.

Along with the velocity $\mathbf{v}(\mathbf{R})$ of this element, we introduce in this equation the number density of molecules $N(\mathbf{R})$. The local flux of molecules is equal $\mathbf{j} = N\mathbf{v}$, and because gaseous molecules do not form and absorb in a space, the continuity equation has the form

$$\frac{\partial N}{\partial t} + \mathrm{div}(N\mathbf{v}) = 0 \tag{3.2.1}$$

The second macroscopic equation of a gas generalizes the motion equation for an individual molecule. If we deal with a collective of molecules, their collisions lead to the action on a gas in the form of the pressure. Let us introduce the pressure as a force that acts on a unit area of an imaginary surface in the frame of reference where a gas as a whole is at rest. We evaluate this force for a uniform gas assuming the axis x to be perpendicular to an imaginary plane. Let us introduce the distribution function $f(v_x)$ over velocities, so that $f(v_x)\mathrm{d}v_x$ is the number density of molecules having velocities between v_x and $v_x + \mathrm{d}v_x$. Elastic reflection of a molecule from the surface leads to inversion of the velocity $v_x \to -v_x$. In turn, one event of molecule reflection is accompanied by the momentum transfer $2mv_x$. Since the force acting on this area is the momentum change per unit time, the gas pressure as the force acting per unit area is equal.

$$p = \int\limits_{v_x>0} 2\,\mathrm{m}v_x f \mathrm{d}v_x = m\int v_x^2 f \mathrm{d}v_x = mN\left\langle v_x^2\right\rangle$$

where the average is made over velocities of molecules. In this operation, we account for the fact that the pressure on both sides of the imaginary plane is identical.

Since the pressure is introduced in the frame of reference where a gas as a whole is at rest, a general expression for the pressure is given by

$$p = mN\,\langle (v_x - \langle v_x \rangle)^2 \rangle = mN[\langle (v_x)^2 \rangle - (\langle v_x \rangle]^2 \tag{3.2.2}$$

In addition note that for an uniform motionless gas the pressure may be represented as

$$p = mN\left\langle (v_x)^2\right\rangle = mN\left\langle (v_y)^2\right\rangle = mN\left\langle (v_z)^2\right\rangle \tag{3.2.3}$$

One can compare the expression for the gas pressure with the definition of the gas temperature that has the form

$$\frac{3T}{2} = \frac{m}{2}\left\langle (\mathbf{v} - \mathbf{w})^2\right\rangle$$

This comparison gives

$$p = NT \tag{3.2.4}$$

This is the equation of state for a gas.

We now derive the momentum equation which follows from the motion equation for individual molecule

$$m\frac{d\mathbf{v}}{dt} = \mathbf{F}$$

One can extract from the consideration a volume V of a gas system, so that the force which acts on a gas inside this volume is

$$-\int p d\mathbf{S} = -\int_V \nabla p,$$

where $d\mathbf{S}$ is the element of a surface which surrounds this volume. Because this equation holds true for any volume, we obtain from this

$$mN\frac{d\mathbf{v}}{dt} = -\nabla p + N\mathbf{F},$$

where $\mathbf{F}$ is the force acted on an individual molecule. Next, transferring from the total derivative to the partial ones, one can transform this equation to the following form

$$mN(\frac{\partial \mathbf{v}}{\partial t} + (\mathbf{v}\nabla)\mathbf{v}) = -\nabla p + N\mathbf{F} \tag{3.2.5}$$

If an external field is absent, this equation is transferred into the Euler equation [17–19]

$$\frac{\partial \mathbf{v}}{\partial t} + (\mathbf{v}\nabla)\mathbf{v} = -\frac{1}{\rho}\nabla p + \frac{\mathbf{F}}{m}, \tag{3.2.6}$$

where $\rho = mN$ is the mass density of a gas.

We now represent this equation for atmospheric air near the Earth's surface taking into account that the gravitation force acts on a gas in the form $\mathbf{F}/m = \mathbf{g}$, where $\mathbf{g}$ is the free fall acceleration. We also take into account the viscosity of the gas. It is convenient to represent the shift viscosity in the tensor form. Basing on formula (3.1.21), we represent the pressure tensor p_{ik} connected with the gas viscosity, in the symmetric form as [17]

$$p_{ik} = \eta \left(\frac{\partial v_i}{\partial x_k} + \frac{\partial v_k}{\partial x_i} - \frac{2}{3}\delta_{ik}\frac{\partial v_i}{\partial x_k} \right), \tag{3.2.7}$$

Here δ_{ik} is the Kronecker symbol, i.e., $\delta_{ik} = 1$ if $i = k$, and $\delta_{ik} = 0$ if $i \neq k$. The summation in this expression is implied over twice repeating symbols. This form of the viscosity pressure tensor allows one to exclude the viscosity part due to gas compression and rarefication. Accounting for this equation, one can obtain equation of the momentum transfer in the form

$$\frac{\partial v_i}{\partial t} + v_k \frac{\partial v_i}{\partial x_k} = -\frac{1}{\rho}\frac{\partial p}{\partial x_i} + g_i + \frac{\eta}{\rho}\frac{\partial}{\partial x_k}\left(\frac{\partial v_i}{\partial x_k} + \frac{\partial v_k}{\partial x_i} - \frac{2}{3}\delta_{ik}\frac{\partial v_i}{\partial x_k}\right) \tag{3.2.8}$$

This is the Navier–Stokes equation [17, 20–22] for a gas. In a usual frame of reference, it takes the form of the Navier–Stokes equation

$$\frac{\partial \mathbf{v}}{\partial t} + (\mathbf{v}\nabla)\mathbf{v} = -\frac{1}{\rho}\nabla p + \mathbf{g} + \frac{\eta}{\rho}\Delta \mathbf{v} + \frac{\eta}{3\rho}\nabla(\mathrm{div}\mathbf{v}) \tag{3.2.9}$$

The third macroscopic equation for a gas relates to heat transport, where we use the continuity equation (3.2.1). In this equation, we use the energy density of molecules $\overline{\varepsilon}N$ instead of the number density of molecules N, where $\overline{\varepsilon}$ is the average energy of the molecule. This equation takes the form

$$\frac{\partial}{\partial t}(\overline{\varepsilon}N) + \mathrm{div}\mathbf{q} = 0$$

We assume the gas to be located in a fixed volume and take into account that the derivative $\partial\overline{\varepsilon}/\partial T = c_\mathrm{p}$ is the heat capacity per molecule. One can represent the heat flux $\mathbf{q}$ in this case as

$$\mathbf{q} = \mathbf{v}c_\mathrm{p}\mathrm{TN} - \kappa\nabla T,$$

where $\mathbf{v}$ is the average velocity of molecules. Substituting this expression into equation of heat transport and excluding the continuity equation from the obtained equation, one can obtain

$$\frac{\partial T}{\partial T} + \mathbf{v}\cdot\nabla T = \frac{\kappa}{c_\mathrm{p}N}\Delta T \tag{3.2.10}$$

For a motionless gas, this equation is an analog of that for diffusion motion (3.1.7).

3.2.2 *The Rayleigh Problem*

In order to understand the nature of convective transport of a gas, we consider below the Rayleigh problem [23, 24] as the simplest problem of this type, where a motionless gas is located between two plates of a different temperature, and an additional force acts on gaseous molecules. Being guided by the atmosphere, we take this force as the gravitation one, i.e., the force per molecule is mg, where m is the molecule mass, and $g = 980\,\mathrm{cm/s^2}$ is the free fall acceleration. This system may be unstable with respect to small perturbations which are able to cause a slow movement of the gas. Development of such instability leads to a convective motion of a gas. Our goal is to find the threshold for this process and to analyze its character being guided by the classical description of this problem (for example, [25–28]).

In this problem, a two-dimensional gas is located between two parallel plates of different temperatures. Let us denote the temperature of a lower plate as T_1 and the temperature of the upper plate as T_2 ($T_2 > T_1$), and the distance between plates to be L. The state of a gas between plates we characterize by the number density N of molecules, the gaseous temperature T which coincides with the plate temperatures near them, the pressure p which is connected with the number density of molecules N and the gas temperature T by the state equation (3.2.4), and the velocity $\mathbf{v}$ of gas motion that in the first approximation is zero. We consider the stationary gas state, and in the first approximation, we have a motionless gas with the parameters

$$N = N_o, \mathbf{v} = 0, T(z) = T_1 + \frac{(T_2 - T_1)}{L}x, \tag{3.2.11}$$

where the axis z is perpendicular to plates, and $z = 0$ at the lower plate.

Since we consider convection as an instability of a motionless gas, one can represent gaseous parameters as the sum of two terms, so that the first term refers to the gas at rest, while the second term corresponds to a small perturbation due to the convective gas motion. In this consideration, the number density of gas molecules is $N_o + N'$, the gas pressure is $p_o + p'$, the gas temperature is $T + T'$, and the gas velocity $\mathbf{v}$ is zero if convection is absent. We insert these parameters into the stationary equations for continuity (3.2.1), momentum transport (3.2.6) and heat transport (3.2.10) and use them as an expansion over a small parameter. Then we obtain in the zero-order approximation

$$\nabla p_o = -\mathbf{F}N, \ \ \Delta T = 0, \ \ \mathbf{v} = 0,$$

where $\mathbf{F}$ is the force acting on a single gas molecule. In the next, first-order approximation these equations give

$$\frac{\nabla(p_o + p')}{N_o + N'} + \frac{\eta\Delta\mathbf{v}}{N_o + N'} + \mathbf{F} = 0, \ \ v_z\frac{T_2 - T_1}{L} = \frac{\kappa}{c_p N_o}\Delta T' \tag{3.2.12}$$

The parameters of the problem under consideration are included in the last equation.

Let us transform the first term in the first equation of (3.2.12) with first-order approximation. This gives

$$\frac{\nabla(p_o + p')}{N_o + N'} \equiv \frac{\nabla p_o}{N_o} + \frac{\nabla p'}{N_o} - \frac{\nabla p_o}{N_o}\frac{N'}{N_o} = \mathbf{F}\left(1 - \frac{N'}{N_o}\right) + \frac{\nabla p'}{N_o}$$

On the basis of equation (3.2.4) of the gas state we have $N' == -N_o T'/T$. Substituting this in the second equation of set (3.2.12), one can represent this set of equations in the form

$$\mathrm{div}\mathbf{v} = 0,\ -\frac{\nabla p'}{N_o} - \mathbf{F}\frac{T'}{T} - \frac{\eta\Delta\mathbf{v}}{N_o} = 0,\ v_z = \frac{\kappa L}{c_p N_o(T_2 - T_1)}\Delta T' \tag{3.2.13}$$

Let us reduce the set of equations (3.2.13) to an equation of one variable. In fulfilling this operation, one can apply the *div* operator to the second equation of (3.2.12) and take into account the first equation of (3.2.12). This gives

$$\frac{\Delta p'}{N_o} - \frac{F\partial T'}{T\partial z} = 0 \tag{3.2.14}$$

Here we assume that $(T_2 - T_1) \ll T_1$. Therefore, the unperturbed gas parameters do not vary significantly within the gas volume. We can neglect their variation and assume the unperturbed gas parameters to be almost spatially constant.

We now take the quantity v_z from the third equation of (3.2.12) and insert it into the $z-$ component of the second equation (3.2.12). Applying the operator Δ to the result, we obtain

$$\frac{1}{N_o}\frac{\partial}{\partial z}\Delta p' - \frac{F\Delta T'}{T} + \frac{\eta\kappa L}{c_p N_o^2(T_2 - T_1)}\Delta^2 T' = 0$$

Using the relation (3.2.14)which establishes the connection between $\Delta p'$ and T', one can obtain finally

$$\Delta^3 T' = -\frac{Ra}{L^4}\left(\Delta - \frac{\partial^2}{\partial z^2}\right)T', \tag{3.2.15}$$

where the dimensionless combination of parameters

$$Ra = \frac{(T_2 - T_1)c_p\, F N_o^2 L^3}{\eta\kappa T} \tag{3.2.16}$$

is called the Rayleigh number [17, 29, 30].

The Rayleigh number is the fundamental parameter of the problem under consideration. One can rewrite the expression for this parameter in the form that conveys clearly its physical meaning. Let us introduce the kinetic viscosity $\nu = \eta/\rho = \eta/(N_o m)$, where $\rho = N_o m$ is the gas density, the thermal diffusivity coefficient is $\chi = \kappa/(N c_p)$ and the force per unit mass is $g = F/m$. In these variables, the Rayleigh number takes the form

$$\mathrm{Ra} = \frac{(T_2 - T_1)}{T}\frac{gL^3}{\nu\chi} \tag{3.2.17}$$

As a matter, the Rayleigh number is the ratio of the rate of heat transfer by convection to that from thermal conduction [30].

According to equation (3.2.15), the Rayleigh number characterizes the possibility of the development of convection. We now determine the threshold of this process with the boundary conditions at which a perturbed gas temperature and gas velocity is zero at the plates, i.e., $T' = 0$ and $v_z = 0$. In addition, the tangential forces $\eta(\partial v_x/\partial z)$ and $\eta(\partial v_y/\partial z)$ are zero at the plates. Differentiating the equation $\mathrm{div}\mathbf{v} = 0$ over z and using the conditions for the continuity of the tangential forces, we find that at the plates $\partial^2 v_z/\partial z^2 = 0$. Thus, we are based on the boundary conditions

$$T' = 0, \ \ w_z = 0, \ \ \frac{\partial^2 v_z}{\partial z^2} = 0$$

We now solve equation (3.2.15). Let us represent its solution with the boundary conditions at $z = 0$ in the form

$$T' = C \exp[i(k_x x + k_y y)] \sin k_z z \tag{3.2.18}$$

and use the boundary conditions at the second plate $z = L$ for determination of the parameters of this formula. The boundary condition $T' = 0$ at $z = L$ gives $k_z L = \pi n$, where n is an integer. Inserting the solution (3.2.18) into (3.2.15), one can obtain

$$\mathrm{Ra} = \frac{(k^2L^2 + \pi^2 n^2)^3}{k^2 L^2}, \tag{3.2.19}$$

where $k^2 = k_x^2 + k_y^2$. Note that the solution (3.2.18) satisfies to all indicated boundary conditions.

Basing on formula (3.2.19), one can determine the minimal value of the Rayleigh number $Ra_{\min}$ which corresponds to $n = 1$ and is given by

$$\mathrm{Ra}_{\min} = \frac{27}{4}(\pi^2 n^2)^2, \ \ k_{\min} = \frac{\pi n}{\sqrt{2}L} \tag{3.2.20}$$

For the lowest excitation $n = 1$ this expression gives $k_{\min} = \pi/(L\sqrt{2})$ and the minimal Rayleigh number at which a gas moves is equal

$$\mathrm{Ra}_{\min} = 27\pi^4/4 = 658 \tag{3.2.21}$$

From this it follows that convection as a form of gas transport is realized at large sizes of a gaseous system and large number density of molecules. Besides that, the threshold of convective transport depends on the geometry of a gaseous system.

The transition from a motionless gas to a convective gas motion is represented often as Rayleigh–Taylor instability [31, 32] which is realized for various systems in the different manner [17, 21, 33–37]. In atmospheric air, it is realized as a result of the atmospheric temperature gradient and gravitation field. The threshold of this instability is described by the Rayleigh number in accordance with formula (3.2.21) in this geometry of a region where a gas is located. For another geometry of this region, other critical values of the Rayleigh number are realized.

Note that above we are considering the gas movement within the framework of the simplest set of the Rayleigh problem if a gas is located between two parallel horizontal plates and the force from a field directs perpendicular to these plates. Because in the case of the atmosphere it is the gravitation field, we take the force acted on a single molecule to be $m\mathbf{g}$, where m is the molecule mass, and the free fall acceleration directs vertical (axis z). Assuming that the motion proceeds in the plane xz, we have the first equation (3.2.13) in the form

$$\mathrm{div}\mathbf{v} = \frac{\partial v_x}{\partial x} + \frac{\partial v_z}{\partial z} = 0$$

Inserting the solution (3.2.18) into this equation, one can obtain for components of the gas velocity

$$v_z = v_o \cos(kx) \sin\frac{\pi n z}{L}, \quad v_x = -v_o \frac{\pi n}{kL} \sin(kx) \cos\frac{\pi n z}{L}, \tag{3.2.22}$$

where n is an integer, and v_o is small compared with a thermal velocity of gas molecules.

Equations of motion of a gaseous element $\mathrm{d}x/\mathrm{d}t = v_x$ and $\mathrm{d}z/\mathrm{d}t = v_z$ on the basis of equations (3.2.21) allow one to find the trajectory of motion of a gas element. This trajectory takes the form

$$\frac{\mathrm{d}x}{\mathrm{d}z} = -\frac{\pi n}{kL} \frac{\tan(kx)}{\cos(\pi n z/L)}$$

From this one can determine the trajectory of a gas element

$$\sin(kx)\sin(\pi n z/L) = C, \tag{3.2.23}$$

where the constant C is determined by the initial position of a given gas element and its value ranges between -1 and $+1$. The specific role play lines with $C = 0$ which are determined by equations

$$z = Lp_1/n, \; x = Lp_2/n \tag{3.2.24}$$

Here p_1 and p_2 are nonnegative integers, and straight lines described by this equation separate the space between plates into some cells. Molecules of each cell can travel only within own cell and cannot leave it. Figure 3.1 shows an example of trajectories of gas elements in the Rayleigh problem for parameters $n = 1$ and $k = \pi/L$, that corresponds to the Rayleigh number $\mathrm{Ra} = 8\pi^4 = 779$. As is seen, at the boundaries of cells the tangential velocity component is zero. This confirms the state that molecules cannot leave an own cell. These cells are known as Benard cells [33, 37, 38].

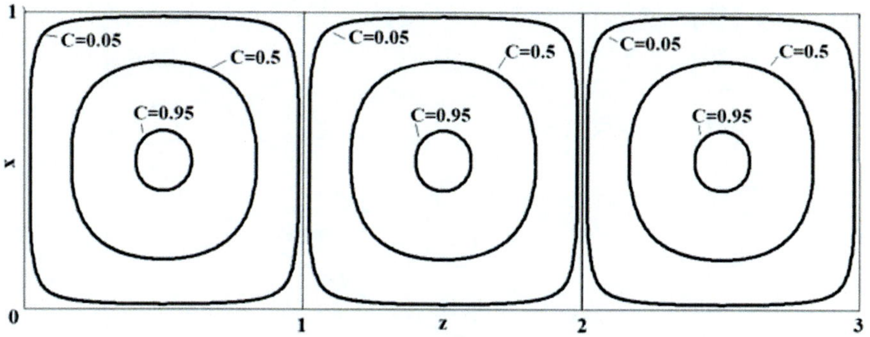

Fig. 3.1 Trajectory of movement of gas elements for Ra $= 779$ in accordance with formula (3.2.23)

Thus, the Rayleigh problem allows us to understand the nature of transition from a motionless gas to its convective motion. Restricting by a simple geometry of a space of gas location, this problem shows that the threshold of this transition and the character of gas movement not far from this threshold is governed by the Rayleigh number which must exceed a certain value for the convective motion of a gas. The Rayleigh number increases with increasing the number density of molecules, a size of a space which is occupied by a gas and also with an increasing of the temperature gradient. In addition, the convective motion as an instability of a motionless gas is caused by an external field, and the Rayleigh is proportional to the specific force g from this field acted on an individual molecule. Besides that, the Rayleigh number is inversely proportional to the thermal conductivity coefficient of a gas and its viscosity coefficient, that is the possibility of development of the convective instability of a motionless gas decreases with an increasing heat transfer and friction in the gas.

Note that the trajectory of a gas movement given in Fig. 3.1 is a closed vortex, and therefore, one can consider the convective gas motion as a system of vortices. Therefore, one can compose the gas movement not far from the convection threshold to be consisted of vortices located in a confined space. In particular, at low excitations, as we have in Fig. 3.2, the gas movement inside one Benard cell corresponds to one vortex. As the Rayleigh number increases, new modes of vortices for gas movement arise. This violates the order in gas movement. Below, we consider the character of this change.

Note that values of the Rayleigh number where only one mode of gas movement is excited, its elements move along closed trajectories, i.e., each vortex remains in a restricted region. This corresponds to an order convection. If the second mode is excited, motion of gas elements becomes random. According to formula (3.2.19), the threshold of excitation of two vortices $n = 2$ corresponds to the Rayleigh number $\mathrm{Ra} = 108\pi^4 \approx 10520$ according to formula (3.2.20). We note that this Rayleigh number is 16 times higher than that for the threshold of excitation of the convective movement, i.e., the range of Rayleigh numbers for the order convection is enough wide.

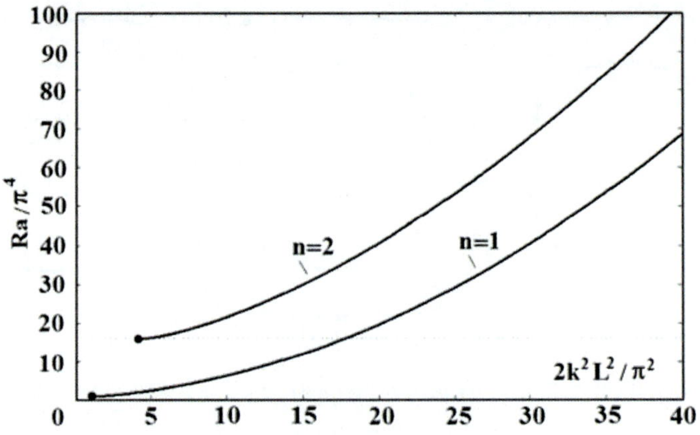

Fig. 3.2 Dependence of the reduced Rayleigh number on the reduced square wave number for two lowest modes. The threshold of each mode is drawn by closed circles

If one convection mode is excited and the Rayleigh number increases, a region of a gas movement is compressed in horizontal. One closed trajectory in vertical corresponds to the first mode, and two closed trajectories relate to the second convection mode in vertical. Above the threshold of excitation of the second convection mode $\mathrm{Ra} > 108\pi^4$, a gas movement includes two modes. Therefore, a test gas element can be shifted in horizontal in this case. As a result, above the threshold of excitation of the second convection mode a random convection takes place.

3.2.3 Dynamics of Atmospheric Air

Above on the basis of the simple Rayleigh model, we consider the nature of convective motion of a gas which allows one to understand the character of transition from a motionless gas to the order convection with formation of the Benard cells and also the transition to the random convection which corresponds to turbulent gas motion. Using this understanding, we below apply it to description of the behavior of real atmospheric air. In this analysis, we take into account the qualitative character of the Rayleigh problem for real gaseous systems. Moreover, we combine this model with the model of standard atmosphere that strengthens the qualitative character of the analysis.

Basing on the above model results, we represent the convective air motion as a combination of individual vortices. The character of motion of atmospheric air is described by the Rayleigh number (3.2.17) which may be represented as

$$\mathrm{Ra} = \frac{gL^4}{\Lambda_{\mathrm{T}}\nu\chi}, \quad \Lambda_{\mathrm{T}} = \frac{T}{dT/dh} \tag{3.2.25}$$

where for the model of standard atmosphere near the Earth surface ($T = 288\,\mathrm{K}$, $\mathrm{d}T/\mathrm{d}z = 6.5\,\mathrm{K/km}$) we have the following value $\Lambda_T = 44\,\mathrm{km}$ near the Earth's surface. Since the mass of an air molecule is $m_a = 4.8 \times 10^{-23} g$ and the number density near the Earth's surface is equal $N = 2.55 \times 10^{19}\mathrm{cm}^{-3}$, one can obtain near the ground for the kinematic viscosity $\nu = \eta/(Nm_a) = 0.15\,\mathrm{cm}^2/\mathrm{s}$ and for the air thermal diffusivity coefficient $\chi = \kappa/(Nc_\mathrm{p}) = 0.2\mathrm{cm}^2/\mathrm{s}$ ($c_\mathrm{p} = 7/2$). The threshold value (3.2.21) is attained at $L \approx 20$ cm. Correspondingly, random convection in atmospheric air relates to a vortex size $L > 40$cm.

Representing the convective motion of atmospheric air as a combination of vortices of various sizes, one can estimate parameters of partial vortices. A typical velocity v_l of the vortex of a size l follows from the Navier–Stokes equation (3.2.8)

$$\eta\frac{\partial^2 v}{\partial h^2} \sim \eta\frac{v_l}{l^2} \sim \mathrm{PN}', \qquad (3.2.26)$$

where η is the air viscosity, $P = mg$ is the molecule weight (m is the molecule mass, g is the free fall acceleration), $N' \sim Nl/\Lambda_\mathrm{T}$ is the difference of the molecule number density at a distance l, N is the number density of air molecules ($N' \ll N$). From this we have for the vortex velocity v_l of a size l

$$v_l \sim \frac{mgNl^3}{\Lambda_\mathrm{T}\eta} = \frac{gl^3}{\Lambda_\mathrm{T} \cdot \nu}, \qquad (3.2.27)$$

From this it follows that because in this consideration the vortex velocity is small compared to the sound speed or a thermal velocity of molecules ($v_l \ll v_\mathrm{T}$), a vortex size l is restricted. In the case of atmospheric air near the Earth's surface this criterion gives $l \ll 3$ m.

From this one can find the contribution to the diffusion coefficient of atmospheric air from vortices of a size l on the basis of formula (3.1.5)

$$D_l \approx 0.3 v_l l = 0.3\nu\mathrm{Re} = 0.3\chi\mathrm{Ra} \qquad (3.2.28)$$

We assume here the character of the diffusion motion to be similar to that in a motionless gas.

We also estimate the value of the Reynolds number [39] which is introduced for a given vortex size l as

$$\mathrm{Re}_l = \frac{v_l l}{\nu} = \mathrm{Ra}_l\frac{\chi}{\nu}, \qquad (3.2.29)$$

and because the kinematic viscosity ν is close to the thermal diffusivity coefficient χ, the values of the Rayleigh number Ra_l and Reynolds numbers Re_l are nearby for gases [36] including atmospheric air

$$\mathrm{Re}_l \approx \mathrm{Ra}_l \qquad (3.2.30)$$

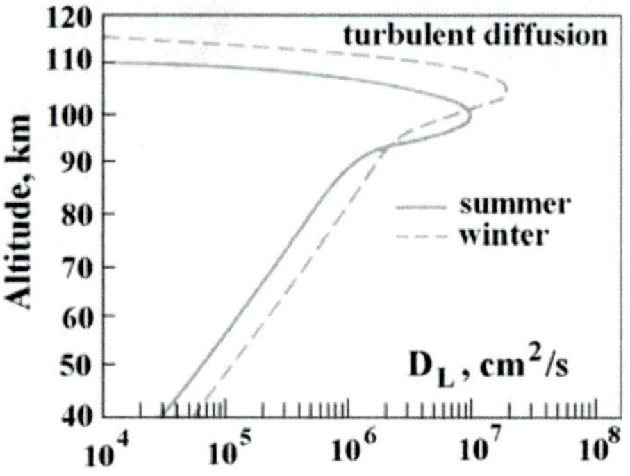

Fig. 3.3 Altitude dependence for the coefficient of turbulent diffusion in atmospheric air [40, 41]

At large Reynolds numbers the turbulent character of gas motion is realized, starting from the Reynolds numbers $\mathrm{Re}_{cr} = v_L L/\nu = 2 \times 10^5$ [17]. From this one can obtain for typical sizes of vortices which give the main contribution to propagation of atmospheric air $L \sim 1$ m. Correspondingly, a typical diffusion coefficient for air propagation through these vortices is estimated as

$$D_{\mathrm{L}} \sim 10^4 \mathrm{cm}^2/\mathrm{s} \quad (3.2.31)$$

A typical velocity in these vortices is $v_{\mathrm{L}} \sim 4\,\mathrm{m/s}$.

Thus, transport of atmospheric air for large distances has the random character [42–44] that results from collisions of air fluxes resulted from mixing of air layers [45, 46]. Because of the random character of this transport process through vertices of restricted sizes, the propagation at distances above a size of a maximal vortex has the diffusion character and is described through the diffusion coefficient D_{L} for convective motion. This diffusion coefficient is estimated according to formula (3.2.31) near the Earth's surface and varies in the atmosphere depending on its altitude. Figure 3.3 gives the convective diffusion coefficient for the standard atmosphere, where observational data for air transport are treated on the basis of the diffusion character of its motion. As it follows from this figure, the diffusion coefficient of tropospheric air from the indicated experiment corresponds to an estimation (3.2.31).

We demonstrate above within the framework of the Rayleigh problem that convection is a more effective mechanism of heat transport than that due to thermal conduction. If the gas flux propagates near a solid surface, a boundary layer of thickness δ is formed near this surface, where the gas velocity varies from zero at the surface up to the flux velocity. The thickness of the boundary layer is determined by the gas viscosity, and the heat transport in the boundary layer is accomplished by thermal conduction, so that the heat flux can be estimated as $q = -\kappa \nabla T \sim \kappa (T_2 - T_1)/\delta$. Applying the Navier–Stokes equation (3.2.9) to the boundary region, one can estimate its thickness.

This equation describes a continuous transition for a gas flow in the vicinity of the solid surface to the region far from the boundary. The second term of the Navier–Stokes equation (3.2.9), $\mathbf{v} \cdot \nabla \mathbf{v}$, must be taken into account in spite of a velocity smallness. Comparing by an order of magnitude some terms of the z-component of the Navier–Stokes equation (3.2.9), one can estimate the boundary layer thickness δ. Indeed, we have $v_z^2\delta \sim F(T_2 - T_1)/T \sim \eta v_z/(N\delta^2)$, and accounting for $T_2 - T_1) \ll T_1,\ T_2$, one can obtain for the boundary layer thickness

$$\delta \sim \left[\frac{\eta^2 T}{N^2 F m^3 (T_2 - T_2)}\right]^{1/3} \tag{3.2.32}$$

In the context of the Rayleigh problem, one can compare the heat flux transported by a gas due to convection (q) and due to thermal conduction ($q_{\rm cond}$). The thermal heat flux is $q_{\rm cond} = \kappa(T_2 - T_2)/L$, and the ratio of the fluxes is equal

$$\frac{q}{q_{\rm cond}} \sim \frac{L}{\delta} \sim \left[\frac{N^2 F m L^3 (T_2 - T_1)}{\eta^2 T}\right]^{1/3} \sim {\rm Gr}^{1/3} \tag{3.2.33}$$

This relation is the definition of the Grashof number [47] as the following dimensionless combination of parameters

$$G = \frac{N^2 F m L^3 (T_2 - T_1)}{\eta^2 T} = \frac{\Delta T}{T}\frac{g L^3}{\nu^2} \tag{3.2.34}$$

A comparison of the definitions of the Rayleigh number (3.2.17) and the Grashof number (3.2.34) gives for their ratio

$$\frac{R}{G} = \frac{c_V \eta}{m\kappa} = \frac{\nu}{\chi}.$$

In the case of gases, the kinematic viscosity ν is close to the thermal diffusivity coefficient χ. Hence, the Rayleigh number Ra has the same order of magnitude as the Grashof number G for a gas. Since convection develops at high Rayleigh numbers, we find that convection proceeds at large Grashof numbers $G \gg 1$. Correspondingly, heat transport via convection is considerably more effective than heat transport in a motionless gas via thermal conduction.

3.2.4 Water Circulation in the Atmosphere

Thus, we consider the motion of atmospheric air as a sum of vortices of various sizes that leads finally to a random motion. We obtain that if a vortex velocity is less than the sound speed, a vortex size l is restricted $l < 3\,{\rm m}$. In atmospheric air this is fulfilled, and propagation on large distances compared with a vortex size has the diffusion character with diffusion coefficient $D_{\rm L}$. Its value is estimated by formula (3.2.31). Air vortices move together with admixture molecules. Hence, motion of water molecules in the atmosphere at large distances proceeds also due to the diffusion law with the diffusion coefficient D_L of air molecules. Correspondingly, the flux of water molecules to the Earth's surface is equal

$$j_w = -wN(H_2O) = \frac{D_L}{\Lambda} N(H_2O), \tag{3.2.35}$$

where w is the drift velocity of air molecules toward the Earth's surface and the number density of air molecules is given by formula (2.1.1), $N(H_2O)$ is the number density of water molecules near the Earth's surface.

In the simplest form [48], water circulation through the atmosphere includes several stages such, that the first stage is evaporation of water from the Earth's surface in the form of water molecules, and then the basic water part is converted into rain drops or snow particles which fall on the Earth's surface. Falling of water molecules to the Earth's surface gives a small contribution to the falling water flux. Let us determine first the flux j_{ev} of evaporated water molecules. Assume first that formation of the condensed water phase is absent in the atmosphere, so that the equilibrium results in equality evaporated and incident fluxes, and the flux of evaporated water molecules for the altitude distribution (2.1.19) is equal

$$j_{ev} = -D_L \frac{dN}{dh} = \frac{D_L}{\lambda} N(H_2O) \tag{3.2.36}$$

Since $\Lambda = 10.4$km and $\lambda = 2$km, the ratio of fluxes toward the Earth in the form of water drops or particles j_c and in the form of water molecules j_w is equal

$$\frac{j_d}{j_w} = \frac{\Lambda - \lambda}{\lambda} \approx 4 \tag{3.2.37}$$

This confirms in principle the scheme circulation of water through the atmosphere [48, 49] according to which water evaporates from the surface in the form of free water molecules, whereas returns back in the form of the condensed phase, i.e., as rain or snow.

One can compare expression (3.2.36) for the average flux of evaporate water with its value (2.1.8) which follows from measured parameters. This allows one to obtain a more precise value of the effective diffusion coefficient D_L due to air convection that is equal

$$D_L = (5 \pm 1) \times 10^4 \text{ cm}^2/\text{s} \tag{3.2.38}$$

The value (3.2.38) of the diffusion coefficient D_L for air convective motion allows one to determine parameters of air vortices which give the main contribution to D_L. According to formulas (3.2.27) and (3.2.28), one can estimate the following vortex parameters which give the main contribution to the diffusion transport in the real atmosphere : $\text{Re}_l = 1 \times 10^6$, $Ra = 8 \times 10^5$, $l = 100$cm, $v_l = 1.5 \times 10^3$ cm/s. As is seen, the Reynolds number exceeds the value which corresponds to the threshold of turbulent motion of a gas $\text{Re}_{cr} = 2 \times 10^5$ [17], i.e., random motion of atmospheric air has the turbulent character and includes several convective modes.

On the basis of this value of the diffusion coefficient D_L for propagation of atmospheric air molecules through the convection motion, one can obtain for an average time of the contact of air and admixture molecules with the Earth's surface that is given by formula

$$\tau_c = \frac{N\Lambda}{D_L|dN/dh|} = \frac{\Lambda^2}{D_L} \approx 0.5\text{year}, \qquad (3.2.39)$$

where $N\Lambda$ is the total number of air molecules per air column of unit area. Let us introduce the residence time of molecules of a given type τ_r in atmospheric air. Since these molecules are captured by vortices of air convective motion, one can determine the probability w that molecules are absorbed by the Earth's surface as a result of a chemical process

$$w_a = \frac{\tau_c}{\tau_r} \qquad (3.2.40)$$

In the case of CO_2 molecules, the residence time in the atmosphere is estimated as 4 year [50, 51]. This gives for the probability of absorption of a molecule as a result of its attachment to the Earth's surface the value $w_a(CO_2) = 0.12$. Absorption of CO_2 molecules results in dissolution of these molecules in oceans, as well as due to photosynthesis process. In the same manner, the probability of absorption of a CH_4 molecule by the Earth surface is $w_a(CH_4) = 0.06$ because the lifetime of methane molecules in the atmosphere is 12 year [52].

References

1. S. Chapman, T.G. Cowling, *The Mathematical Theory of Non-Uniform Gases* (Cambridge University Press, Cambridge, 1952)
2. J.H. Ferziger, H.G. Kaper, *Mathematical Theory of Transport Processes in Gases* (North Holland, Amsterdam, 1972)
3. M. Capitelli, D. Bruno, A. Laricchiuta, in *Fundamental Aspects of Plasma Chemical Physics. Transport* (New York, Springer, 2013)
4. N.B. Vargaftic, *Tables of Thermophysical Properties of Liquids and Gases* (Halsted Press, New York, 1975)
5. B.M. Smirnov, *Reference Data on Atomic Physics and Atomic Processes* (Springer, Heidelberg, 2008)
6. B.M. Smirnov, *Physics of Ionized Gases* (Wiley, New York, 2001)
7. M.V. Smolukhowski. Zs. Phys. **17**, 585 (1916)
8. A. Einstein, Ann. Phys. **17**, 549 (1905)
9. A. Einstein, Ann. Phys. **19**, 371 (1906)
10. A. Einstein. Zs.für Electrochem. **14**, 235 (1908)
11. B.M. Smirnov, *Clusters and Small Particles in Gases and Plasmas* (Springer NY, New York, 1999)

12. E.R.G. Eckert, R.M. Drake, *Analysis of Heat and Mass Transfer* (McGraw Hill, New York, 1972)
13. J.R. Welty, C.E. Wicks, R.E. Wilson, in *Fundamentals of Momentum, Heat, and Mass Transfer.* (New York, Wiley 1984)
14. R.B. Bird, W.E. Stewart, E.N. Lightfoot, *Transport Phenomena* (Wiley, New York, 2002)
15. Y.A. Cengel, *Heat and Mass Transfer: A Practical Approach* (McGraw Hill, Boston, 2003)
16. F. Incropera, *Fundamentals of Heat and Mass Transfer* (Wiley, Hoboken, 2007)
17. L.D. Landau, E.M. Lifshits, *Fluid Mechanics* (Pergamon Press, London, 1959)
18. A.J. Chorin, J.E. Marsden, *A Mathematical Introduction to Fluid Mechanics* (Springer, New York, 1993)
19. J.A. Fay, *Introduction to Fluid Mechanics* (MIT Press, Cambridge, 1994)
20. https://en.wikipedia.org/wiki/Navier-Stokes-equations
21. G.K. Batchelor, *An Introduction to Fluid Dynamics* (Cambridge University Press, New York, 2000)
22. F.M. White, *Viscous Fluid Flow* (McGraw Hill, New York, 2006)
23. G.G. Stokes. Trans. Cambridge Phil. Soc. **9**, Part II, 8 (1851)
24. Lord Rayleigh, Phil. Mag. **21**, 697 (1911)
25. G.A. Bird, *Molecular Gas Dynamics* (Clarendon Press, Oxford, 1976)
26. I.G. Currie, *Fundamental Mechanics of Fluids* (Dekker, New York, 2003)
27. J.H. Spurk, N. Aksel, *Fluid Mechanics* (Springer Verlag, Berlin, 2008)
28. P.K. Kundu, I.M. Cohen, *Fluid Mechanics* (Academic Press, New York, 2008)
29. https://en.wikipedia.org/wiki/Rayleigh-number
30. https://en.wikipedia.org/wiki/Heat-transfer
31. Proc. Lond. Math. Soc. **14**, 170 (1883)
32. S.I. Taylor, Proc. Roy. Soc. **201A**, 192 (1950)
33. S. Chandrasekhar, *Hydrodynamic and Hydromagnetic Stability* (Oxford University Press, Oxford, 1961)
34. P. Drazin, W. Reid, *Hydrodynamic Stability* (Cambridge Universiy Press, Cambridge, 1981)
35. D.J. Acheson, *Elementary Fluid Dynamics* (Oxford University Press, Oxford, 1990)
36. V.P. Krainov, *Qualitative Methods in Physical Kinetics and Hydrodynamics* (American Institute of Physics, New York, 1992)
37. P.G. Drazin, W.H. Reid, *Hydrodynamic Stability* (Cambridge University Press, Cambridge, 2004)
38. E.L. Koschmieder, *Benard Cells and Taylor Vortices* (Cambridge University Press, Cambridge, 1993)
39. https://en.wikipedia.org/wiki/Reynolds-number
40. T. Shimazaki, A.R. Laird, Radio Sci. **7**, 23 (1972)
41. M.J. McEwan, L.N. Phillips, *Chemistry of the Atmosphere* (Edward Arnold, London, 1975)
42. E. Lorenz, *The Nature and Theory of General Circulation of the Atmosphere* (World Meteorological Organization, Geneva, 1967)
43. D.G. Andrews, J.R. Holton, C.B. Leovy, in *Middle Atmospheric Dynamics* (Academy Press, 1987)
44. D.G. Andrews, *An Introduction to Atmospheric Physics* (Cambridge University Press, Cambridge, 1999)
45. G.T. Csanady, *Turbuletnt Diffusion in Environment* (Holland, Reidel, Dordrecht, 1973)
46. D.J. Tritton, *Physical Fluid Dynamics* (Claredon Press, Oxford, 1988)
47. https://en.wikipedia.org/wiki/Grashof-number
48. R. Braham, J. Meteorol. **9**, 227 (1952)
49. https://en.wikipedia.org/wiki/Atmospheric-circulation
50. https://en.wikipedia.org/wiki/Atmospheric-carbon-cycle
51. H. Schmidt, Glob. Planet Change. **152**, 19 (2017)
52. F. Keppler, J.T.G. Hamilton, M. Brass, T. Rackmann, Nature **439**, 187 (2006)

Chapter 4
Condensation Processes in Atmosphere

Abstract Water condensation in the atmosphere is possible in a supersaturated vapor that results from mixing atmospheric parcels under the action of vertical winds. The first parcel of the mixture contains wet air from near-surface altitudes and the second one consists of cold air from high altitudes. The threshold of this process and optimal conditions are analyzed. Atmospheric winds are of importance both for formation of a supersaturated water vapor in the atmosphere and stave off them from washing out. As a result, clouds formed from a supersaturated water vapor occupy a restricted part of the atmosphere and the stability of clouds is determined by thermal effects. Transformation of a water excess of a supersaturated air into condensed phase in the form of micron-size droplets proceeds due to condensation nuclei which are present in the atmosphere in the form of ions, radicals and active molecules. Subsequent growth of microdroplets up to rain drops takes place through the Ostwald ripening process for small droplet sizes and the gravitation mechanism of growth for large droplet sizes. In order to satisfy the observed fact of a large lifetime of clouds, it is required that water microdroplets in dense (cumulus) clouds must be charged of the same charge sign, i.e., cumulus clouds have an electric nature.

4.1 Supersaturation Atmospheric Air

4.1.1 Formation of Supersaturated Air in Mixing of Atmospheric Layers

Clouds which play the important role in the human life consist of water microdroplets, i.e., of droplets of micron sizes. Since water evaporates from the Earth's surface in the form of molecules, water condensation proceeds in the atmosphere. But condensation is possible in points with a supersaturated water vapor, whereas the model of standard atmosphere deals with average atmospheric parameters, leads to a nonsaturated water vapor at any altitudes. In other words, condensation of a water vapor is impossible within the framework of the model of standard atmosphere. This means that the partial pressure of atmospheric water at any altitude is lower than the water saturated vapor

B. M. Smirnov, *Global Energetics of the Atmosphere*, Springer Atmospheric Sciences, https://doi.org/10.1007/978-3-030-90008-3_4

pressure. Moreover, an altitude increase leads to a decreasing average atmospheric moisture. This fact follows both from data of Fig. 2.2 and from formula (2.1.20) for the altitude dependence of the average moisture of atmospheric air. Hence, condensation processes proceed outside the model of standard atmosphere.

Evidently, a supersaturated vapor may be formed in the atmosphere if a small amount of air from low atmospheric layers is transported to a higher layer and mixes with it. Then due to a decreasing temperature, the air moisture increases and air may become supersaturated. Therefore, we below consider adiabatic mixing of two volumes of atmospheric air taken from different layers with parameters corresponding to the model of the standard atmosphere. Our goal is to determine the conditions at this operation under which the dew point can be reached, i.e., the air humidity reaches 100%. Along with this, we consider optimal conditions for this air mixing.

A vertical displacement of a test atmospheric region, which is called as a parcel, may be caused by external forces, and also by motion of parcels due to atmospheric nonuniformities. In particular, we find above that the lapse rate in a dry atmosphere (2.1.5) differs from that in atmospheric saturated air (Table 2.2). If these regions are located close, mixing of nearby parcels takes place. This mixing of two parcels of a different temperature and moisture leads to formation of a joined air region. Our goal is to determine these parameters for a unit parcel.

In this analysis, we use the adiabatic character of mixing, i.e., the thermal energy of a unit parcel is determined by those of mixed parcels. In addition, the water amount in the unit parcel is a sum of those of mixed parcels. Under these conditions we have for the temperature T of a united parcel with accounting for the adiabatic character of the mixing process

$$T = \frac{n_1 T_1 + n_2 T_2}{n_1 + n_2} = T_1 - (T_1 - T_2)x, \quad x = \frac{n_2}{n_1 + n_2}, \tag{4.1.1}$$

where n_1, n_2 are the numbers of air molecules of each parcel, T_1, T_2 are the initial temperatures of these parcels, and x is the concentration of air molecules resulted from the second parcel. At the same time, we assume the air heat capacity to be independent of the temperature, and take into account the adiabatic character of parcel mixing.

One can represent the mixture temperature T in the form

$$T = \frac{n_1 T_1 + n_2 T_2}{n_1 + n_2} = T_1 + hx\frac{dT}{dh}, \tag{4.1.2}$$

where h is the difference of altitudes for mixing parcels. Taking the first parcel to be located near the Earth's surface, we use the parameters $T_1 = 288$ K and $dT/dh = 6.5$ K/km in formula (4.1.2) for the model of standard atmosphere.

Because the first parcel is located near the ground, h is the altitude of the second parcel, and the number density of water molecules for this altitude is approximated by formula (2.1.19). Then the number density of water molecules in a united parcel is equal

$$N_w = \eta N_{\text{sat}}(T_E)(1-x) + xN_o \exp(-h/\lambda), \tag{4.1.3}$$

where η is the air moisture near the Earth's surface. If the number density of formula (4.1.3) exceeds $N_{\text{sat}}(T)$ that is the number density at the saturated vapor pressure for an indicated temperature, the water vapor becomes a supersaturated one. As a result, a part of water molecules of a united parcel may be transferred into the condensed phase.

Let us introduce the supersaturation degree $S(h,x)$ in the case where the number density of water molecules exceeds that at the saturated vapor pressure $N_w > N_{\text{sat}}(T)$. This quantity is the characteristic of a supersaturated vapor and is given by [1, 2]

$$S(h,x) = \frac{N_w}{N_{\text{sat}}(T)} - 1 = \frac{\eta N_{\text{sat}}(T_E) - N_{\text{sat}}(T) + x[N_o \exp(-h/\lambda) - \eta N_{\text{sat}}(T_E)]}{N_{\text{sat}}(T_E - h(1-x)\frac{\mathrm{d}T}{\mathrm{d}h})} \tag{4.1.4}$$

Formulas (4.1.2) and (4.1.4) allow one to determine the supersaturation degree $S(h,x)$ depending of the initial altitude h and the concentration of molecules from different parcels. The threshold of the condensation process is described by equation

$$S(h,x) = 0$$

Figure 4.1 shows the position of this threshold that follows from this equation and expression (4.1.4) for the degree of supersaturation. As is seen, the minimal altitude of condensation grows both with a decreasing moisture and with an increasing concentration of air molecule from the lower parcel in a united parcel. In addition, a nonsupersaturated vapor corresponds to a range of parameters below a given curve, while above this curve a united parcel is supersaturated one.

In addition, the maximal degree of supersaturation $S(h,x)$ is evaluated on the basis of formula (4.1.4) and the results are represented in Fig. 4.2. This value grows with an increasing altitude of the second parcel at a given moisture of air in the first

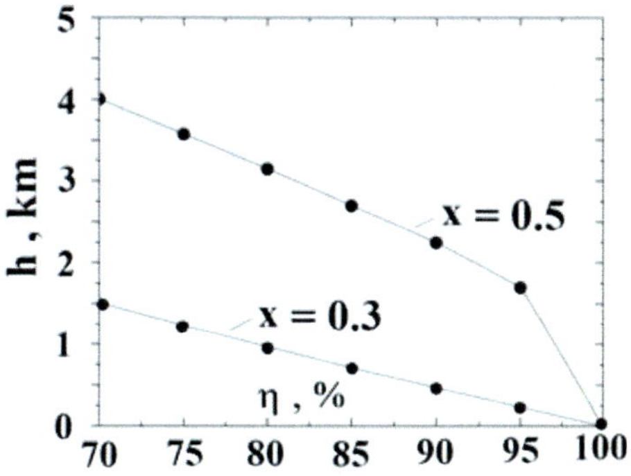

Fig. 4.1 Threshold altitude at which the dew point is attained in adiabatic mixing of two parcels for an indicated concentration of air molecules taking from the near-surface parcel. This altitude is evaluated on the basis of formula (4.1.4) within the framework of the model of standard atmosphere [1, 2]

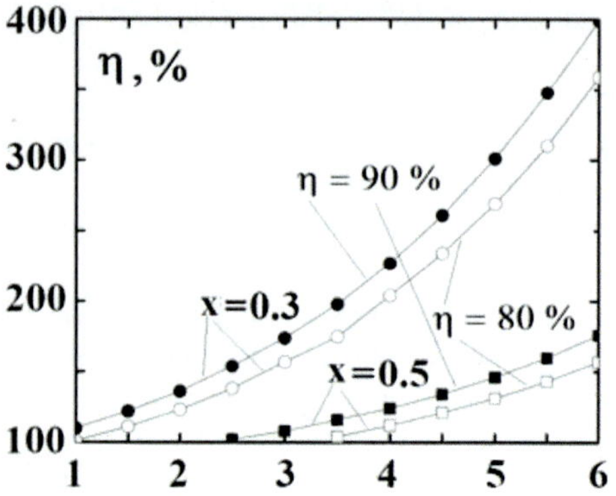

Fig. 4.2 Maximums of the supersaturation degree for adiabatic mixing of two identical parcels that follows from expression (4.1.4), so that the first parcel is taken from the near-surface layer, and the altitude of the second parcel is given. Parameters of atmospheric air correspond to the model of standard atmosphere

parcel, as well as at a given concentration of molecules in the first parcel. From the above analysis it follows that clouds are formed starting from some altitudes above the Earth in accordance with observations. Of course, a fog may be formed near the Earth's surface if the temperature of near-surface layers is low. Such conditions may be realized, for example, at autumn tomorrow, but formed condensed water in this case does not partake in circulation of atmospheric water, as well as in electric processes of the atmosphere.

Therefore, we restrict now by condensation processes in a cumulus cloud which proceed as a result of mixing of air parcels under the action of vertical winds. In this case, condensation processes proceed starting from some altitudes and include restricted portions of the atmosphere. Therefore, the amount of condensed water in the atmosphere is less compared to that consisting of free molecules. Indeed, let us take a cumulus cloud with average parameters (2.2.12) where the number density of bound water molecules in microdroplets is $nN_d = 7 \times 10^{16}$ cm^{-3}. If this cloud is located at the altitude of 3 km, where the temperature is 268 K for standard atmosphere, we have for the number density of free water molecules $N_w = 1.1 \times 10^{17}$ cm^{-3} according to Table 2.2. As is seen, in this the number density of free water molecules exceeds that of bound ones. In addition, regions with such condensed water include a small atmosphere part. All this leads to a small amount of condensed atmospheric water compared to the amount of water in the form of free molecules.

One can expect, that the system consisting of air with water microdroplets and free water molecules at the saturated vapor pressure is unstable in the atmosphere. Indeed, let us take an element of a cumulus cloud with average parameters (2.2.12) of water microdroplets which are typical for a cumulus cloud. If we displace it down, water microdroplets are evaporated in lower layers of standard atmosphere due to the equilibrium between free and bound water molecules. According to data of Table 2.2, at the temperature 274 K condensed water disappears under equilibrium conditions. Evaporation of condensed water in this case requires an energy which is able to heat air contained in this water by 6 K. Because the process of microdroplet evaporation requires an additional energy, this increases the stability of the mixture of free and bound water in atmospheric air.

Convective motion of air is accompanied by the drift of molecules and particles of air under the action of gravitation forces and temperature gradients. We now analyze this drift of atmospheric molecules and particles. In the above consideration, the convection motion is created by air molecules as a mixture of nitrogen and oxygen molecules with the average mass $m = 29$ of molecules in units of atomic mass. Other components and admixtures do not contribute to the convection motion of air, and the mean free path of molecules is small compared with a vortex size. Therefore, components of air are captured by vortices in the course of a convective motion. We below consider the air motion as a result of convection as in diffusion motion. Hence, this corresponds to large distances of propagation compared to a size of larger vortices that is $\sim$1 m. In the course of the convection motion, molecules and small particles are captured by air vortices. This creates a force acted on an individual molecule which exceeds remarkably its weight.

4.1.2 Peculiarities of Clouds

Clouds are ensembles of water microparticles, mostly microdroplets. Therefore, analyzing properties of water microdroplets and processes with their participation, we are guided by clouds. Transition from microdroplets to clouds as systems of microdroplets, some questions arise with respect to the nature of clouds. First of this is why water microdroplets are grouped in clouds, rather they are distributed uniformly over a space. Indeed, forces which can compel the microdroplets to be concentrated in restricted space regions are absent. Therefore, the only reason of joining of free water microdroplets in groups results from the fact that atmospheric regions contained a supersaturated water vapor are restricted. This follows from the character of mixing of atmospheric parcels that was analyzed above.

Note that various properties of clouds and processes in them are analyzed in a lot of books, in particular, in [3–21]. Basing on data from these books, we represent shortly main features of clouds as a system of free water microdroplets as the most spread form of water particles of clouds. There are various types of clouds with specific processes of their formation, existence and degradation. We now focus on cumulus clouds which contains the main part of condensed atmospheric water and are responsible for atmospheric electricity as a global phenomena. Moreover, because these clouds consist of microdroplets, we below analyze processes in atmospheric air involving water microdroplets. Hence, representing a cloud as an ensemble of water microdroplets located in air, one can explain some cloud properties and estimate their parameters.

In analyzing the character of mixing of atmospheric parcels in the course of origin of a supersaturated vapor, we convince in the role of winds in this process. But winds are important also in conservation of cloud boundaries. Let us compare a cloudy atmosphere with a suspension, i.e., a liquid contained nonsaturated microparticles which example is milk. In suspensions, microparticles are distributed over a liquid more or less uniformly, and a special separation is required in order to extract

Fig. 4.3 Clouds in the painting "The rape of Europa" by Swiss artist F. Vallotton (1908) [22, 23]

a substance of particles from the liquid. Another situation is realized in a cloudy atmosphere which is a nonuniform system and this is created under the action of winds. As an example, we give in Fig. 4.3 [22, 23] the painting of the Swiss artist F. Vallotton. Leaving aside the main topic of this painting, we note that clouds have sharply boundaries here.

Above we convince that the formation of clouds is impossible in a motionless air and mixing of atmospheric layers—parcels which are located at the beginning at different altitudes can lead to the origin of supersaturated air that subsequently is converted in air contained water microdroplets. One can expect that formed clouds are washing out, and this degradation proceeds as a result of a random motion of individual microdroplets. As it follows from formula (3.1.14), the product of the diffusion coefficient for a microparticle in atmospheric air D_L and the radius of a microdroplet r is equal $D = 7.5 \times 10^{-8}\ \mathrm{cm}^3/\mathrm{s}$. Correspondingly, a displacement Δ of a microdroplet of a radius $r = 1\,\mu\mathrm{m}$ during a day as a result of diffusion is equal according to formula (3.1.8) $\Delta \approx 0.2$ cm. Thus, washing out of clouds is impossible due to a random motion of microdroplets and degradation of clouds proceeds under the action of winds.

We now consider penetration of an external air stream inside a cloud under conditions given in Fig. 4.4. Figure 4.4 represents the scheme of this process where an air flux moves toward a cloud. This flux transfers the momentum to motionless air prompt, but water microdroplets acquire a momentum with a delay. The relaxation time τ_{rel} for microdroplets is given by formula (2.2.15), and for microdroplet parameters (2.2.12) corresponded to cumulus clouds this time equals 8×10^{-4} s. Taking a typical velocity as $l \sim 10$ m/s, one can obtain a way of a microdroplet with respect to the flux $l \sim 1$ cm. As a result, this value is a thickness of the boundary of a cumulus cloud.

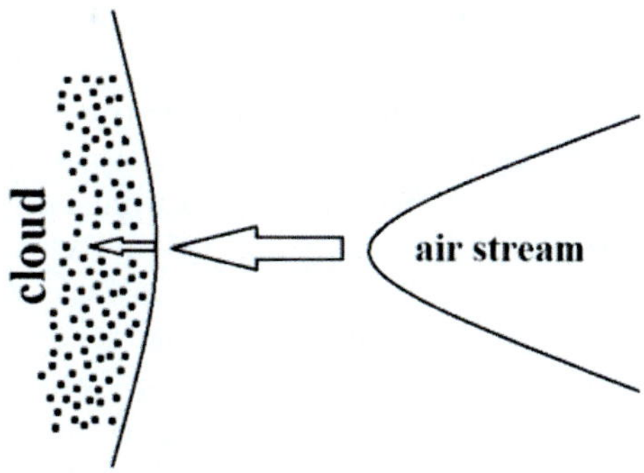

Fig. 4.4 Character of interaction of the air stream with a cumulus cloud

This estimation allows one to estimate the character of interaction of air flows with clouds. Note that water microdroplets of clouds do not influence on air flows. Indeed, the mass density of condensed water in a cumulus clouds, as it follows from formula (2.2.12), is three orders of magnitude below the mass density of air inside which microdroplets are located. Therefore, water microdroplets are not obstacles in air fluxes, whereas microdroplets react on air motion with a delay. These air displacements are joined in air convection, and, if clouds are large compared to vortex sizes, their interaction with air vortices leads to fastening of cloud boundaries. In this case, the force atmospheric flows on clouds is directed inside them. For this reason, the character of the distribution for water microdroplets in the atmosphere differs from that for solid particles in suspensions.

Along with the above mechanism, the cloud stability is determined by thermal processes. Indeed, evaporation of microdroplets requires of energy which is compensated by air cooling. Let us determine the temperature decrease ΔT which follows from total evaporation of water microdroplets for a typical cumulus cloud which parameters are given by formula (2.2.12). This value follows from equation of the heat balance for the evaporation process

$$c_p N_a \Delta T = \varepsilon_o N_b, \tag{4.1.5}$$

where $c_p = 7/2$ is the heat capacity per air molecule, N_a is the number density of air molecules, $N_b = 7.2 \times 10^{16}\,\mathrm{cm}^{-3}$ is the number density of bound water molecules in microdroplets of a typical cumulus cloud, $\varepsilon_o = 0.44$ eV is the binding energy of a water molecule in a liquid droplet according to formula (2.1.13). Let us apply this formula to the standard atmosphere at an altitude $h = 3$ km, where the air temperature equals $T = 273$ K, the air pressure is $p = 0.7$ atm, and the number density of water molecules is equal $N_w = N_{\mathrm{sat}} = 1.6 \times 10^{17}\,\mathrm{cm}^{-3}$. From this, we have for the temperature change ΔT as a result of microdroplet evaporation of a typical cumulus cloud

$$\Delta T = \frac{\varepsilon_o N_b}{c_p N_a} \approx 6\,\mathrm{K} \tag{4.1.6}$$

Therefore, for degradation of a cumulus cloud it is necessary to heat it with this temperature.

The parameter which characterizes the degree of covering of the sky with clouds is the cloudiness [24] that is the sky part which is covered by clouds. The cloudiness may be described by the optical thickness of a cloudy atmosphere in the visible spectrum range. The lower limit of a cloudy atmosphere corresponds to its optical thickness of 0.1, the average optical thickness of a cloudy atmosphere is equal 0.68 [24]. One can divide a cloudy atmosphere in three tiers, and their typical altitudes are represented in Fig. 2.2. In addition, the optical thickness of a cloudy atmosphere over a land is 0.10–0.15 less than that over oceans [24]. Of course, cloudiness depends on the latitude and seasons. But in the above analysis we deal with global parameters, i.e., with averaged atmospheric parameters over the globe.

Thus, from the physical standpoint, clouds are regions of the atmosphere with a low temperature, and water condensation proceeds there. As a result of water condensation, the air temperature in such a region increases. Winds lead to equalizing the cloud temperature with that of a surrounding regions of the atmosphere. Winds also provide a slow degradation of the clouds and their disappearance.

4.2 Kinetics of Growth of Water Microdroplets in Atmospheric Air

4.2.1 Droplet Growth in Supersaturated Air

In the analysis of evolution of wet atmospheric air, we assume above that transformation of the excess part of a supersaturated water vapor into condensed phase proceeds prompt. We now consider the character of this process on the basis of simple models, as well as analyze the growth of forming water droplets. Let us start from supersaturated atmospheric air where the excess of a water vapor inside air is converted in the condensed phase in the form of microdroplets or microparticles under thermodynamic equilibrium. For simplicity, we below restrict ourselves by water microdroplets.

The first stage of new phase formation is the origin of an embryo which becomes subsequently the condensation nucleus. In uniform air according to the classical theory [25–28], this process is long because under thermodynamic equilibrium small clusters consisting of bonded molecules decay until these reach a critical size. Such an event lasts long and the rate of growth of clusters consisting of bound molecules in a supersaturated gas is determined by reaching of a critical size [25, 28, 29]. In this consideration of growth of a water microdroplet in supersaturated atmospheric air we assume that nuclei of condensation exist in it in the form of ions, radicals, dust. In this case growth of a new phase proceeds through attachment of water molecules to condensation nuclei.

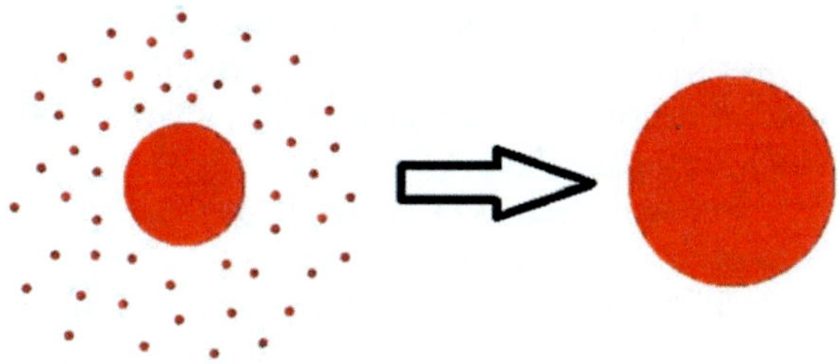

Fig. 4.5 Growth of water droplets in supersaturated atmospheric air as a result of attachment of water molecules (small circles) to forming clusters (a system of bound molecules) which are nuclei of condensation in this process [35–37]

Nuclei of condensation are of importance for growth of water droplets under usual atmospheric conditions. In reality, attachment of water molecules to nuclei of condensation is a chemical reaction between water molecules and radicals in the form of nitrogen oxides, ammonia and sulfur compounds [30–34]. Attachment of water molecules to microdroplets proceeds with participation of some acids and their salts. The rate of subsequent growth of water droplets decreases with an increasing size. Hence, the first stage of transformation of a water excess in the condensed phase in supersaturated air gives a small contribution to a time of the condensation process.

Because growth of a new phase in a supersaturated gas proceeds through formation and growth of condensation nuclei, the total process of a new phase growth is called the nucleation process. There are various mechanisms for this process [38–42] depending on conditions, as well as specifics of a gas and a type of nanometer-sized particles-aerosols formed in the first stage of this process. Below we concentrate only on nucleation processes in atmospheric air involving water that decreases a number of nucleation mechanisms. In particular, Fig. 4.5 describes the character of transformation of supersaturated air in saturated one contained water microdroplets.

In considering kinetics of growth of microdroplets, we note that this process depends on the character of motion of nucleating molecules in a gas [43–45]. If the mean free path of associating molecules in a gas is large compared to a size of the growing particle, we deal with the kinetic regime of growth of a new phase, whereas if another relation between these parameters is fulfilled, the diffusion regime of growth is realized with the diffusion character of motion for attaching molecules. We are guided below by the diffusion regime of growth of water microdroplets in atmospheric air because this mechanism of growth takes place under normal conditions and micron sizes of a growing particle. This means that in the course of conversion of a supersaturated atmospheric air in saturated one, the basic time of this process proceeds when conditions of the diffusion regime of droplet growth are realized.

We now analyze the rate of the first stage of the nucleation process where supersaturated vapor are transformed in saturated one and water microdroplets are formed. In the diffusion regime of growth, evolution of the number of molecules n for a test microdroplet is given by

$$\frac{dn}{dt} = 4\pi D_w r N_w - 4\pi D_w r N_{\rm sat}, \tag{4.2.1}$$

where D_w is the diffusion coefficient of molecules in surrounding air, r is a microdroplet radius, N_w is a current number density of free water molecules, N_{sat} is the number density of water molecules at the saturated vapor pressure. We write this rate by using the Smolukhowski formula (2.2.11) [46] for the rate of attachment of surrounding molecules to a microdroplet under the condition that the droplet radius is large compared to the mean free path of molecules, so that motion of attaching molecules near the microdroplet has the diffusion character. In addition, we take into account in formula (4.2.1) evaporation of molecules from the microdroplet surface, so that at the saturated vapor pressure, the rates of molecule attachment to the microdroplet surface and their evaporation are equal.

Let us consider a nonsaturated vapor in the limit of a low number density $N_w \ll N_{sat}$ of water molecules. In this limit water microdroplets evaporate, and equation (4.2.1) takes the form

$$\frac{r\mathrm{d}r}{\mathrm{d}t} = -D_w r_W^3 N_{sat},$$

where $r_W = 1.92$ Å[35] is the Wigner–Seitz radius for water. Above we use formula (2.2.3) which connects a number of molecules in a microdroplet n and its radius r. From this, one can define the total evaporation time τ_{ev} of a water microdroplet if its radius is small compared to the critical one in accordance with this equation

$$\tau_{ev} = \frac{r^2}{2D_w N_{sat} r_W^3}, \tag{4.2.2}$$

We now make estimations for atmospheric air with using that the diffusion coefficient of water molecules in the air at atmospheric pressure is $D_w = 0.22\ \mathrm{cm^2/s}$ [47, 48]. In these estimations, we are guided by clouds which are located at an altitude $h = 3$ km, as a typical altitude of clouds. Within the framework of the model of standard atmosphere, the temperature at this altitude is $T = 268$ K which corresponds to the number density of water molecules at the saturated vapor pressure $N_{sat} = 1.1 \times 10^{17}\ \mathrm{cm^{-3}}$. This number density of air molecules at this altitude corresponds to the diffusion coefficient of water molecules in air $D_w = 0.31\ \mathrm{cm^2/s}$. Then formula (4.2.2) for a typical microdroplet radius $r_o = 8\,\mu\mathrm{m}$ (2.2.12) in a cumulus cloud gives $\tau_{ev} \approx 0.6$ s. This value is a typical time of equilibrium establishment between water molecules and water microdroplets in a cumulus cloud.

The value (4.2.2) can be considered as a typical time of establishment of equilibrium between a water vapor in the atmosphere and free water atmospheric microdroplets. This value is small compared to the lifetime of clouds which is a few of hours. In the course of subsequent evolution of clouds, the equilibrium is supported, so that the rates of molecule attachment to water microdroplets and of molecule evaporation from their surface are equal, so that the number density of free water molecules equals to the saturated vapor pressure of atmospheric water. Subsequent processes in the atmosphere involving water microdroplets proceed under such conditions.

Let us consider the first stage of the nucleation process from another standpoint expressing the rate of growth of the condensed phase through a large number density

$N_{\rm nc}$ of condensation nuclei. Denoting a current number density of excess molecules by Δ_w, and its initial value by Δ_o, we have for the current average number n of molecules in one microdroplet after finishing the process of total transformation of excess water molecules into microdroplets

$$n = \frac{\Delta_w}{N_{\rm nc}} = \left(\frac{r_f}{r_W}\right)^3, \qquad (4.2.3)$$

where r is the current average radius of growing microdroplets and we use formula (2.2.3) which connects the average number n of molecules of one microdroplet and its average radius r. The average final droplet radius r_f is equal

$$r_f = r_W \left(\frac{\Delta_o}{N_{nc}}\right)^{1/3} \qquad (4.2.4)$$

Equation (4.2.1) for growth of forming microdroplets has the form

$$\frac{d\Delta_w}{N_{\rm nc} dt} = 4\pi D_w r \Delta_w$$

One can rewrite this equation as

$$\frac{d\Delta_w}{dt} = -\frac{\Delta_w}{\tau_{\rm gr}} \left(\frac{\Delta_o - \Delta_w}{\Delta_o}\right)^{1/3}, \quad \frac{1}{\tau_{\rm gr}} = 4\pi D_w r \Delta_o^{1/3} N_{\rm nc}^{2/3} \qquad (4.2.5)$$

As is seen, $\tau_{\rm gr}$ is a typical time of microdroplet growth as a result of transformation of excess water molecules in microdroplets. Being guided by a cumulus cloud, we have according to (2.2.12) $\Delta_o \sim 10^{16}$ cm^{-3}. Because the number density of condensation nuclei is $N_{\rm nc} > 10^3$cm^{-3}, this gives $\tau_{\rm gr} < 1$s. Thus, from the standpoint of the concept of condensation through condensation nuclei, we obtain again that a small time of establishment of equilibrium between free water molecules of an atmospheric water vapor and water microdroplets located there compared to typical times of cloud evolution.

4.2.2 Droplet Growth Through Ostwald Ripening

We now consider the next stage of evolution of atmospheric air with water microdroplets that proceeds in clouds. Under conditions in atmospheric air, the basic mechanism of growth is Ostwald ripening [49, 50] or coalescence. The character of this growth mechanism is represented in Fig. 4.6 and results from a specific interaction between microdroplets and free water molecules. Indeed, let us divide the system of water molecules of a certain atmosphere region in two subsystem, so that the first one includes free water molecules, whereas the second one consists of bound molecules.

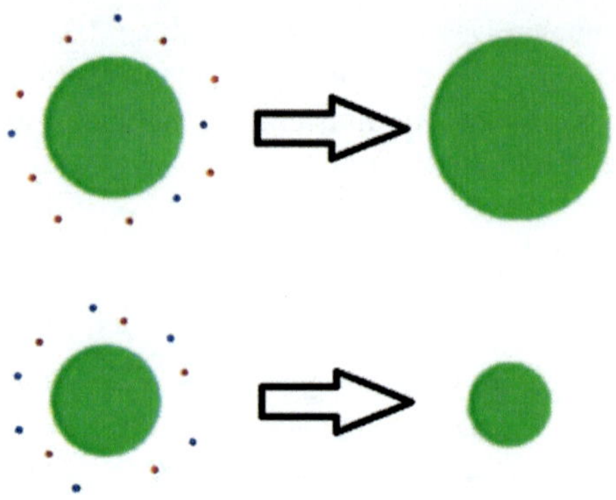

Fig. 4.6 Mechanism of Ostwald ripening (coalescence) for microdroplet growth in atmospheric air. Small red circles are attached molecules, small blue circles are evaporated molecules. A number of attaching molecules exceeds that of evaporating ones for a large microdroplet, and the reciprocal relation between these rates takes place for small microdroplets. As a result, a large microdroplet (large green circle) grows, whereas a small one disappears

Within the framework of this consideration, the above first stage of growth of the water condensed phase in atmospheric air consists in transition of water molecules from the first to the second subsystem. In contrast to this, the mechanism of growth of water microdroplets for Ostwald ripening proceeds without transition between these subsystems, but interaction between indicated subsystems is of importance for droplet growth.

In this process, each microdroplet participates in processes of molecule attachment to the microdroplet and evaporation from its surface, but the equilibrium number densities of free molecules at which the rates of attachment and evaporation are equalized and depend weakly on the microdroplet size. As a result, large microdroplets grow while small ones evaporate, and the average droplet size increases. The character of this mechanism of droplet growth is presented in Fig. 4.6. The theory of this growth mechanism for the diffusion character of molecule motion was developed in [51–54].

Following to this theory, we define the critical droplet size $n_{\rm cr}$ for which the rates of molecule attachment and molecule evaporation are equal. Let us use equation (4.2.1) for the balance between attached and evaporated molecules

$$\frac{dn}{dt} = -4\pi D_w r_W n^{1/3} N_{\rm sat} \exp\left(\frac{2A}{3n^{1/3}T} - \frac{2A}{3n_{\rm cr}^{1/3}T}\right), \qquad (4.2.6)$$

Here we use expression (2.2.1) for the binding energy of a bound molecule in a microdroplet, and the critical droplet size $n_{\rm cr}$ is determined from the relation

$$N_w = N_{\rm sat} \exp\left(-\frac{2A}{3n_{\rm cr}^{1/3}T}\right)$$

For considering droplet sizes $n \gg A^3/T^3$, the balance Eq. (4.2.6 for a test microdroplet of a current size n) takes the form

$$\frac{dn}{dt} = \frac{8\pi D_w r_W A N_w}{3T}\left[\left(\frac{n}{n_{cr}}\right)^3 - 1\right] \tag{4.2.7}$$

We use that under these conditions $N_w = N_{sat}$. From Eq. (4.2.7) it follows also that microdroplets of sizes $n > n_{cr}$ grow, whereas in the case $n < n_{cr}$ microdroplets decrease.

This process has a self-consistent character with the automodel form of the size distribution function of microdroplets. Equation (4.2.7) of microdroplet growth shows that the rate of growth for the average microdroplet size $\overline{n}$ has the form

$$\frac{d\overline{n}}{dt} = J, \tag{4.2.8}$$

and the rate of growth J is independent of a size. Formula (4.2.7) gives the dependence of the growth rate J on problem parameters, and solution of the problem allows one to determine the size distribution function $f(n)$ [51–55]. Then Eq. 4.2.8 has the form [37, 56]

$$\frac{d\overline{n}}{dt} = \frac{1.4 D_w r_W A N_{sat}}{T} = 1/\tau_{Ost} \tag{4.2.9}$$

Note that for the average size of a water microdroplet in cumulus clouds $r = 8\mu m$ according to formula (2.2.12) a small parameter of this problem is estimated as $A/(T\overline{n}^{1/3}) \sim 10^{-4}$.

Because the automodel character of droplet growth for the Ostwald ripening mechanism of growth, all the parameters depend on the parameter

$$u = \frac{r}{r_{cr}} = \left(\frac{n}{n_{cr}}\right)^{1/3}$$

Figure 4.7 gives the size distribution function of droplets for the Ostwald ripening mechanism of growth. In addition, as it follows from [37, 56]

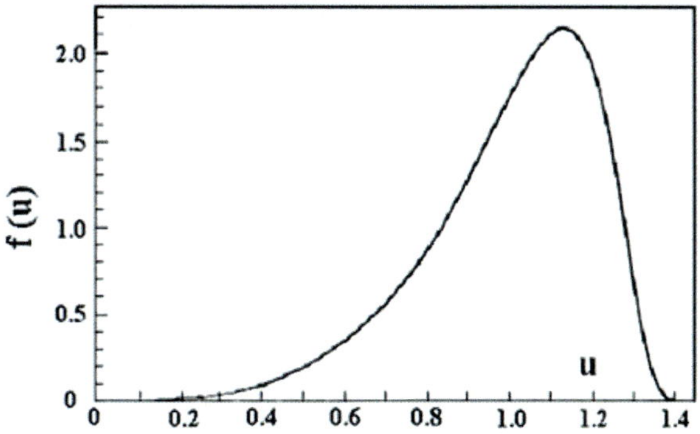

Fig. 4.7 Size distribution function of droplets for the Ostwald ripening mechanism of droplet growth as a function of a reduced microdroplet radius [53–55]

$$\overline{n} = 1.123 n_{cr} \tag{4.2.10}$$

Let us determine on the basis of formula (4.2.9) a growth time τ_{Ost} up to average size $r = 8$ μm of a microdroplet in a cumulus cloud assuming the process proceeds at the altitude $h = 3$ km, where the temperature is $T = 268$ K and the air pressure is $p = 0.7$ atm, and, correspondingly, the diffusion coefficient of water molecules in atmospheric air under indicated conditions is $D_w = 0.31\ \mathrm{cm}^2/\mathrm{s}$. Next, the number density of water molecules at the saturated vapor pressure for this temperature is equal $N_w = N_{sat}(T) = 1.6 \times 10^{17}\mathrm{cm}^{-3}$ according to data of Table 2.2. On the basis of these parameters and formula (4.2.9), one can obtain for the growth time

$$\tau_{Ost} = \frac{T}{1.4 D_{wr_W} A N_{sat}} = 2.3 \times 10^{-10}\ \mathrm{s} \tag{4.2.11}$$

From the growth Eq. (4.2.9) one can find for the time growth τ_g up to a microdroplet radius $r = 8$μm under considering conditions ($T = 273$ K, $p = 0.7$ atm)

$$\tau_c = n\tau_{Ost} \approx 4.7\ \mathrm{h} \tag{4.2.12}$$

4.2.3 Gravitation Mechanism of Growth of Water Droplets

The Ostwald mechanism of microdroplet growth leads to an increase of the growth time with an increasing size. Therefore, at a certain size, this mechanism is replaced by the mechanism of gravitation growth which is of importance for large drops let up to formation of rain drops. The character of gravitation growth is presented in Fig. 4.8, so that in the course of droplet falling in atmospheric air under the action of their weight, droplets are joined as a result of their contact.

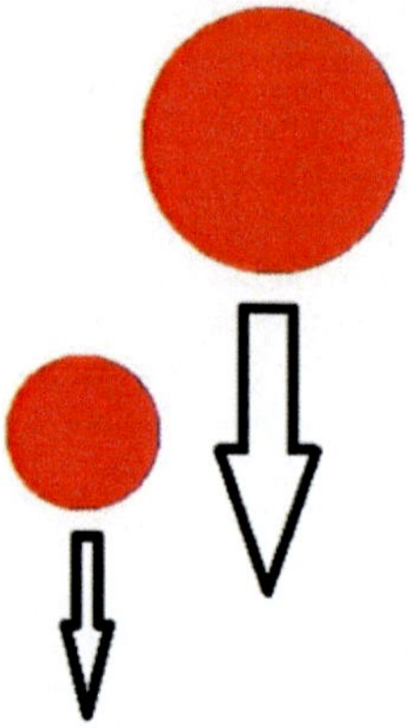

Fig. 4.8 Gravitation mechanism of droplet growth. A large droplet catches up with a small one in the course of their falling and joins with it

Evidently, this mechanism of droplet association is stronger than that due to Ostwald ripening, if the velocity of gravitation fall v_g exceeds significantly a thermal velocity of droplets, i.e., if the following criterion holds true

$$v_g \gg \sqrt{\frac{8T}{\pi M}}, \qquad (4.2.13)$$

where M is the droplet mass, T is the gas temperature expressed in energetic units. This criterion gives

$$r \gg 0.7\left(\frac{\eta^2 T}{\rho^3 g^2}\right)^{1/7}$$

In particular, for water droplets located in atmospheric air at the temperature $T = 268$ K this criterion takes the form

$$r \gg 1\ \mu\text{m} \qquad (4.2.14)$$

On the basis of this mechanism of droplet association as a result of merging of droplets which move in atmospheric air under the action of their weight, one can estimate the rate constant of gravitational association of droplets k_{as} assuming a typical difference of falling velocity Δv_g to be of the order of this velocity v_g given by formula (2.2.9)

$$k_{\text{as}} \sim \Delta v_g \cdot r^2 \sim \frac{\rho g r^4}{\eta},$$

The accurate evaluation of this rate constant with accounting for the automodel form of the size distribution function of microdroplets give [37]

$$k_{\text{as}}(r) = \frac{2g\rho r^4}{\eta} = \chi_o\left(\frac{r}{r_W}\right)^4 = \chi_o n^{4/3}, \ \chi_o = \frac{2\rho g r_W^4}{\eta}, \qquad (4.2.15)$$

where r is a droplet radius. In particular, in air at atmospheric pressure and temperature $T = 268$ K ($\eta = 1.7 \times 10^{-4}$ g/(cm · s)) the parameter χ_o is equal $\chi_o = 1.6 \times 10^{-24}$ cm^3/s. For a typical size of water microdroplets in cumulus clouds $r = 8\mu$m according to (2.2.12) we have from this

$$k_{\text{as}} \approx 5 \times 10^{-6}\ \text{cm}^3/\text{s}$$

Correspondingly, a typical time of association τ_{as} of two water microdroplets of a cumulus cloud with droplet parameters according to (2.2.12) is

$$\tau_{\text{as}} \sim \frac{1}{k_{\text{as}} N_d} \sim 3\ \text{min}$$

As a matter, τ_{as} is of the order of the doubling time for droplet growth.

Equation of droplet growth for the average number $\overline{n}$ of droplet molecules is expressed through the total number density N_b of bound molecules of a microdroplet has the form [36]

$$\frac{d\overline{n}}{dt} = \frac{1}{2} k_{\text{as}} N_b,$$

where N_b is the total number density of bound molecules. Note that this growth equation relates to the association rate constant which is independent of the droplet size. In addition, this equation assumes a uniform distribution of droplets in a space. Because these assumptions are not fulfilled for gravitation droplet growth, subsequent results can be considered as estimations.

Let us extract the size dependence in formula (4.2.15) for the rate constant of gravitation association, and representing it in the form $k_{\text{as}} = \chi_o \overline{n}^{4/3}$. Taking into account that large droplet sizes do not give a contribution to the total growth time t_{gr} of a droplet for the gravitation mechanism of droplet growth, we then obtain for the total growth time τ_{gr} of a group of microdroplets with an average droplet size $\overline{n}$

$$\tau_g = 2 \int \frac{d\overline{n}}{k_{\text{as}} N_b} = \frac{6}{\chi_o \overline{n}^{1/3} \cdot N_b} = 6\tau_{\text{as}}, \tag{4.2.16}$$

and the total time of growth of water droplets up to large drops in cumulus clouds is estimated as $\tau_g \approx 20$ min.

The time growth of microdroplets τ_{gr} characterizes transition from clouds consisting water microdroplets to drops of rain, that we consider as the lifetime of clouds. Joining the of Ostwald ripening and gravitation mechanisms, one can obtain the following equation growth of microdroplets

$$\frac{d\overline{n}}{dt} = \frac{1}{\tau_{\text{Ost}}} + \frac{\chi_o n^{4/3} N_b}{2}, \tag{4.2.17}$$

where τ_{Ost} is the growth time for the mechanism of Ostwald ripening in accordance with formula (4.2.11). Then the total growth time due to both mechanisms of growth in a supersaturated water vapor up to formation of rain drops is given by

$$\tau_{\text{gr}} = \int_0^\infty \frac{dn}{1/\tau_{\text{Ost}} + \chi_{\text{on}}^{4/3} N_b/2} = \frac{2.7\tau_{\text{Ost}}^{1/4}}{(2N_b\chi_o)^{3/4}} \tag{4.2.18}$$

Taking the values of times for coalescence ($\tau_{\text{Ost}} = 2.3 \times 10^{-10}$ s) and gravitation growth $t_{\text{grav}} = 1.5 \times 10^7$ s according to formulas (4.2.11) and (4.2.18), one can obtain for the total growth time $\tau_{\text{gr}} = 25$ min.

One can see that a growth time of the mechanism of Ostwald ripening (4.2.11) is independent of a droplet size, whereas a typical growth time decreases sharply (as $\sim n^{-4/3}$) for the gravitation mechanism of droplet growth. Then on the basis of formulas (4.2.11) and (4.2.15) one can find a typical number of droplet molecules n_t at which typical growth times for these growth mechanisms are comparable

$$n_t \approx \frac{2}{(N_b \chi_o \tau_{\rm Ost})^{3/4}} = 2.5 \times 10^{12} \qquad (4.2.19)$$

We apply this formula to air conditions at the altitude 3 km as usually, where $T = 268$ K, $p = 0.7$ atn. Then we obtain that rates of these growth processes are comparable at the droplet radius $r_t = 2.6\ \mu$m. The growth time from zero up to this is equal due to the process of Ostwald ripening $\tau_t = n_t \cdot \tau_{\rm Ost} = 9$ min. In addition, the growth time due to the gravitational mechanism of droplet growth from an indicated size up to infinite one is equal

$$\tau_g' = \frac{6}{N_b \chi_o n_t^{1/3}} = 28\ \text{min}$$

The combination of these two growth processes in accordance with formula (4.2.18) gives the total growth time $\tau_{\rm gr} = 25$ min.

4.2.4 Growth of Water Microdroplets in Clouds

In spite of the roughness of the above analysis, one can formulate a general character of droplet growth in supersaturated air under real conditions. We are guided mostly by the case of formation of a cumulus cloud where the number density of bound water molecules is enough large and is comparable with the number density of free water molecules. Starting from a supersaturated water vapor in atmospheric air which is formed as a result of atmosphere parcel mixing from different altitudes, one can analyze the first stage of the condensation process. In this process, an excess of the water vapor is transformed in microdroplets by attachment to condensation nuclei at the beginning. As a result, all the excess of a supersaturated water vapor is transformed into microdroplets.

When microdroplets are formed, they grow such that the amount of the number density of water molecules is not changed practically. On the stage of submicron-sized droplets grow through the process of Ostwald ripening, which nature is determined by a different size dependence for evaporation and attachment processes. As a result of the sum of these processes, small droplets evaporate, whereas large droplets increase, so that the average droplet size grows. In addition, the dependence of a time $t(r)$ of reaching of a given radius r for a droplet has the form $t(r) \sim r^3$, i.e., the growth rate decreases with an increasing size.

On contrary, the rate constant of gravitation growth according to formula (4.2.15) increases sharply with an increasing droplet size. Hence, joining these mechanisms of droplet growth, one can find a growth time from supersaturated air to its condensation up to rain drops proceeds during ten minutes. Comparison of results of this analysis with evolution of the condensation process in atmospheric air with observed data leads to two contradictions. First, in this analysis the development of atmospheric water proceeds continuously in contrast to a large lifetime of clouds. Second, an observed lifetime of clouds consisting of water microdroplets exceeds by an order of magnitude that followed from this analysis.

In order to overcome these contradictions, one can assume the electric nature of clouds, namely, water droplets of clouds are charged. Therefore, we arrive at the concept of charged microdroplets. Then we assume as early that water droplets have the same radius which increases in the course of the condensation process. In the same manner, we assume an identical charge of individual microdroplets which grow with an increasing droplet radius. In addition, the number density of water molecules, as well as the charge number density, do not vary in the course of evolution of this system. In particular, this means that joining of two microdroplets with n_1 and n_2 numbers of molecules, as well as Z_1 and Z_2 charges leads to formation of a droplet with $n_1 + n_2$ number of molecules and with $Z_1 + Z_2$ charge. Keeping this growth scheme, we consider below the character of evolution of atmospheric air with water microdroplets.

Assuming an identical charge Z of joining microdroplets, we have for the repulsion potential of interaction for two droplets of a radius r and charge Z during their contact

$$U(r) = \frac{Z^2 e^2}{2r}$$

Correspondingly, one can obtain for the rate constant of gravitation association of charged droplets instead of formula (4.2.15)

$$k_{\mathrm{as}} = \frac{2g\rho r^4}{\eta} \exp\left(-\frac{Z^2 e^2}{2rT}\right) \tag{4.2.20}$$

We are based on the equation of growth which takes the form

$$\frac{d\overline{n}}{dt} = \frac{1}{\tau_{\mathrm{ost}}} + \frac{\chi_o N_b \overline{n}^{4/3}}{2\exp\left(-\frac{Z^2 e^2}{2r_W T \overline{n}^{4/3}}\right)} \tag{4.2.21}$$

Then a current time $t(r)$ through which a given droplet radius is attained an indicated value is determined by the expression

$$t(\overline{n}) = \int_0^{\overline{n}} dn \left[\frac{1}{\tau_{\mathrm{ost}}} + \frac{\chi_o N_b n^{4/3}}{2\exp\left(-\frac{Z^2 e^2}{2r_W T n^{1/3}}\right)}\right]^{-1} \tag{4.2.22}$$

It is convenient to use a current droplet radius as a variable. Then Eq. (4.2.22) takes the form

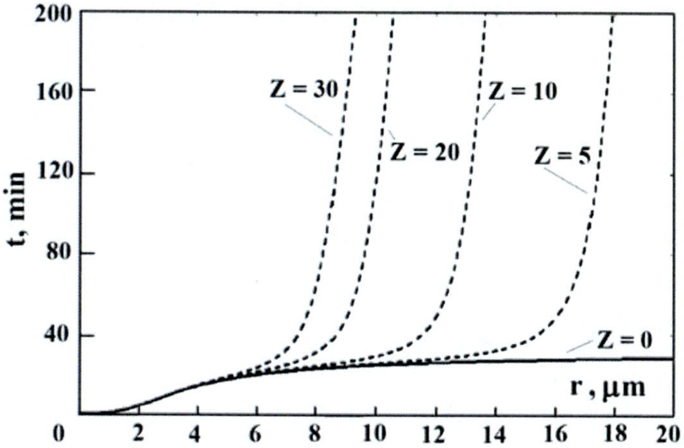

Fig. 4.9 Time t during which a growing water droplet reaches an indicated radius r according to formula (4.2.23). Average atmospheric parameters are used which are realized at an altitude 3 km (the temperature is $T = 268$ K, the air pressure is $p = 0.7$ atm). An indicated average droplet charge Z is attained at the droplet radius $r = 8$ μm

Table 4.1 Radius of growing water microdroplets under parameters of Fig. 4.9 through 2 hours from the growth beginning where atmospheric air contains a supersaturated water vapor

Z	5	10	20	30
r, μm	17	13	10	8.7

$$t(r) = \int_0^r \frac{3r^2 dr}{r_W^3}\left[\frac{1}{\tau_{ost}} + \frac{\chi_o N_b r^4}{2r_W^4 \exp\left(-\frac{Z^2e^2}{2r_W T n^{1/3}}\right)}\right]^{-1} \qquad (4.2.23)$$

Fig. 4.9 contains this dependence .

Results of Fig. 4.9 confirm conclusions which follow from observations according to which cumulus clouds are stable through a large time, and we explain it by the droplet charge. This means that charged microdroplets grow in time slow, i.e., cumulus clouds become stable in spite a large droplet density. This property is used in formula (2.2.12) where certain average parameters belong to cumulus clouds. Table 4.1 presents microdroplet radii which are attained through 2 h of droplet evolution under conditions of Fig. 4.9. In addition, a slow variation of the microdroplet radius is observed on this stage of microdroplet evolution. For example, for a charge $Z = 20$ at the microdroplet radius $r = 8$ μm the microdroplet radius varies by 4% during time from 2 hours up to 3 hours from the beginning.

We now consider growth of microdroplets in a saturated vapor from another standpoint, basing on the doubling time of droplets and with using the above growth mechanisms [1, 2]. The doubling time is a time during which the number of droplet

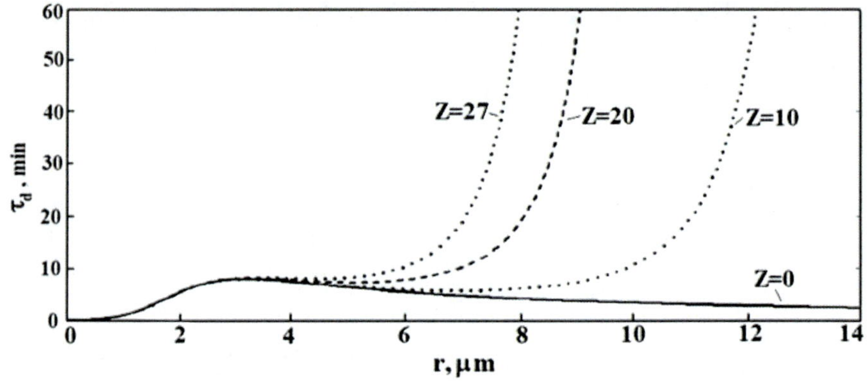

Fig. 4.10 Doubling time for the number of molecules of water droplets at the temperature is $T = 273$ K in accordance with formula (4.2.24). An indicated droplet charge Z corresponds to a droplet radius $r = 8$ μm

molecules is doubled. Correspondingly, a droplet radius varies from $2^{-1/6}r$ up to $2^{1/6}r$, i.e., the radius change Δr is

$$\Delta r = (2^{1/6} - 2^{-1/6})r = 0.23r$$

As a result, we have for the doubling time of the number molecules of a neutral droplet

$$\tau_d(r) = \frac{0.7r^3}{r_W^3/\tau_{Ost} + r^4\chi_o N_b/(2r_W)} \tag{4.2.24}$$

Figure 4.10 contains the doubling time of microdroplet growth according to formula (4.2.24) for combination of two growth mechanisms. The maximum doubling time $\tau_d = 7.9min$ corresponds to a microdroplet radius of 3.1 μm, where both mechanisms of droplet growth give comparable contributions to this value. For the droplet radius of an average cumulus cloud (2.2.12) $r = 8$ μm the doubling time is equal $\tau_d = 4.1$ min.

Next, the doubling time for charged droplets $\tau_d(r)$ is equal [1, 2]

$$\tau_d(r) = \frac{0.7r^3}{r_W^3 + \chi_o N_b + r^4/[2r_W \exp[-Z^2e^2/(2rT)]}, \tag{4.2.25}$$

where as early $Z(r) = z_o r^3$ and z_o is independent on a droplet radius r. Figure 4.10 contains the doubling time of charged microdroplets according to this formula. In addition, Table 4.2 contains values of the doubling time when it reaches a radius $r = 8$ μm.

Table 4.2 Doubling times of growth of water microdroplets when they attain the radius $r = 8\ \mu m$ in the course of their growth in atmospheric air at the air temperature $T = 273$ K

Z, e	τ_d, min
0	4.1
2	4.2
5	4.6
10	6.1
15	9.7
20	19
25	42
30	104
40	400

This table contains the charge Z which has a droplet when its radius in atmospheric air is $r = 8\ \mu m$ [1]

Table 4.2 contains the values of growth times for charged droplets. For the growth mechanism under consideration, the specific droplet charge, i.e., the ratio of the charge to the droplet mass, does not vary in the course of droplet growth. Thus, in this analysis of droplet growth, we account for the Ostwald ripening and gravitation growth mechanisms in accordance with Fig. 4.10. From this it follows that growth of charged droplets proceeds slowly at large droplet sizes, so that growth of droplets up to rain drops exceeds the doubling time for this microdroplet by one order of magnitude.

Note that above we use parameters of cumulus clouds according to formula (2.2.12) which are based on the assumption that parameters of a cumulus cloud varies slightly during a long time. This assumption is fulfilled if microdroplets carry a charge of the same sign. Then Coulomb repulsion of colliding microdroplets hampers their approach, and therefore their parameters vary slowly.

Along with clouds which are formed at some altitudes over the ground due to a low atmospheric temperatures at such altitudes, at night or tomorrow when the ground temperature is low, condensation may proceed neat the ground. Indeed, if humid air near the ground cools below its dew point, fog is formed, as it is demonstrated in Fig. 4.11. Especially, fog formation is possible in autumn after cold nights and at clear sky. Then the Earth's surface cools sharply, and air of near-surface layers become supersaturated. Condensation in this air proceeds during cold time, and a subsequent heating the ground tomorrow leads to evaporation of water droplets. Because air contained a fog is heavier than dry air, a fog is located longer in low places of a landscape [57, 58]. Figure 4.11 represents an example of fog formation in nature on lowlands.

Fig. 4.11 Fog formation of in lowlands during an early tomorrow at autumn as a result of a temperature decrease at night [57]

References

1. B.M. Smirnov, *Global Atmospheric Phenomena Involving Water* (Switzerland, Springer Atmospheric Series, 2020)
2. B.M. Smirnov, Phys. Sol. State **62**, 24 (2020)
3. R.G. Fleagle, J.A. Businger, *Introduction to Atmospheric Physics* (Acad. Press, San Diego, 1980)
4. H.L. Green, W.R. Lane, *Particulate Clouds: Dust, Smokes and Mists* (Princeton, Van Nostrand, 1964)
5. H.R. Byers, *Elements of Cloud Physics* (Univ. Chicago, Chicago, 1965)
6. N.H. Fletcher, *The Physics of Rainclouds* (Cambridge Univ. Press, London, 1969)
7. B.J. Mason, *Clouds, Rain and Rainmaking* (Cambr. Univ. Press, Cambridge, 1975)
8. S. Twomey, *Atmospheric Aerosols* (Elsevier, Amsterdam, 1977)
9. P.C. Reist, *Introduction to Aerosol Science* (Macmillan Publ. Comp, New York, 1984)
10. R.R. Rogers, M.K. Yau, *A Short Course in Cloud Physics* (Pergamon Press, Oxford, 1989)
11. F.H. Ludlam, *Clouds and Storms: The Behavior and Effect of Water in the Atmosphere* (University Park, Penn State University Press, 1990)
12. K. Young, *Microphysical Processes in Clouds* (Oxford Univ. Press, New York, 1993)
13. K. Friedlander, *Smoke, Dust, and Haze* (Oxford Univ. Press, Fundamentals of Aerosol Dynamics (Oxford, 2000)
14. H.R. Pruppacher, J.D. Klett, *Microphysics of Clouds and Precipitation* (Kluwer, New York, 2004)
15. M. Satoh, *Atmospheric Circulation Dynamics and General Circulation Models* (Springer-Praxis, Chichester, 2004)

16. J. Straka, *Clouds and Precipitation Physics* (Cambridge Univ. Press, Cambridge, 2009)
17. B.J. Mason, *The Physics of Clouds* (Oxford Univ. Press, Oxford, 2010)
18. H. Pruppacher, J. Klett, *Microphysics of Clouds and Precipitation* (Springer, Dordrecht, 2010)
19. D. Lamb, J. Verlinde, *Physics and Chemistry of Clouds* (Cambr. Univ. Press, Cambridge, 2011)
20. P.K. Wang, *Physics and Dynamics of Clouds and Precipitation* (Cambr. Univ. Press, Cambridge, 2013)
21. A.P. Khain, M. Pinsky, *Physical Processes in Clouds and Cloud Modeling* (Cambr. Univ. Press, Cambridge, 2018)
22. https://www.handmadepiece.com/the-rape-of-europa-handmade-oil-painting-reproduction-on-canvas-by-artist-felix-vallotton.html
23. https://www.soho-art.com/oil-painting/1277087106/Felix-Vallotton/The-Rape-of-Europa-1908.html
24. https://en.wikipedia.org/wiki/Cloud-cover
25. L.D. Landau, E.M. Lifshitz, *Statistical Physics*, vol. 1 (Pergamon Press, Oxford, 1980)
26. Ya..B.. Zeldovich, ZhETF **12**, 525 (1942)
27. F.F. Abraham, *Homogeneous Nucleation Theory* (Acad.Press, New York, 1974)
28. I. Gutzow, J. Schmelzer, *The Vitreous State* (Springer, Berlin, 1995)
29. B.M. Smirnov, *Principles of Statistical Physics* (Wiley VCH, Berlin, 2006)
30. J.H. Seinfeld, S.N. Pandis, *Atmospheric Chemistry and Physics* (Wiley, Hoboken, 2006)
31. M.L. Salby, *Physics of the Atmosphere and Climate* (Cambr. Univ. Press, Cambridge, 2012)
32. B.M. Smirnov, *Microphysics of Atmospheric Phenomena* (Switzerland, Springer Atmospheric Series, 2017)
33. K.S.W.Champion, A.E.Cole, A.J.Kantor, in *Chemical dynamics in extreme environments* ed by R.A. Dressler. (Singapore, World Sci.Publ., 2001)
34. B. Sportisse, *Fundamentals in Air Pollution* (Springer Science, Dordrecht, 2010)
35. B.M. Smirnov, *Clusters and Small Particles in Gases and Plasmas* (Springer NY, New York, 1999)
36. B.M. Smirnov, *Cluster Processes in Gases and Plasmas* (Wiley, Berlin, 2010)
37. B.M. Smirnov, *Nanoclusters and Microparticles in Gases and Vapors* (De Gruyter, Berlin, 2012)
38. M.M.R. Williams, S.K. Loyalka, *Aerosol Science Theory and Practice* (Pergamon, Oxford, 1991)
39. A.A. Lushnikov, *Introduction to aerosols* in *Aerosols Science and Technology* ed by I. Agranovski (Weinheim, Wiley, 2010), p. 142
40. A.A. Lushnikov, *Condensation, evaporation, nucleation* in *Aerosols Science and Technology* ed by I. Agranovski (Weinheim, Wiley, 2010), pp. 91126
41. A.A. Lushnikov, *Nanoaerosols in the atmosphere* in *The Atmosphere and Ionosphere, Physics of Earth and Space Environments* ed by V.L. Bychkov et al. (Dordrecht, Springer, 2012), pp. 79–164
42. A.A. Lushnikov, V.A. Zagaynov, Yu.S. Lyubovtseva, *Formation of Aerosols in the Atmosphere* in *The Atmosphere and Ionosphere, Physics of Earth and Space Environments* ed by V.L. Bychkov et al. (Dordrecht, Springer Science, 2012), pp. 69–95
43. N.A. Fuchs, *Evaporation and Growth of Drops in a Gas* (Moscow, Izd.AN SSSR, 1958; in Russian)
44. N.A. Fuchs, A.G. Sutugin, *Highly Dispersed Aerosols* (London, Ann Arbor, 1971)
45. N.A. Fuchs, *Mechanics of Aerosols* (Pergamon, New York, 2002)
46. M.V.Smolukhowski. *Zs.Phys.* **17**, 585(1916)
47. N.B. Vargaftic, *Tables of Thermophysical Properties of Liquids and Gases* (Halsted Press, New York, 1975)
48. B.M. Smirnov, *Reference Data on Atomic Physics and Atomic Processes* (Springer, Heidelberg, 2008)
49. W.Ostwald, Zs.Phys.Chem. **22**, 289 (1897)
50. W. Ostwald, Zs. Phys. Chem. **34**, 495 (1900)
51. I.M. Lifshitz, V.V. Slezov, JETP **35**, 331 (1958)

52. I.M. Lifshitz, V.V. Slezov. **1**, 1401 (1959)
53. I.M. Lifshitz, V.V. Slezov, J. Phys. Chem. Sol. **19**, 35 (1961)
54. E.M. Lifshitz, L.P. Pitaevskii, *Physical Kinetics* (Pergamon Press, Oxford, 1981)
55. V.V. Slezov, V.V. Sagalovitch, Sov. Phys. Usp. **30**, 23 (1987)
56. B.M. Smirnov, Phys. Usp. **57**, 1041 (2014)
57. http://wallpaperson.net/-ph/22/234421342.jpg
58. https://en.wikipedia.org/wiki/Fog

Chapter 5
Energetics of the Earth

Abstract Energetic balance of the Earth and its atmosphere is determined by three groups of processes, namely absorption and scattering of solar radiation by the atmosphere and Earth's surface, absorption and emission of infrared radiation by the Earth and atmosphere, and heat transport from the Earth to the atmosphere through air convection and due to water evaporation from the Earth's surface and its condensation in the atmosphere. Global energetic fluxes due to these channels from various sources are compared. Heat fluxes due to convection and water condensation in the atmosphere are evaluated directly and differ from accepted ones. Correction of parameters of heat transfer from the Earth to atmosphere is made on the basis of the nature of these processes. The photosynthesis process results in transformation of atmospheric carbon dioxide in a biomass, i.e., in organic compounds of plants located at land and in oceans. The process of degradation of the biomass results from its oxidation with formation of carbon dioxide which subsequently is injected into the atmosphere. These two processes establish the equilibrium between the biomass and atmospheric carbon dioxide. Atmospheric carbon dioxide is mixed with air because of a large residence time. Extraction of fossil fuels from the Earth's interior and their combustion shifts this equilibrium. From the middle of eighteenth century up to now as a result of this human activity, the biomass increases by approximately 40% and the amount of atmospheric carbon dioxide increases by 50%. The mass of atmospheric carbon is 870 GtC now (1 GtC is 10^9 tons of carbon in carbon dioxide or in organic compounds), and the biomass is correspondingly 685 GtC and 530 GtC for land and oceans. The rate of the global photosynthesis process, as well as that of the opposite process of oxidation of surface carbon, is estimated as 220 GtC/year, whereas this value due to the human activity is approximately 10 GtC/year. The photosynthesis process provides production of food. The Malthusian concept that the Earth is not capable to provide the growing population by food is not valid, at least, from sixties of twentieth century because of the green revolution. The land surface is exhausted more or less for agriculture, whereas the contemporary efficiency of food production is very low and may be significantly increased due to new technologies. An increase of the global temperature during last 150 years is about 1 °C, whereas temperature variations in past exceeds 10 °C. The contemporary warming of the Earth causes the rise of sea level approximately 3.3 mm/year.

B. M. Smirnov, *Global Energetics of the Atmosphere*, Springer Atmospheric Sciences, https://doi.org/10.1007/978-3-030-90008-3_5

5.1 Energetic Balance of the Earth and Atmosphere

5.1.1 Channels of Energetic Balance

We now scrutinize the energetic balance of the Earth and its atmosphere or the global energy budget. This means the analysis of the channels of energy gain and loss both by the Earth and its atmosphere. Because the heat state of the Earth follows from this balance, the global energy budget is examined more century (for example [1]). The basis of all these energetic processes is solar radiation which is penetrated in the Earth's atmosphere and is absorbed by it, as well as by the Earth's surface. One can divide global energetics processes involving the Earth and its atmosphere in three parts. The first one includes penetration of solar radiation in the Earth's atmosphere, so that the energy of solar radiation is absorbed partialy by the atmosphere and Earth and partially is reflected by them. The second channel is connected with emission and absorption of infrared (long wave) radiation by the Earth and atmosphere. The third group of global energetic processes corresponds to heat exchange between the Earth's surface and atmosphere as a result of air convection and water evaporation from the Earth surface in the form of free water molecules and subsequent water return to the Earth's surface in the condensed state, i.e., in the form of liquid drops or snow. The energy resulted from water condensation in the atmosphere remains there, and in this way, the energy transfers from the Earth to the atmosphere.

The first energetic channel of the global energetic balance results from penetration of solar radiation in the atmosphere. The power of Sun radiation is 3.86×10^{26} W on average [2] that corresponds to the energy flux or to the solar irradiance [3] of 1365 W/m^2 [4–8]. Approximating the solar emission by radiation of a blackbody, one can obtain the effective temperature $T_{ef} = 5777$ K of solar radiation. The main contribution to the solar radiation gives the visible and infrared spectrum range. Correspondingly, ultraviolet radiation, radio waves and X-rays give a contribution below 1% to the total solar power. Time oscillations of the solar irradiance which early was named by the solar constant are small [6, 7]. The canonical solar irradiance on the basis of data of 1990s was accepted as 1365 W/m^2 and during the minimum during 2008 the solar irradiance minimum was 1361 W/m^2. This value is accepted now as the solar irradiance [9, 10] and corresponds to the average flux of solar radiation approximately 340 W/m^2. In addition, the most likely drift in quiet Sun irradiance since 1700 up to now is between +0.07 W/m^2 and −0.13 W/m^2 [11].

From the standpoint of solar radiation as the energy source which induces energetic processes in the Earth and atmosphere, it is of importance for its stability. Early the solar irradiancy was named the solar constant, until the accuracy of its measurement was remarkably worse than 10^{-3} [12]. As it follows from Fig. 5.1, the solar irradiance correlates with the magnetic activity, and hence, evolution of the solar irradiance subjects to the solar cycle [13] which is characterized by the quasi-periodicity of 11 year observed in oscillations in the number of solar spots. Solar spots are a dark area on the Sun surface. They are formed as a result of processes involving motion

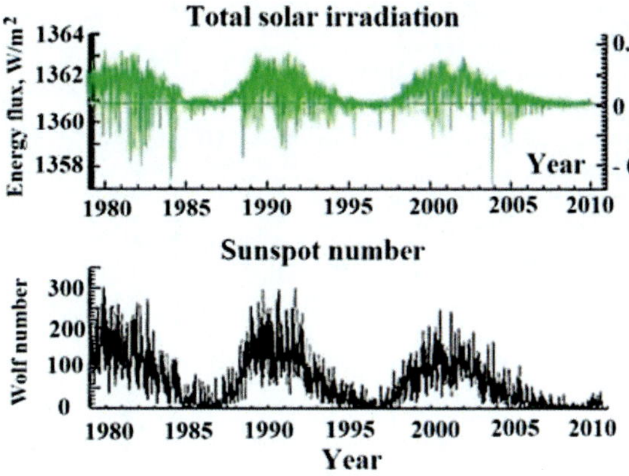

Fig. 5.1 Evolution of the solar irradiance as the average energy flux at the Earth's atmosphere top expressed in W/m^2 and evolution of the Wolf number [11]

of the solar plasma in a magnetic field, so that instabilities in convective motion of this plasma lead to solar flares in the form of plasma explosions as a result of reconnection of magnetic lines of force.

This periodicity is determined by the Wolf number [14] which accounts for both a number of solar spots, as well as a number of groups of solar spots, and also accepts the parameter of a measuring device. The Wolf number as an accepted characteristic of solar activity was introduced by Switzerland scientist R. Wolf in 1848 [15]. The periodicity in evolution of the number of solar spots was discovered by S.H. Schwabe [16] in 1843, and the solar cycle is named sometimes as the Schwabe cycle. Note that, the solar cycle period is not constant, though it is close to 11 years. In spite of small variations of the solar radiance, in considering the energetic balance of the Earth and its atmosphere, we will take the solar irradiance to be constant in time.

The characteristic of interaction between solar radiation and the system consisting of the Earth and atmosphere is albedo, that is the ratio of the solar flux reflected from the total system Earth–atmosphere [17]. Note that we consider solar radiation as a sum of electromagnetic waves of wavelengths $(0.4 - 0.7)\,\mu$m. In reality, the albedo depends on a material of the Earth's surface and is 0.84 if the Earth is covered by snow or ice [18]. In the opposite case, if the Earth's surface is covered by green forrest, the albedo is equal approximately 0.14 [18]. The average Earth's albedo is accepted approximately 0.3 [18]. In addition, the albedo depends on an angle of incidence of solar radiation in the atmosphere. In particular, according to [19], the average Earth's albedo is estimated as 0.34 with a minimum of 0.28 in the subtropics and a maximum of 0.67 at the poles. This value depends only on presence of clouds, and without them the atmosphere is transparent practically for visible light. Thus, the albedo depends on certain conditions of the atmosphere, as well as on covering of the Earth's surface.

The second channel of the global energy budget is infrared radiation because just such radiation may be emitted at room temperature of a surface. Interaction of infrared radiation with atmosphere is stronger than that for visible radiation. In particular, propagation of radiation through clouds which consist of water microdroplets has

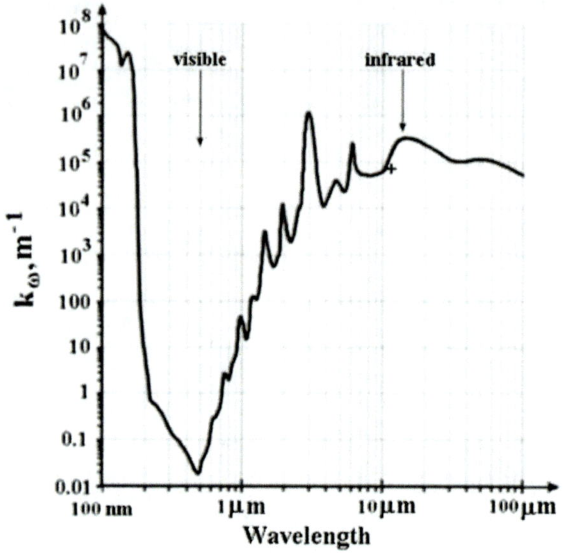

Fig. 5.2 Absorption coefficient of liquid water under normal conditions [20, 21]. Arrows show the wavelength for solar radiation and infrared (longwave) one, a cross is a result of atmospheric measurements [22]

a different character. Fig. 5.2 contains the absorption coefficient of liquid water depending of the wavelength [20]. As is seen, the absorption coefficient for liquid water and for the infrared spectrum range exceeds that for visible light almost seven orders of magnitude. As a result, water microdroplets absorb effectively infrared radiation, while it is possible only elastic scattering of visible photons on water droplets.

This difference in the absorption coefficients for infrared and visible spectrum ranges determines the Twomey effect [23, 24]. According to this effect, aerosols as small particles consisting of various chemical compounds are nuclei of condensation in formation water microdroplets. Pure water microdroplets, as well as clouds consisting of them, are transparent for visible radiation and hence cannot be observed by eye. But aerosols dissolved in water microdroplets can absorb visible radiation, and water microdroplets become visible for light.

One more peculiarity of interaction of infrared radiation with the ground and atmosphere is that emission of the ground in the infrared spectrum range is close to that of a blackbody. Indeed, the equilibrium radiative flux J_E from a surface of a temperature T_E is equal

$$J_E = \gamma J_{bl}, \quad J_{\rm bl} = \sigma T_E^4, \tag{5.1.1}$$

where γ is the gray coefficient of this surface, $J_{\rm bl}$ is the radiative flux from the blackbody surface which is determined by the Stephan–Boltzmann law and $\sigma = 5.67 \times 10^{-8}\,\mathrm{W/(m^2 \times K^4)}$ is the Stephan–Boltzmann constant. The emissivity of various objects which are located at the ground is close to one, the radiative flux from the Earth surface is close to that according to a blackbody. Indeed, Fig. 5.3

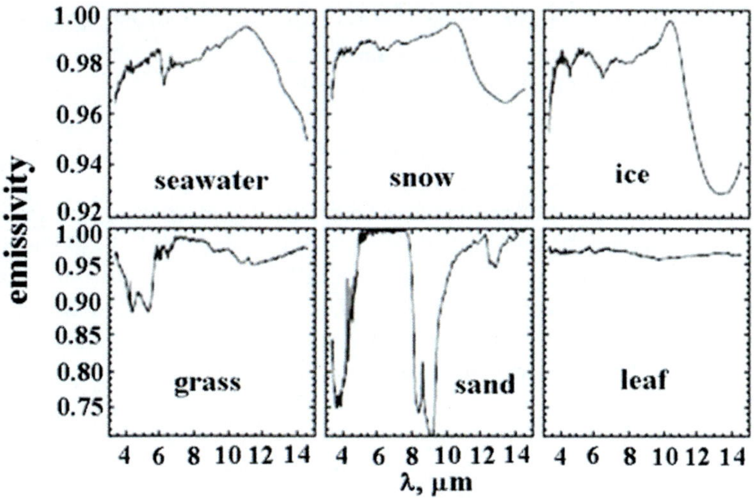

Fig. 5.3 Gray coefficient of various surfaces in the infrared spectrum range [25]

which contains values of the gray coefficient in the infrared spectrum range for various natural objects confirms this fact. This allows one to consider emission of the Earth's surface similar to that of a blackbody.

Let us use this for the model of standard atmosphere according to which the temperature of the Earth's surface is $T_E = 288\,\mathrm{K}$. According to the Wien law [26, 27], the radiation wavelength λ_T that corresponds to the maximum of the partial radiative flux for the temperature of standard atmosphere is equal $\lambda_T \approx 10\,\mu\mathrm{m}$. In addition, this temperature gives for the radiative flux from the blackbody surface at the indicated temperature

$$J_{\mathrm{bl}}(T_E) = 390\ \mathrm{W/m^2} \tag{5.1.2}$$

This value may be considered as the upper limit for the average radiative flux J_E emitted from the Earth's surface that must be close to this value.

Note that infrared radiation at wavelengths of the order of $10\,\mu\mathrm{m}$ is created by vibration–rotation and rotation transitions in molecules, as well as by microdroplets and microparticles of the order of this size. At wavelengths with a large optical thickness emission of the atmosphere toward the Earth is created by its low layers, the atmospheric regions which are responsible for atmosphere emission toward the Earth and outside are separated. Subsequently, in analyzing the greenhouse atmospheric effect, we consider methods of determination of the radiative temperature at a given frequency toward the Earth and outside depending on the density of optically active components of the atmosphere.

The third group of energetic processes includes heat transfer from the Earth's surface to the atmosphere as a result of air convection and transport of latent heat as a result of evaporation of water molecules from the Earth's surface with subsequent condensation of these molecules in the atmosphere. The energy flux J_h as a result of heat transport from the Earth to the atmosphere is equal

$$J_h = J_{\rm con} + J_{\rm ev}, \tag{5.1.3}$$

where $J_{\rm con}$ is the energy flux as a result of heat transport through the atmosphere due to air convection and $J_{\rm ev}$ is the energy flux due to condensation of evaporated water molecules in the atmosphere. The later flux may be expressed through the rate of evaporation of water molecules in the atmosphere. Let dM/dt be the total water mass reached the Earth's surface per unit time in the form of the condensed phase (rain or snow). We have for the energy flux due to condensation of evaporated molecules

$$J_{\rm cond} = \frac{\varepsilon}{m_o S}\frac{{\rm d}M}{{\rm d}t} = \varepsilon j, \tag{5.1.4}$$

where $\varepsilon_o = 0.44\,{\rm eV}$ is the binding energy per one molecule in liquid water, $m_o = 3 \times 10^{-23} g$ is the mass of a water molecule, $S = 5.1 \times 10^{14}$ m^2 is the area of the Earth's surface and j is the flux of precipitated condensed water. Taking $j = 1 \times 10^{17}$ cm$^{-2}s^{-1}$ according to formula (2.1.8) or dM/dt $= 1.5 \times 10^{13}$ g/s [28–32], one can obtain from formula (5.1.4) for the energy flux due to water evaporation from the Earth's surface and its subsequent condensation in the atmosphere

$$J_{\rm ev} = 69 \text{ W/m}^2 \tag{5.1.5}$$

Note that this value is based on the scheme of water circulation through the atmosphere where evaporated water returns to the Earth's surface in the condensed form only that leads to a heightened energy flux due to this channel.

5.1.2 Global Energy Budget

Let us analyze the energy balance for the Earth and its atmosphere. Evidently, in this analysis, it is correctly to operate with powers P which the Earth and atmosphere obtain or loss due to each channel. In this case, the role of some regions of the globe is not remarkable. The average fluxes from various channels are used, so that the power of a certain channel is divided to the area of the Earth's surface $S = 5.1 \times 10^{14}$ m^2, and the average fluxes are measured in units W/m^2. Table 5.1 represents the Earth's energy balance in the form of the average energy fluxes for basic channels of energy exchange involving the Earth's surface and the atmosphere.

We use in Table 5.1 five sources of the used information. Data marked by the cipher 1 include the information from National Academy of Science (USA) [33] within the framework of the NASA program and was included in the author's books [34–36]. The cipher 2 gives information from the Salby book [37] which along with

Table 5.1 Global energy budget according to different sources

Channel, process	1	2	3	4	5	Average
1. Solar radiation	–	–	–	–	–	–
Incident solar flux	340	343	341	340	340	341 ± 1
Absorbed by A*	57	68	78	75	79	72 ± 9
Absorbed by E	163	164	161	165	161	163 ± 2
Reflected by E	14	16	23	23	24	19 ± 4
Reflected by A	106	90	79	77	76	86 ± 10
2. Infrared radiation	–	–	–	–	–	
Emission by E	386	390	396	398	397	393 ± 5
Emission by A toward E	327	327	333	346	342	335 ± 7
Emission by A outside	200	215	217	220	218	215 ± 8
Outgoing radiation from E	20	22	22	20	21	21 ± 1
Absorbed by E	327	327	333	346	342	335 ± 8
Absorbed by A	366	368	374	379	375	372 ± 6
3. Heat transport from E to A	104	106	97	112	106	105 ± 6
Evaporation and condensation	(69)	90	80	88	86	86 ± 4
Convection	(46)	16	17	24	20	19 ± 4

*A—the atmosphere, E—the Earth
The energy fluxes for an indicated process are expressed in W/m^2

data of National Academy of Science (USA) [33] contain satellite measurements of atmospheric emission and their analysis [38–43]. Other sources of indicated data for the global energy budget occur later. Namely, the cipher 3 relates to data of [44–47], the cipher 4 corresponds to [10] data, and information for the cipher 5 is taken from [48]. The above data were included in monographs for atmospheric physics (for example, [49–53]). Data of the subsequent analysis for the global energetic balance of the Earth and its atmosphere [54, 55] support the data of Table 5.1.

In considering the energy fluxes for the Earth and its atmosphere, we assume an equilibrium between the Earth, atmosphere and their environment. Therefore, we use three laws of the energy conservation, namely for the Earth, atmosphere and environment. Hence, in constructing Table 5.1, we correct the used data such that the above laws of energy conservation were fulfilled. Results of the statistical average of the Table 5.1 data are given in Fig. 5.4.

One can summarize data for various sources given in Table 5.1 for the total average energy fluxes separately for the Earth and atmosphere. The average energy fluxes for emission and absorption for the Earth and its atmosphere are represented in Table 5.2. The part of emission in the form of infrared radiation directed to the Earth is called the greenhouse phenomenon or greenhouse effect.

In addition, Table 5.3 contains the conservation laws of energy for three objects, namely for the Earth, its atmosphere and environment, or the sum of the Earth and atmosphere. The values of Table 5.3 are taken from Table 5.1 as the average ones for a corresponding process in accordance with their position in Table 5.1.

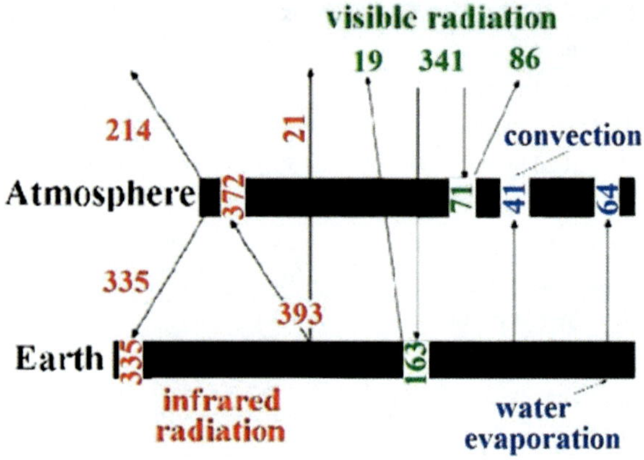

Fig. 5.4 Energetic balance of the Earth and atmosphere in the form of average energy fluxes expressed in W/m^2 for indicated channels [34]. Absorbed energy fluxes are given inside corresponding rectangulars, consumed energy fluxes are indicated near arrows

Table 5.2 Average summarized energy fluxes for gain and loss of energy by the Earth and atmosphere

Channel, process	1	2	3	4	5	Average
Absorbed by E	490	491	494	509	503	497 ± 8
Emitted by E	386	390	396	398	397	393 ± 5
Absorbed by A	423	436	433	454	454	440 ± 14
Emitted by A	527	542	532	566	560	545 ± 16

The energy fluxes are given in W/m^2. E denotes the Earth, A means the atmosphere

Table 5.3 Conservation laws for energy fluxes of indicated processes

System	Income	Loss	Total flux
The Earth and atmosphere	341	19+86+215+21	341
The Earth	163+335	393+105	498
The atmosphere	72+372+105	335+214	549

Energy fluxes are measured in W/m^2 and are given in accordance with data of Table 5.1 for the Earth and the Earth's atmosphere as a whole

Moreover, we correct the average parameters of Table 5.1 with fulfilling of the energy conservation laws. This leads to a small shifts of some average values compared with their statistical average ones. The total flux means the summarized energy flux which is gained or is lost by an indicated object.

Data of Table 5.1 testify about separation the longwave energy flux toward the Earth and outside. This means that different atmospheric regions create this radiation. Let us introduce the effective regions of the atmosphere which are responsible for emission of longwave radiation toward the Earth's surface and outside. Indeed, let us take the energy fluxes for infrared radiation from the atmosphere toward the Earth $J_\downarrow$ and outside $J_\uparrow$ according to data of Table 5.1 as

$$J_\downarrow = 335\ \mathrm{W/m^2},\ \ J_\uparrow = 215\ \mathrm{W/m^2} \tag{5.1.6}$$

From this, one can obtain the temperatures $T_\downarrow$ and $T_\uparrow$ for the atmosphere layers which are responsible for atmospheric emission toward the Earth $T_\downarrow$ and outside it $T_\uparrow$ on the basis of the Stephan-Boltzmann law

$$J_\downarrow = \sigma T_\downarrow^4, \quad J_\uparrow = \sigma T_\uparrow^4, \tag{5.1.7}$$

Correspondingly, within the framework of this model [34, 56] which are based on a large optical thickness of the atmosphere, one can introduce the effective regions of emission. These effective regions emit as a blackbody that gives for radiative temperatures or for effective temperatures of emission to the Earth $T_\downarrow$ and outside $T_\uparrow$ as

$$T_\downarrow = 277\text{ K}, \quad T_\uparrow = 248\text{ K} \tag{5.1.8}$$

The altitudes of effective regions $h_\downarrow$ and $h_\uparrow$ for infrared emission to the Earth and outside follow from relations $T_\downarrow = T[h_\downarrow]$ and $T_\uparrow = T([h_\uparrow])$ that have the form

$$T_\downarrow = T_E - h_\downarrow \frac{\mathrm{d}T}{\mathrm{d}h}, \quad T_\uparrow = T_E - h_\uparrow \frac{\mathrm{d}T}{\mathrm{d}h}, \tag{5.1.9}$$

where within the framework the standard atmosphere model, the temperature of the Earth's surface is the $T_E = 288\,\text{K}$ and the average temperature gradient is $\mathrm{d}T/\mathrm{d}h = 6.5\,\text{K/km}$ [57]. Thus, from equation (5.1.9), one can obtain

$$h_\downarrow = 1.6\text{ km}, \quad h_\uparrow = 6.1\text{ km} \tag{5.1.10}$$

Values (5.1.8) and (5.1.10) describe the character of longwave atmospheric emission under the condition of a large optical thickness of the atmosphere.

In analyzing energetic processes of third group of Table 5.1, one can see nearby total heat fluxes from the Earth's surface to the atmosphere. But partial energy fluxes due to air convection and water condensation in the atmosphere are different for parameters marked by cipher 1 and other ones. Hence, we separate these data and the average values correspond to averaging of parameters from sources 2–5.

Table 5.1 takes into account basic channels of energetic processes involving the Earth's surface and atmosphere which determine the global energy budget. Now we also consider weak atmospheric processes. These processes do not give the contribution to the energy balance of the Earth and atmosphere, but they may be important for other aspects of atmospheric physics including the influence of the human activity on the environment. Some processes of such a type are represented in Table 5.4. For comparison, the total power of incident solar radiation penetrated in the Earth's atmosphere is 1.7×10^{17} W.

Let us analyze briefly data of Table 5.4. The main contribution to the human energetics goes from combustion of fossil fuels (oil, methane, coal) which power is 85% of the total one [58, 59]. A small part of solar radiation includes hard radiation of the wavelengths below 100 nm, and the photon flux at these wavelengths is equal approximately to $2.4 \times 10^{10}\ \text{cm}^{-2}\,\text{s}^{-1}$. Though this radiation does not reach the

Table 5.4 Powers of the total energetic processes P expressed in W

Type of energy	Process	Power, P, W
Human energetics	Combustion of fossil fuels, etc.	2×10^{13}
Ionization by solar radiation	Ionization of molecules and atoms of ionosphere	7×10^{10}
Tropospheric ionization	Ionization of air molecules by cosmic rays	1×10^{9}
Kinetic energy of particles	Solar wind outside atmosphere	5×10^{10}
Electric energy	Lightnings	1×10^{11}
Electric energy	Ion drift in troposphere	6×10^{8}

Earth's surface, absorption of these photons leads to ionization and dissociation of molecules and atoms of the upper atmosphere. As a result of these processes, the ionosphere is created in the range of altitudes of (100–200) km, and its existence is of importance for applications because short-length electromagnet waves reflect from it. Cosmic rays penetrate in lower layers of the atmosphere and cause ionization there. The rate of ion formation per unit atmospheric column under the action of cosmic rays is 4.5×10^7 cm^{-2} s^{-1} [60]. Taking the energy of 30 eV that is spent per one ionization event, one can obtain the value given in Table 5.4.

The solar wind is a plasma flux created in the solar corona [61–63] contains electrons and protons of energy of the order of 1 keV. Except polar regions, the solar wind cannot penetrate in the Earth's atmosphere due to the action of the Earth's magnetic field. A typical velocity of the solar wind is approximately $v = 4 \times 10^7$ cm/s, and the number density of electrons or ions is $N_i = 7$ cm^{-3}. From this, one can estimate the power of the solar wind for the Earth's cross section $\pi R_{\oplus}^2$, where $R_{\oplus}$ is the Earth's radius. From this, one can estimate the power of this process given in Table 5.4.

In considering energetic processes of atmospheric electricity, we have that the average electric current from clouds to the Earth's surface is 1700 A [64]. The electric potential of cumulus clouds with respect to the Earth's surface is 20–100 MV [65], and taking its average value $U = 60$ MV, one can obtain for the power of this process given in Table 5.4. Another electric process in the clear-sky atmosphere is ion drift in the atmosphere under the action of the electric potential of the tropospheric upper layers that is equal $U_o = 240-300$ kV [66–68]. Taking the number density of ions as $N_i \sim 10^3$ cm^{-3}, both positive and negative ones, one can obtain the power of this process represented in Table 5.4. Note that atmospheric electric processes are secondary ones with respect to water circulation processes though the atmosphere, the power of these processes is small compared to that for processes involving water.

5.1.3 Heat Transfer from Ground to Atmosphere

Examining the heat transfer between the Earth and atmosphere, we conserve in Table 5.1 only total average energy flux. According to data which denote by ciphers 2–5 of Table 5.1, the statistically averaging energy flux due to condensation of evaporated water in the atmosphere is equal $J_{ev} = (86 \pm 4)$ W/m^2, and the average energy flux due to air convection is $J_c = (19 \pm 4)$ W/m^2 that also is found in accordance with [69]. But using the rates of water evaporation from the Earth's surface and the part of atmospheric water which returns to the ground in the condensed form (rain or snow) and is equal $(\Lambda - \lambda)/\Lambda = 0.77$, where parameters λ and Λ characterize the altitude distribution for the number density of air molecules (2.1.1) and water molecules (2.1.19). In this case, we obtain on the basis of formula (5.1.5) for the energy flux due to condensation of atmospheric water the value $J_c = 69/$ m^2 which differs from the above one.

Our goal now is to determine which of these values is more reliable. For this aim, we analyze first the energy flux due to air convection. In this analysis, we use the analogy in transfer of mass and heat both in motionless air and in the case of air convection. In both cases, heat and mass are transported simultaneously. This leads to identical ratios for the coefficient of thermal conductivity κ to the diffusion coefficient D_L. In the case of motionless at atmospheric pressure and temperature $T = 288$ K, the diffusion coefficient of air molecules is equal $D_m = 0.18\,\mathrm{cm}^2/\mathrm{s}$ under normal conditions, and the thermal conductivity coefficient is $\kappa_m = 2.51 \times 10^{-4}\,\mathrm{W}/(\mathrm{cm}\cdot\mathrm{K})$ [70, 71]. Thus, we have for the thermal conductivity coefficient from comparison with that of motionless air

$$\kappa_{\mathrm{con}} = \mathrm{N}D_L\xi, \ \xi = \frac{\kappa_m}{ND_m} = 3.8, \tag{5.1.11}$$

where $N = 2.55 \times 10^{19}\,\mathrm{cm}^{-3}$ is the number density of molecules under normal conditions. According to formula (3.2.38), the diffusion coefficient in convective air near the air surface is equal $D_L = 5 \times 10^4\,\mathrm{cm}^2/\mathrm{s}$. From this, one can find the coefficient of thermal conductivity for convective motion of air near the Earth's surface $\kappa_{\mathrm{con}} = 70\,\mathrm{W}/(\mathrm{cm} \times \mathrm{K})$. Note that this value is obtained on the basis of the assumption which accounts for the transfer of air molecules that cause transport of energy. This character of mass and energy transport is identical in motionless air and air which partakes in convective motion of the atmosphere.

Let us determine the energy flux due to heat transfer from the Earth to atmosphere due to air convection according to formula

$$J_{\mathrm{con}} = -\kappa_{\mathrm{con}}\frac{\mathrm{d}T}{\mathrm{d}h}, \tag{5.1.12}$$

where the average observed temperature gradient for a low atmosphere is $dT/dh = 6.5K/km$. This gives for the heat flux due to air convection

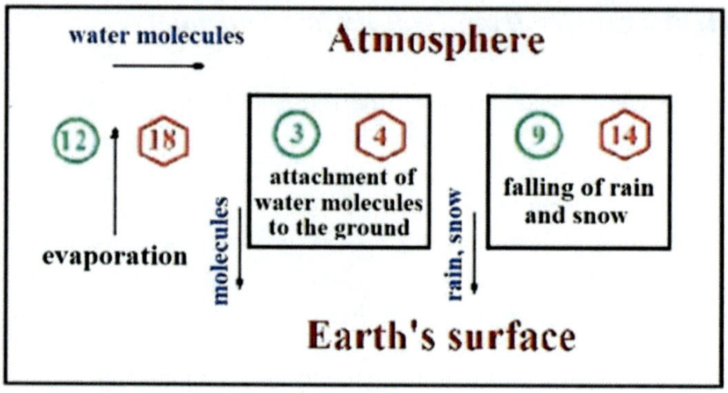

Fig. 5.5 Circulation of water through the atmosphere. Three channels of this circulation include evaporation of water molecules from the Earth's surface, return of water back in the form of water molecules, as well as in the form of the condensed phase, i.e., in the form of rain and snow. Values of rates of global water masses $\mathrm{d}M/\mathrm{d}t$ are given in red inside hexagons and are expressed in units 10^{12} g/s, as well as values of average fluxes of water molecules are drawn in green inside circles and are given in units $10^{16}\ \mathrm{cm}^{-2}\ \mathrm{s}^{-1}$

$$J_{\mathrm{con}} = 45\ \mathrm{W/m^2} \tag{5.1.13}$$

We expect the accuracy of energy fluxes as approximately 20%. As is seen, the energy flux (5.1.13) due to air convection differs from that of Table 5.1 due to sources given by ciphers 2–5 approximately in two times. This is outside the accuracy of the value (5.1.13). Because evaluation of the value (5.1.13) is grounded, we accept this value, rather than those of Table 5.1 due to sources given by ciphers 2–5.

The total heat flux (5.1.3) that is the sum of energy fluxes due to air convection and water condensation in the atmosphere according to formulas (5.1.5) and (5.1.13) is $J_h = 114\ \mathrm{W/m^2}$. This is close to that of Table 5.1 which is equal $(105 \pm 6)\ \mathrm{W/m^2}$. Correcting data (5.1.5) and (5.1.13) on the basis of the value of the total heat flux, we obtain

$$J_c = 64\ \mathrm{W/m^2},\ \ J_{\mathrm{con}} = 41\ \mathrm{W/m^2} \tag{5.1.14}$$

Just these values are included in Fig. 5.4. In addition, this gives for the thermal conductivity coefficient related to convective motion of atmospheric air

$$\kappa_{\mathrm{con}} \approx 60\ \mathrm{W/(cm \cdot K)} \tag{5.1.15}$$

The energy flux J_c due to water condensation of evaporated water molecules in the atmosphere given in formula (5.1.14) allows us to recalculate water fluxes between the Earth and atmosphere, and we give in Fig. 5.5 corrected values of water circulation through the atmosphere. Indeed, the energy flux J_c in formula (5.1.14) relates only to such evaporated molecules which return to the ground in the condensed state, i.e., in the form of rain or snow. Taking the energy of release of one water molecule from the liquid surface as $\varepsilon_o = 0.44$ eV according to formula (2.1.13), one can obtain for this part of the flux $j_c = j_{\rm ev} - j_w$, where $j_{\rm ev}$ is the flux of evaporated water molecules from the Earth's surface, j_c is the part of this flux which is transformed in the condensed state and returns to the Earth's surface in the form of rain or snow and j_w is the part of the flux of evaporated molecules which return back in the form of free water molecules. Thus, we have

$$j_c = \frac{J_{\rm ev}}{\varepsilon_o} = 9 \times 10^{16}\ {\rm cm}^{-2}{\rm s}^{-1}$$

Another part of evaporated water molecules forms the flux j_w which corresponds to water molecules which return to the ground in the form of free molecules. On the basis of the value of formula (3.2.37), one can find for this flux

$$j_w = 3 \times 10^{16}\ {\rm cm}^{-2}{\rm s}^{-1}$$

From this, one can obtain for the total average flux j_{ev} of evaporated water molecules from the Earth's surface

$$j_{\rm ev} = j_c + j_c == 12 \times 10^{16}\ {\rm cm}^{-2}{\rm s}^{-1}$$

The above analysis allows us to determine the total water mass $\mathrm{d}M/\mathrm{d}t$ evaporated from the Earth's surface per unit time. Evidently, the partial value for $\mathrm{d}M/\mathrm{d}t$ is connected with the corresponding flux j of evaporated water molecules through the relation

$$\frac{\mathrm{d}M}{\mathrm{d}t} = jSm_o,$$

where $m_o = 3 \times 10^{-23}$g is the mass of the water molecule. Fig. 5.5 contains the rates of global water masses dM/dt evaporated from the Earth's surface.

One can treat the above results from another standpoint. Let us consider the energy flux to the Earth's surface as the reason of water evaporation from the Earth's surface. Let us determine the energy cost of one water molecule $\varepsilon(H_2O)$ which is introduced as the ratio of the total energy flux J_E absorbed by the Earth's surface to to flux $j_{\rm ev}$ of evaporated water molecules, that is

$$\varepsilon(H_2O) = \varepsilon_o \frac{J_E}{J_c} = 3.4\ {\rm eV}, \qquad (5.1.16)$$

where we use the value of the total energy flux $J_E = 498$ W/m^2 according to Tables 5.2, 5.3 and the energy consumed on evaporation of water molecules is $J_c = 64$ W/m^2 according to formula (5.1.14). We also take the binding energy $\varepsilon_o = 0.44$ eV per one water molecule that is the energy consumed for a release of one molecule. This means that the absorbed energy is consumed for other channels.

On the basis of the above results, one can estimate a typical size l of an air vortex which is responsible for propagation of water molecules through the atmosphere. Indeed, we have the following estimation for the energy flux J_{ev} due to the formula

$$J_{ev} \sim \varepsilon_o \times D_L \cdot \frac{N(H_2O)}{l},$$

where $D = 5 \times 10^4\ cm^2/s$ is the diffusion coefficient due to air convection according to formula (3.2.38), $N(H_2O)$ is the number density of water molecules near the Earth's surface, and $\varepsilon_o = 0.44$ eV is the binding energy of a water molecule with the liquid water surface. From this, one can obtain for a typical vortex length l

$$l \sim 2\,m \tag{5.1.17}$$

The effective coefficient of thermal conductivity in atmospheric air, which is given by formula (5.1.15), allows one to determine the thermal diffusivity coefficient χ

$$\chi = \frac{\kappa_{con}}{C_p \rho} = 5 \times 10^4\,cm^2/s, \tag{5.1.18}$$

where the heat capacity is $C_p = 1.005 J/(g \times K)$ and the mass density equals $\rho = 1.225 \times 10^{-3}\ g/cm^3$ for atmospheric air near the Earth's surface within the framework of the model of standard atmosphere [57]. In particular, this allows one to determine the average altitude $\overline{h}$ of heat propagation through time t

$$\overline{h} = \sqrt{2\chi t} \tag{5.1.19}$$

In particular, from this, it follows that during a day or a night, the Earth's surface can change by heat as a result of convection with air layers located at altitudes up to 500m.

5.2 Photosynthesis

5.2.1 *Character of Photosynthesis*

Photosynthesis is a process of transformation of atmospheric carbon dioxide in a solid carbon under the action of solar radiation in the visible spectrum range. Solid carbon means carbon atoms which are included in an organic compound of growing plants. As a result of this process, carbon dioxide of atmospheric air and water located inside a plant produce various forms of solid carbon. Photosynthesis starts from absorption of a solar photon by a chlorophyll molecule [72] as a green pigment molecule. The mechanism of the absorption event consists in displacement of an electron, so that an excited state of a chlorophyll molecule has an electrochemical nature [73]. In particular, intermediate compounds of transformation of water in the course of photon absorption are molecules of nicotinamide adenine dinucleotide phosphate (NADPH) and adenosine triphosphate (ATP). The energy of absorbed photons is kept in these compounds during a short time, and then, various stable carbon-contained molecules are formed as a result of interaction of chlorophyll molecules with absorbed carbon dioxide and water molecules. In this manner, an excitation energy from absorption of solar photons is used in transformation of CO_2 and H_2O molecules in various organic molecules.

Chlorophyll molecules are located inside cells of plants–chloroplasts or plactids [74] which play an important role in this process [75, 76]. Typical chloroplasts have the form of lenses of diameter $(3 - 10)\mu$m and thickness $(1 - 3)\mu$m. Concentrating on energetics of the photosynthesis process, we leave aside growth, replication and dividing of plactids, as well as the multistage character of photosynthesis and various chemistry of these processes [77–82]. Below we take into account that plactids as cells with chlorophyll molecules inside them are contained in plants at lands and in phytoplankton of oceans, and they are an elementary structure where the photosynthesis takes place.

Let us focus on the summarized photosynthesis process which proceeds according to the scheme

$$6H_2O + 6CO_2 + n\hbar\omega \to C_6H_{12}O_6 + 6O_2, \tag{5.2.1}$$

The photosynthesis process proceeds in several stages, and the scheme (5.2.1) summarizes some stages of this process. Here $\hbar\omega$ is the photon, and n is the number of photons which is required for the photosynthesis process, so that $n\hbar\omega$ is the energy used for this process. According to the scheme (5.2.1), glucose $C_6H_{12}O_6$ and oxygen molecules are formed as a result of this multistage process. One can consider this process as a model one for the photosynthesis.

We give in Fig. 5.6 the chemical structure of glucose as a chemical compound. Comparing it with the chemical structure of molecules of carbon dioxide O=C=O and water H–O–H, one can expect that the glucose molecule is constructed by replication of its elements on some matrix. Subsequently, glucose may be transformed in various forms of solid carbon [78, 83]. This process is complex, proceeds in several stages and depends on a matter where it is realized. Taking the process (5.2.1) as the model photosynthesis process, we have for the enthalpy of this process $\Delta H = 468$ kJ/mol [84], and this enthalpy refers to one carbon atom.

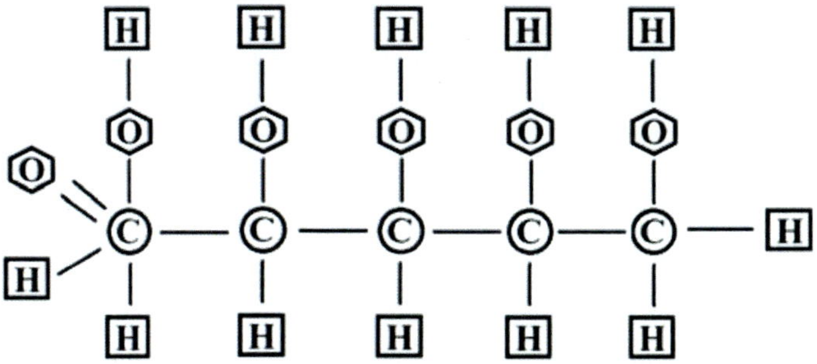

Fig. 5.6 Character of chemical bonds between atoms in glucose

One can join a series of photochemical processes in photosynthesis in a general model where several carbon atoms of CO_2 molecules are joined in elementary units of forming solid carbon. According to this feature, plants are divided in C3 and C4 groups, so that the character of the photosynthesis process depends on a number of carbon atoms which participate in transformation of carbon atoms from CO_2 molecules in an organic compound that may be modeled as an elementary chemical process [85]. Plants of the group C3 include wheat, rice and beans, and this process proceeds through collisions between the CO_2 molecule and some intermediate products. In this case, a CO_2 molecule collides with so-called rubisco molecule (ribulose bisphosphate carboxylase) (for example, [86–89]) which plays a role of the catalyst in replication of solid carbon. Because in this process excess oxygen is removed from a growing product of the photosynthesis process, this process is called also the photorespiration [90, 91].

Plants of the group C4 [84, 92] include maize, sugarcane and sorghum. In this case, carbon dioxide molecules are captured by the mesophyll cell, and the chemical process involves molecules in the bound state [93]. As a result, dependencies of the photosynthesis rate on parameters of this process, in particular, on the concentrations of CO_2 molecules in atmospheric air, are different for plants of these groups [94]. Comparison of photosynthesis mechanisms for C3 and C4 groups, one can find that the C4 one is more complex chemical process, but it is energetically more profitable under optimal conditions.

C4 is energetically favorable for plants only at certain temperatures and concentrations of atmospheric CO_2. The group C4 prevails at high CO_2 concentrations that taken place in past when the concentration of CO_2 molecules was above 700 ppm. However, at current concentrations of CO_2 molecules approximately 400 pm and at a summer temperature approximately 30 °C, the maximum theoretical efficiency of the photosynthesis process for plants of $C4$ groups is 4.6%, whereas for plants of the group C3 the maximum of the energetic efficiency is 6%. Under optimal conditions, the maximum theoretical efficiency for transformation of the solar energy into the chemical one of bound carbon is approximately 11%, and under real conditions, it is estimated as (3–6)% [95, 96]. Of course, real efficiencies of the photosynthesis process are lower.

According to the mechanisms of the photosynthesis process, it is different at land and in oceans. At land, this process proceeds mostly in forests, whereas grasses and agriculture products as a result of the human activity give a smaller contribution to the global rate of this process. In any case, carbon dioxide in these processes is taken from the atmosphere. In oceans, the photosynthesis process proceeds with participation of dissolved carbon dioxide. The phytoplankton of ocean is an object where assimilation of CO_2 molecules, as well as water molecules, leads to the solid carbon, and phytoplankton of oceans is an analog of plants at land [97, 98]. Subsequently, phytoplankton is used by fishes as a nutrient medium [99]. In this way, a biomass of oceans grows. For such measurements, the isotope method is used on the basis of the radioactive ^{14}C isotope [100, 101]. Along with it, the satellite method is used by measurements of the chlorophyll concentration in oceans [102].

Dissolution of carbon dioxide in oceans is an additional channel of removal of carbon dioxide from the atmosphere along with the photosynthesis process. The subsequent process with participation of dissolving carbon dioxide molecules leads to the chemical process with formation of salts, such as $CaCO_3$ [103]. Destruction of carbonates leads to formation of carbon dioxide molecules which go into the atmosphere. Formation of atmospheric carbon dioxide from land results from plant rot, as well from processes of breathing of plants and microbes.

Let us use that the residence time of CO_2 molecules in the atmosphere is 4 years [104]. Because the total amount of atmospheric carbon dioxide is 870 GtC, one can find from this for the total rate of the photosynthesis process $J_C = 220$ GtC/year as that of carbon absorption by the Earth's surface. Let us determine now on the basis of this rate the efficiency of the photosynthesis process. Take the process (5.2.1) as a model process with the enthalpy $\Delta H = 468$ kJ/mol $= 39$ kJ/gC [84]. From this, we have the power $P = J_C \times \Delta H = 2 \times 10^{14}$W that is consumed on realization of the photosynthesis process. Because the power of solar radiation that penetrates in the Earth's atmosphere is $1.7 \times 10^{17} \times$ W, the average efficiency of the photosynthesis process is approximately 0.1%.

5.2.2 Atmospheric CO_2 Molecules

As a result of photosynthesis, the Earth's atmosphere differs from that of neighboring planets, the Venus and Mars, which atmospheres were formed under nearby conditions. Due to plants located on the Earth's surface, the most part of atmospheric carbon dioxide was transformed in oxygen as a result of the photosynthesis process. Therefore, oxygen is one of the main atmospheric components, whereas the concentration of atmospheric carbon dioxide is relatively small. Therefore, the carbon balance of the Earth and its atmosphere results from transformation of atmospheric carbon dioxide in organic products at the Earth's surface due to the photosynthesis process. Returning of carbon from the Earth's surface into the atmosphere in the form of CO_2 molecules proceeds through oxidation of carbon-contained compounds located at the Earth's surface. We below examine this equilibrium.

In this analysis, we express the carbon amount in carbon-contained compounds in GtC, i.e., in gigatons of carbon (10^{15} gram). In addition, the concentration of CO_2 molecules in the atmosphere is measured in ppm, that is one molecule per million (10^6) air molecules. Then, 2.1 GtC of the atmosphere is equivalent to the concentration of atmospheric CO_2 molecules of 1 ppm. The most reliable monitoring of the concentration for CO_2 molecules at the Mauna Loa Observatory is given in Fig. 2.1, this concentration increases from 316 ppm in 1959 up to 415 ppm in 2020, and the amount of the carbon dioxide mass varies from approximately 660 GtC up to 870 GtC.

Absorption of CO_2 molecules by land is determined by the photosynthesis process, while a sink of CO_2 molecules in oceans has another nature. CO_2 molecules are dissolved in ocean water and then form carbonic acid (H_2CO_3). Its salt transforms

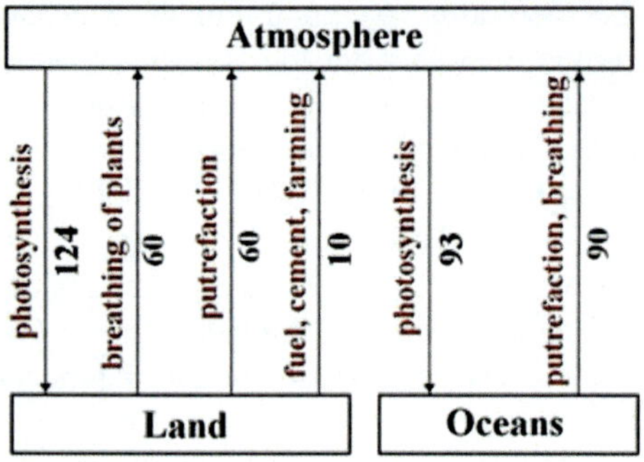

Fig. 5.7 Carbon fluxes between the Earth and its atmosphere which support the carbon equilibrium for the atmosphere [107–110, 110–112]. Fluxes of carbon are given in GtC/year

ocean water into electrolyte, where the positive ion is H_3O^+ and the negative ion is HCO_3^-. Of course, this is a simplified scheme of this electrolyte, whereas according to this so-called dissolved inorganic carbon has the structure of HCO_3 in 90% of cases, 9% of that dissolved inorganic carbon has the structure of CO_3, and only 1% of this matter is a dissolved carbon dioxide CO_2 [105]. Dissolved inorganic carbon forms subsequently some compounds which transit in sediments; a typical time of their formation is 1700 years [106]. In addition, dissolved inorganic carbon is used for formation and growth of algae and phytoplankton as a result of photosynthesis in oceans, and subsequently both formed algae and phytoplankton become a food for fishes and other living organisms located in oceans.

On the basis of this character of dissolution of carbon dioxide in oceans, we consider this as thermodynamic equilibrium between atmospheric and oceanic carbon dioxide [113, 114]. In particular, one can scrutinize this as a result of thermodynamic equilibrium with the salt $CaCO_3$ which proceeds according to the scheme

$$CO_2 + CaO \leftrightarrow CaCO_3 \tag{5.2.2}$$

The enthalpy of this transition at room temperature is equal $\Delta H = 178$ kJ/mol [103]. Since the amount of ocean carbon dioxide exceeds significantly that of atmospheric one, we have for concentration c of atmospheric CO_2 molecules

$$c = \text{Const} \exp\left(-\frac{\Delta H}{T}\right) \tag{5.2.3}$$

If the concentration of atmospheric carbon dioxide is established as a result of this equilibrium, this relation allows one to check the Pauling concept according to which the change of the concentration of atmospheric carbon dioxide results from an increase of the ocean temperature [113, 115]. If the initial concentration of the carbon dioxide molecules is c_1, and their final concentration is c_2, this causes by an increase ΔT in the global temperature or the ocean temperature according to the Pauling concept given by the relationship

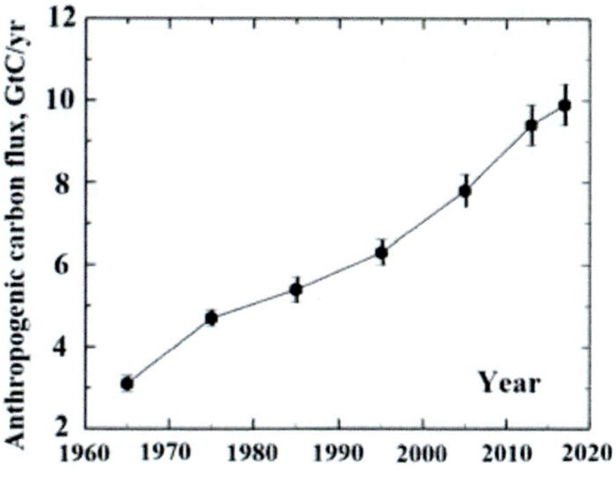

Fig. 5.8 Amount of the industrial carbon which is taken mostly from fossil fuels and subsequently is injected in the atmosphere in the form of CO_2 molecules

$$\Delta T = \frac{T^2}{\Delta H} \cdot \ln\left(\frac{c_2}{c_1}\right) \tag{5.2.4}$$

In particular, as it follows from this formula, the variation of the concentration of CO_2 molecules from 316 ppm in 1959 up to 415 ppm corresponds to the Pauling concept if the global temperature varies during this time by 1.1 °C. One can expect that the Pauling concept gives a certain contribution to the change of the carbon dioxide concentration.

Figure 5.7 presents the carbon balance for the equilibrium between the Earth's surface and atmosphere. We note that the key process in establishment of this equilibrium is the photosynthesis process. Carbon dioxide partakes in this atmospheric process, while in oceans its role is fulfilled by salts of the carbonic acid. One can expect that the accuracy of rates of Fig. 5.7 is restricted and may be estimated by comparison the data from different sources. It may be done also by comparison different aspects of atmospheric processes. In particular, the residence time of CO_2 molecules in the atmosphere is 4 years [104]. Because the total amount of atmospheric carbon dioxide is 870 GtC, one can find from this the total rate of the photosynthesis process 220 GtC/year as the only process of carbon absorption by the Earth's surface. This corresponds more or less to data of Fig. 5.7 and testifies also about its accuracy. In addition, this allows one to increase the accuracy of values of the total photosynthesis rate [116–120] which are found in a wide corridor.

Basic carbon fluxes of Fig. 5.7 has the natural character. The flux of anthropogenic carbon according to Fig. 5.7 is 10 GtC/year and is approximately 5% of the total carbon flux from the Earth to the atmosphere. The anthropogenic flux is analyzed carefully in [58] and is divided into the industrial part and land-use one. In 2017, the industrial part includes (9.9 ± 0.5) GtC/year [58], 40% in this value follows from coal, 35% is from the oil, 20% is due to a gas combustion, and 4% of anthropogenic carbon is extracted in the course of cement production. The land-use CO_2 results from deforestation, as well as from agriculture production. During the decade from 2008–2017, the land-use part of extraction of carbon dioxide is equal (1.5 ± 0.7) GtC/year compared to the industrial one of (9.4 ± 0.5) Gtc/year [58]. Figure 5.8 contains the rate of industrial carbon averaged over the corresponding decade from

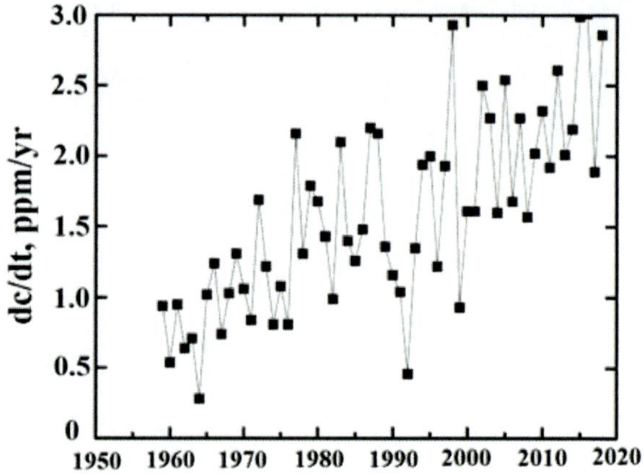

Fig. 5.9 Rate of accumulation of CO_2 molecules in the atmosphere averaged over the year [121]

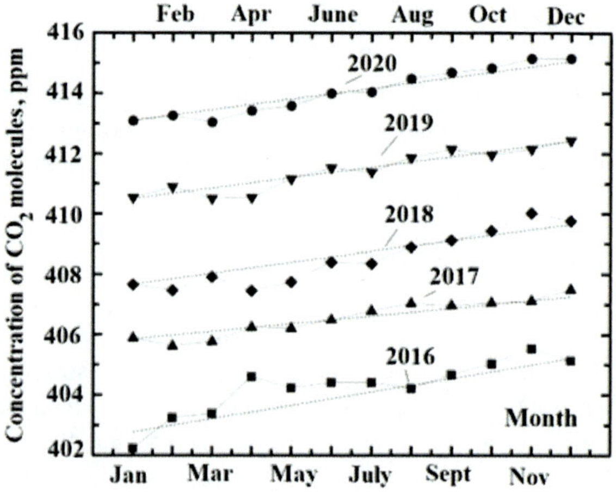

Fig. 5.10 Rate of growth of the amount of atmospheric CO_2 molecules with averaging over year from 2016 up to 2020 [121]. Dot lines corresponds to an average over fluctuations

1960 up to 2017 and also in 2017 according to [58]. Just this carbon is taken from the Earth's interior and is included in the biomass that is found in equilibrium between the Earth's surface and atmosphere.

The photosynthesis process is associated with atmospheric carbon oxide, and therefore, we scrutinize below evolution of the amount of carbon dioxide in the atmosphere. Figure 2.1 represents evolution of carbon dioxide in the atmosphere, and we further consider the peculiarity of this evolution. According to this figure, the concentration of atmospheric CO_2 molecules increases in time, as well as the average rate of this increase. Figure 5.9 gives the rate of this growth (the change of the concentration of atmospheric CO_2 molecules per year) according to measurements in the Mauna Loa Observatory. Though this value slowly increases in time, one can see also large fluctuations of this value. In addition, Fig. 5.10 contains the same quantity for last five years which follows also from [122, 123].

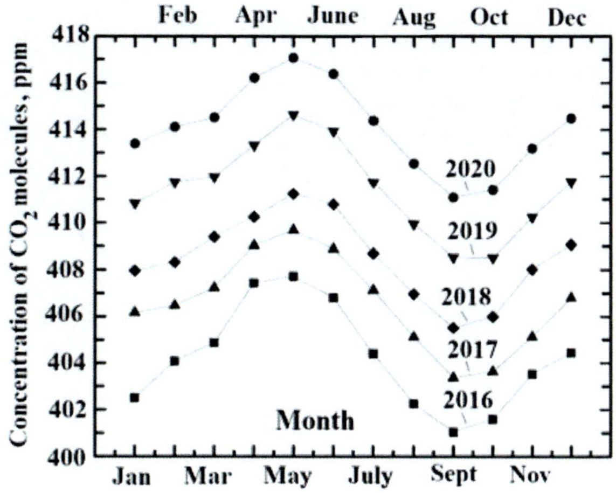

Fig. 5.11 Rate of growth of the amount of atmospheric CO_2 molecules with averaging over month from 2016 up to 2020 [121]

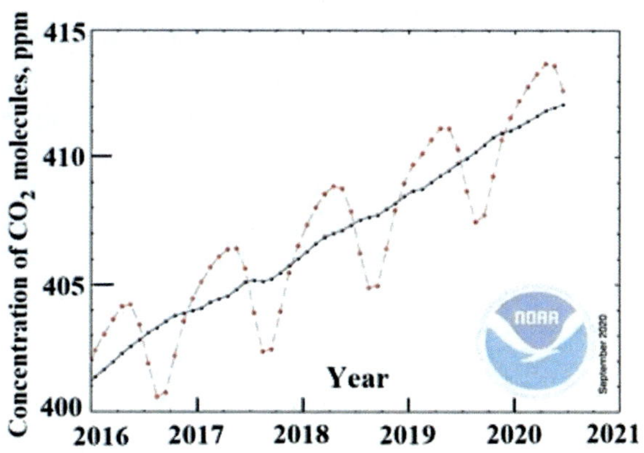

Fig. 5.12 Evolution of the concentration of atmospheric CO_2 molecules with averaging over month and year for last five years [122]

The peculiarity of the photosynthesis process under the action of solar radiation is such that this process at the land has the season character. This follows from Fig. 2.1 where along with a monotonic growth of the carbon dioxide, the season oscillations are observed. Indeed, land gives the basis contribution to the rate of the photosynthesis process, and the most part of land is located in the Northern Hemisphere. Therefore, consumption of atmospheric CO_2 molecules as a result of their accumulation by the Earth's surface proceeds mostly at the vegetation period of the Northern Hemisphere (Fig. 5.12).

Because the amplitude of oscillations exceeds that for a monotonic increase of the concentration of atmospheric CO_2 molecules per year, one can separate the monotonic and oscillation parts. It is clear that the photosynthesis process lasts several months of the year, but for tropical regions, this process can proceed during all the year. In addition, the season dependence for the rate of the photosynthesis process is different for Northern and Southern Hemispheres. In addition, the amplitude of

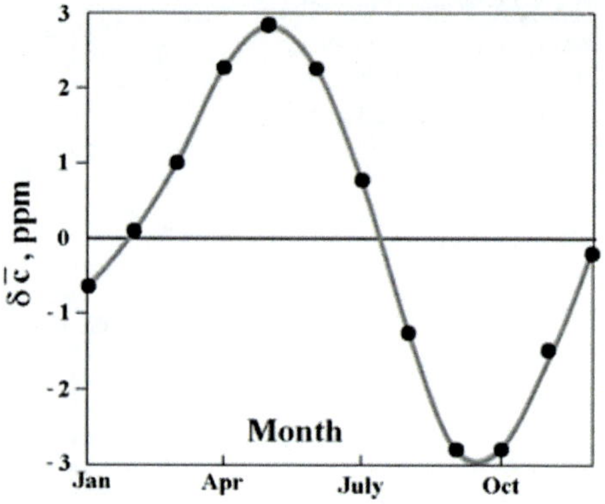

Fig. 5.13 Oscillation of the growth rate for the amount of atmospheric CO_2 molecules during approximately 30 years averaged over month [123]

season oscillations for the concentration of carbon dioxide molecules is different depending on the latitude.

Let us consider limiting cases for evolution of the concentration of CO_2 molecules. In the case of a strong mixing, the rates of biomass degradation at the Earth's surface is equal to the average rate of the photosynthesis process that is equal approximately 220 GtC/year, 100 ppm/year or 9 ppm/month on average, if the concentration of carbon dioxide molecules in the atmosphere does not vary, i.e., season oscillations are absent. In another limiting case, we assume for simplicity that the photosynthesis process proceeds uniformly during 4 months. This means that the degradation of biomass leads to an increase of the concentration of atmospheric carbon dioxide for this time by 33 ppm, while the total decrease of this value with accounting for the photosynthesis process is 70 ppm. This is the maximum amplitude of season oscillations of the concentration of CO_2 molecules in the atmosphere.

We now scrutinize the character of season oscillations under consideration for conditions of the Mauna Loa observation. One can see from Fig. 5.13 that the amplitude of oscillations is 6 ppm for the Mauna Loa region. Indeed, because of a low latitude of the Mauna Loa Observatory (19.5 °N), carbon dioxide from the Southern Hemisphere can penetrate in this region. In addition, the photosynthesis season in nearest regions lasts practically all the year.

Let us treat the data of Fig. 5.11 for evolution of the concentration of atmospheric carbon dioxide during last 5 years within a simple model. Averaging over oscillations, one can approximate the concentration $c(t)$ of atmospheric CO_2 molecules as

$$c(t) = c_o + t\frac{\mathrm{dc}}{\mathrm{dt}} \tag{5.2.5}$$

From data of Fig. 5.11, we have now

$$\frac{\mathrm{d}c}{\mathrm{d}t} = 2.4\ \text{ppm/year} \tag{5.2.6}$$

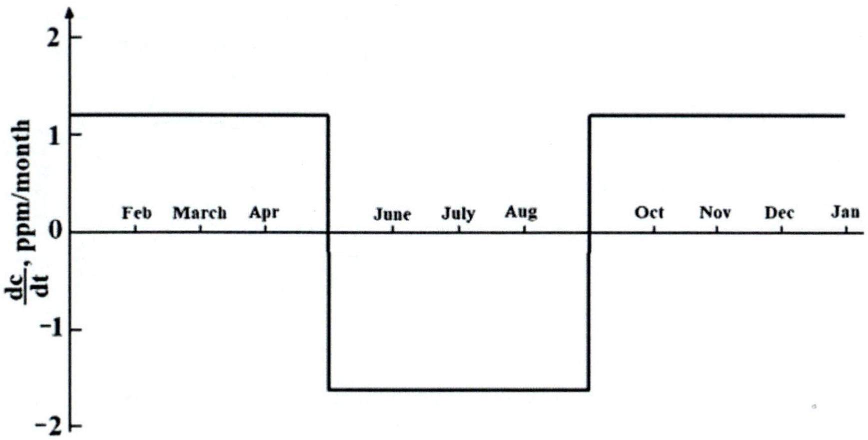

Fig. 5.14 Season evolution of the derivative dc/dt of the concentration of CO_2 molecules in the atmosphere per time. The average concentration of CO_2 molecules assumes to be unvaried, as well as the rate of biomass degradation. This dependence under indicated assumptions follows from data of Fig. 5.11 for last five years

This derivative corresponds to the doubling time of the amount of atmospheric carbon dioxide as 110 years. Figure 5.14 contains the value dc/dt which is obtained on the basis of the above model and using data of Fig. 5.11. In this case, we have for the parameter dc/dt the value -1.8 ppm/month during four months from May up to September, where the photosyntheses process proceeds. During 8 months from September up to May, where the photosynthesis process does not act, this value is 1.2 ppm/month. In addition, from this, we obtain the increase rate for the concentration of atmospheric carbon dioxide molecules as this relation may be represented also in another form

$$\frac{d \ln c}{dt} = 6 \times 10^{-3} \text{year}^{-1} \tag{5.2.7}$$

5.2.3 *Active Carbon of the Earth*

Photosynthesis is a complex biochemical process which leads to partial conversion of the energy of solar radiation in the chemical energy of carbon-contained compounds. Various compounds partake in this process. Such elements as sulfur, chlorine and phosphor are elements of these compounds and are of importance for principal stages of the photosynthesis process. Moreover, at the first stage of development of the Earth, sulfur compounds replace carbon ones in collection of the solar energy. In this analysis, leaving aside the chemical aspect of the photosynthesis process, we concentrate on evolution of the mass of carbon which partakes in the photosynthesis and oxidation processes. These processes establish the carbon equilibrium between atmospheric and oceanic carbon, as well as carbon located at land.

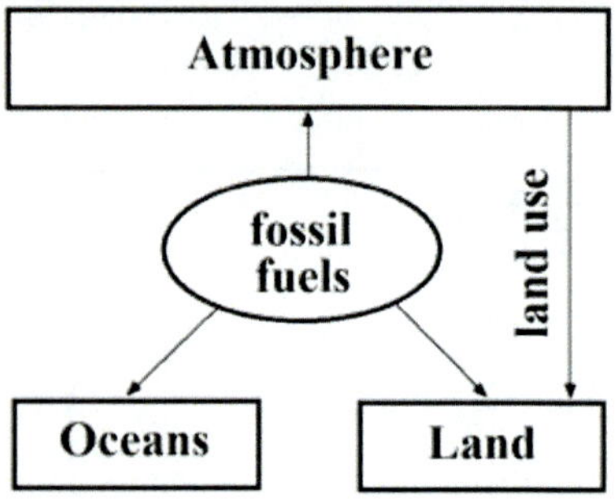

Fig. 5.15 Scheme of carbon fluxes as a result of injection of anthropogenic carbon

Let us define the biomass [124–126] as the mass of organic compounds at land and in oceans which partake in the photosynthesis process. In other words, the chemical energy of these compounds results from the photosynthesis process, and their degradation leads to formation of carbon dioxide and its injection in the atmosphere. The biomass includes living organisms and plants. Its mass is estimated as 500 GtC [102, 117]. In particular, according to [127], the biomass is 550 GtC and includes 450 GtC of plants and 70 GtC of bacteria. According to [126], the rate of biomass production as a result of photosynthesis equals 56 GtC/year at lands and 48 GtC/year in ocean. This coincides with the total rate (123 ± 8) GtC/year of biomass formation [128] and contradicts to the study [118] that gives the estimation 50GtC/year for this value. In addition, the most part of the living biomass (86%) [125] is located at land. But according to methods of determination of these values, they may be considered as estimations which are valid by the order of magnitude. Indeed, subsequent evaluations show that these estimations approximately two times below the values which use more reliable grounding.

We call atmospheric carbon in the form of carbon dioxide, as well as carbon which is located at the Earth's surface and is formed in the photosynthesis process as active carbon. Subsequently oxidation of bound carbon leads to formation of carbon dioxide which goes into the atmosphere. One can determine the amount of active carbon on the basis of data which follows from results of the Global Carbon Project [129] represented in [58, 116, 130–134]. Figure 5.15 gives the scheme for carbon fluxes which follow from the Global Carbon Project. In this consideration, we assume that the mass of active carbon varies very long as a result of transition of active carbon in the inactive phase where it does not partake in the photosynthesis process. Then, variation of the mass of active carbon results from the use of fossil fuels. This means that carbon in the form of fossil fuels is taken from the Earth's interior where it does not interact with solar radiation and then it is transferred at the Earth's surface where this interaction takes place.

According to the scheme of Fig. 5.15, carbon of fossil fuels is distributed subsequently between the atmosphere, oceans and land. Besides that, an increase of the biomass as a result of human activity causes an additional formation of carbon compounds at the Earth's surface in the form of land-use carbon. In this consideration, we assume that an increase of the mass of active carbon is distributed between these objects, i.e. between the atmosphere, land and oceans, and the rate of carbon transition to each of these objects is proportional to the carbon mass which they contains. Then on the basis of results of the Global Carbon Project [129] one can determine the carbon mass located in the atmosphere, oceans and at land. Following to this scheme, on the basis of [116] we have that in 2018 the anthropogenic carbon flux of (5.1 ± 0.2) GtC/year was absorbed by the atmosphere, the flux of (2.6 ± 0.6) GtC/year was absorbed by oceans, and the flux of (3.5 ± 0.7) GtC/year was absorbed by land. This anthropogenic flux of carbon is the sum of the industrial ones, mostly as a result of combustion of fossil fuels, that is (10.0 ± 0.5) GtC/year as well as of the land-use part, which is equal (1.5 ± 0.7) GtC/year and includes mostly the agriculture crop.

Let us treat these data in order to determine the amount of active carbon. Assuming the equilibrium between the atmosphere, oceans and land to be established due to exchange by carbon, we take into account that fossil fuels contain additional carbon to active one, i.e., the mass of active carbon increases by the carbon mass contained in fossil fuels. This means that a typical time of transformation of active carbon into an inactive one is relatively large. The equilibrium under consideration gives that the amount of active carbon in the atmosphere, ocean and land is proportional to carbon fluxes for these channels. Correspondingly, the masses of active carbon in the atmosphere, oceans and at land are 50%, 23% and 27% of its total mass, respectively. Note that the accuracy of these results corresponds to that of used data and is equal approximately 20%.

In addition, one can use variations of the above values during a large time. In particular, from 1750 up to 2018 according to [116] fluxes of anthropogenic carbon emitted in the atmosphere, in oceans and at land are equal (275 ± 5) GtC/year, (170 ± 10) GtC/year, and (220 ± 50) GtC/year correspondingly. According to this, the masses of active carbon in the atmosphere, oceans and land are 41, 26 and 33% of its total mass correspondingly. Comparing these values with the previous ones, one can estimate their accuracy as 20%, as early.

Note also that the concentration of CO_2 molecules is equal (277 ± 3) ppm in 1750, and according to data of Fig. 2.1, the concentration of atmospheric CO_2 molecules was 408 ppm in 2018. Transforming these values into carbon masses, one can find the mass of atmospheric carbon to be 588 GtC in 1750, as well as 866 GtC in 2018, so that their difference is $278 GtC$ in accordance with the above value (275 ± 5) GtC/year [116]. Because the carbon masses and their changes in oceans and at land are proportional to those ones in the atmosphere, we have the mass of active carbon in oceans 360 GtC and 530 GtC in 1750 and in 2018 correspondingly, whereas at land these values are equal to 465 GtC and 685 GtC, respectively. Variations of these values between 1750 and 2018 in oceans and at land are equal 170 GtC and 220 GtC in accordance with [116].

The sum of masses of active carbon in oceans and at land is the biomass. From the above values, it follows that the biomass was 820 GtC in 1750 and 1220 GtC in 2018. These values exceed approximately twice the above biomass value of 500 GtC [102, 117], but the basis [116] for these values is more reliable. Note a restricted accuracy of these data ($\sim$20%) in accordance with the accuracy of initial data. In

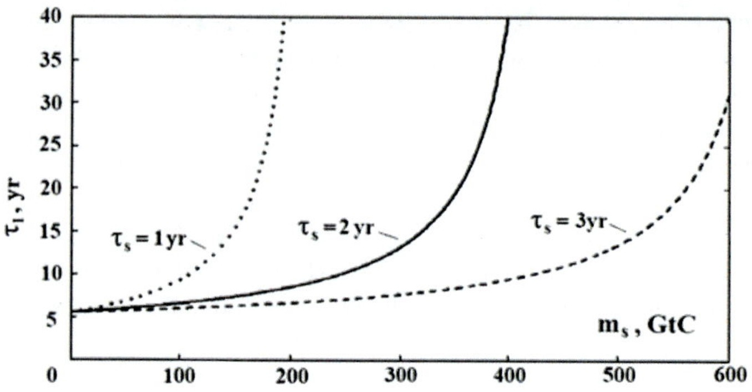

Fig. 5.16 Dependence of the lifetime of a long-lived carbon-containing component τ on the mass m of the short-lived carbon-containing component

addition, on the basis of these data [116], we have for the total mass of active carbon in atmosphere, in oceans and at land as 930 GtC and 1.370 GtC in 1750 and 2018 correspondingly. The difference of masses of active carbon for this time period is equal to the mass of fossil fuels which was used in this time and is equal (440 ± 20) GtC [116]. Thus, from the analysis of data [116] for changes of masses in the atmosphere, oceans and at land, as well as the mass of fossil fuels used during time under consideration allow one to evaluate the evolution of masses of active carbon in the atmosphere, oceans and at land. From these values, one can conclude that extraction of fossil fuels from the Earth's interior and including it in the carbon balance between the atmosphere and the Earth's surface led to an increase of the biomass as a result of the human activity by approximately 40%.

We above consider plants of the photosynthesis process as an uniform system on the basis of their average parameters. Transiting to details of this process, we take into account that plants give the main contribution to the total rate of the photosynthesis process. From this standpoint, we divide plants in two parts, short-living plants and long-living ones. In the first stage of the photosynthesis process with formation of short-lived plants, they are grass and leafs of trunks. They accumulate the biomass during a season of the photosynthesis process and then are degradated. Evidently, the lifetime of these objects is 1–3 years, and below for definiteness, we take it as $\tau_s = 2$ years. Algae assimilates the energy of solar radiation and are an analog of plants at land [97, 98]. The solar energy is fixed in oceans by the phytoplankton that is used subsequently by fishes as a nutrient medium [99]. Indeed, phytoplankton absorbs atmospheric carbon dioxide and transforms it into bonded solid carbon under the action of solar radiation.

One more process of absorption of solar light by leafs is the transfer of its energy to trunks of trees as the transformation of the solar energy in the chemical energy of formed wood. Evidently, a typical lifetime of wood τ in nature is tens of years. Denoting by m_s the total short-lived biomass and by m_l that for its long-lived part, as well as by τ_s, τ_l the lifetime of the short-lived and long-lived components of the biomass correspondingly, we have the following equation for the photosynthesis rate

$$\frac{m_s}{\tau_s} + \frac{m_l}{\tau_l} = \frac{\mathrm{d}M}{\mathrm{d}t}, \quad M = m_s + m_l, \tag{5.2.8}$$

where $M = 1220$ GtC is the contemporary total biomass and $\mathrm{d}M/\mathrm{d}t = 220$ GtC/year is the photosynthesis rate. This equation connects the mass m_s of the short-lived component of the biomass with the lifetime τ_s and long-lived component of the biomass with the lifetime τ_l. Figure 5.16 gives the solutions of Eq. (5.2.8) for different contributions of the short-living biomass. As is seen, the long-lived component gives the main contribution into the biomass amount. This component is concentrated in forest. Because the total area of forest is shorten, this leads to an increase of the amount of atmospheric carbon dioxide.

5.2.4 Food Production in Earth

The photosynthesis process provides the humanity life through products of agriculture production. Let us determine the possibility of the photosynthesis process for this. According to data of UN Food and Agriculture Organization, the energy of food consumption lies between 2500 and 3500 kcal/day per person [135] depending on the country. Being guided by the latter value, one can obtain that the total power of the food consumptioned by the planet population is approximately 1×10^{12} W. One can compare this with the power of the total photosynthesis process that is 2×10^{14} W or the land photosynthesis power that is approximately 1×10^{14} W.

Early, starting from eighteenth century, the concept existed that an increase of humanity may lead to a food lack because the Earth's capacity with respect to food production is restricted. But so called the green revolution proceeded in sixties of twentieth century led to a sharp crop increase both due to use of fertilizes and owing to new technologies. As a result, the problem of humanity growth takes place in poor countries only. Table 5.5 contains the values of the world crop for basic food cultures during a long time starting from the green revolution. The total population of the world was 3.21×10^9 in 1961 and 7.63×10^9 in 2018 [136]. As it follows for Table 11 data, from 1961 up to 2018, growth of the total world crop of indicated cultures varies stronger. Indeed, the ratio of crops for 2018 and 1961 is equal 3.6, while the world population increases in 2.4 times, i.e., the overpopulation is not now sharp.

Let us analyze on the basis of these data the Malthus concept [140, 141] according to which an increase of the Earth's population is limited because of a restricted amount of food that may be obtained at the Earth. According to this concept, the population growth proceeds in the exponential manner, whereas the food supply grows linearly in time [142, 143]. Finally, this leads to the catastrophe because the population cannot be provided by food. These dependencies were valid in a certain range of time of the civilization development, and this concept was discussed seriously in 19th and first half of twentieth century when this problem was sharp. Subsequently, the above

Table 5.5 Total yields of major grain crops across the Earth in millions of tons [137–139]

Culture	Crop, 1961	Crop, 2018
Maize	205	1132
Rice	285	781
Wheat	222	751
Barley	72	141
Sum	784	2805

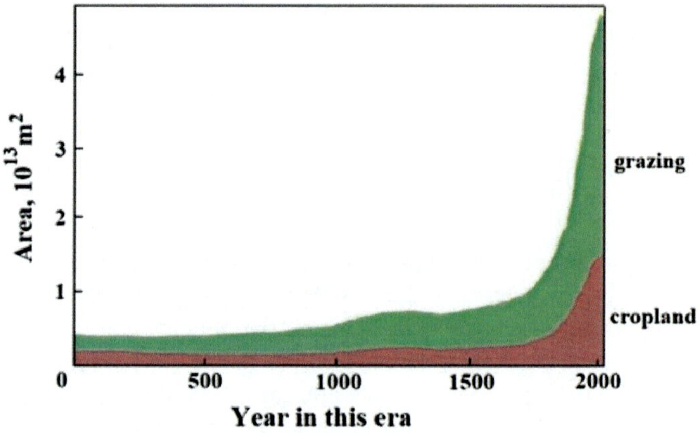

Fig. 5.17 Evolution of the Earth's area which is used in the agriculture for food production [145]

dependencies were violated, especially, starting from sixties of twentieth century (for example, [144]), when the green revolution proceeded. Table 5.5 confirms this.

Nevertheless, it is clear that the Earth's capacity with respect to food is restricted, and our task is to understand what resources may be used in future to provide the population by food. Figure 5.17 contains the Earth's surface which is used for agriculture during last centuries. Because the total area of the Earth's land equals to 1.5×10^{14} m^2, from this, one can conclude that possibility of the Earth's surface for agriculture was exhausted more or less. Because of the efficiency of the food production as an energetic process is very small ($\sim 10^{-5}$), one can expect that this efficiency will be increased significantly due to new technologies that allow one to provide an increasing population by food.

Let us estimate the possibility of the Earth as a source of food from the energetic standpoint. It is clear that the energetic estimation for food is overstated, since along with plant food, a person consumes animal food, which is several times more expensive energetically. To take this factor into account, we will analyze the current situation. Indeed, the value of 3000 kcal/day when converted to consumed carbon in the case of the process (5.2.1) corresponds to the use of the carbon mass of the order of 100 kg/year per person if we transform it to glucose. If the total mass of the crop, given in Table 5.5, is divided by the number of people currently living, we determine the mass of grain per person, approximately 360 kg per year. Taking this into account, we will get as an upper estimate the number of people of the order of 300 billion, whom our planet is able to provide with food of Table 5.5. At the same time, if we assume that food is created only at land, which accounts for about half of the planet's photosynthetic power, the resulting upper estimate for the Earth's capacity as a food source should be lowered by half, so that the Earth is able to feed about the habitation of the order of 100 billions.

Thus, the Earth is in principle able to provide by food for a population that exceeds the contemporary one by at least one order of magnitude. But in order to increase the yield of products, it is necessary to increase the land area used for agriculture. But Fig. 5.17 represents the areas on the Earth's surface used for agriculture, and from this, the land area is 1.5×10^{14} m^2. In preindustrial 1750, 8–9 million km^2 or approximately 6% of the global land area were occupied for cultivation and pasture, mostly in Europe, South-East Asia, China and India. But croplands and pasture expanded over last centuries and occupy now all the appropriate area practically. Thus, the possibilities of the photosynthesis process for the Earth can provide with food significantly larger number of people than the contemporary one. But this requires to create more efficient technologies for food production.

5.3 Thermal State of the Earth

5.3.1 Evolution of the Earth's Global Temperature

The main parameter that characterizes the thermal state of the Earth, is the global temperature, i.e., the temperature averaged over the globe and time. Evidently, the local temperature of the Earth's surface as a function of time experiences large oscillations, as well as the temperature of the Earth's surface depends greatly on a geographical point. Therefore, the global temperature defined as the average temperature of the Earth's surface is characterized by large fluctuations in time.

One can determine the global temperature only for last time, starting from the middle of the nineteenth century, when meteorological stations appeared in different points of the globe. Then, the global temperature follows from averaging of local temperatures at different points of the Earth. Then, statistical error for the global temperature in this definition is estimated as several degrees of Celsius, and it is impossible to describe evolution of the Earth's thermal state during a century. To overcome this contradiction, Hansen et al. [146] proposed the method based on determination the variation of local temperatures taken from measurements of meteorological stations. In this method, the difference of local temperatures is taken at a given geographical point and at a certain time (day and season), but in different years. Averaging the change in local temperature for a year, for two years, etc., over time, and then averaging this value over geographical points of the globe, one can find the average change in the global temperature ΔT correspondingly for a year,

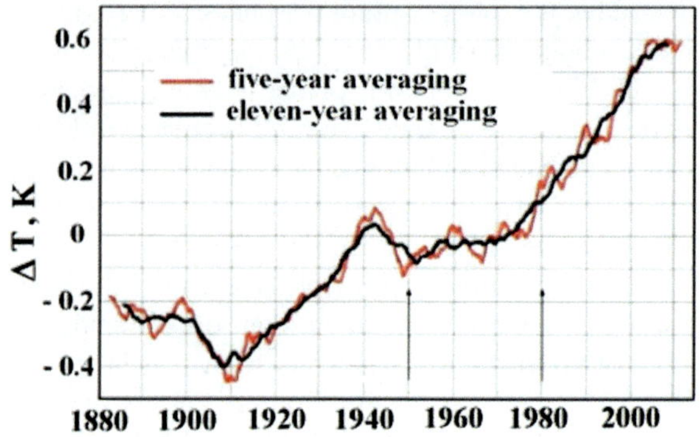

Fig. 5.18 Evolution of the global temperature averaged over five and fifteen years [147]. The temperature change is calculated from the average value in 1951–1980

two years and several years. At the same time, if the global temperature fluctuation (i.e., the statistical error for this average temperature) is of the order of several °C, then the fluctuation in the global temperature change ΔT is measured in tenths of °C. Hence, this method allows one to analyze the nature of changes in the global temperature for a very short period of time on the basis of data from meteorological stations.

This method [146] was carried out within the framework of the NASA [Gottard Institute for Space Studies (GISS)] project, and its results are published, in particular, in papers [147–151]. The number of meteorological stations which information was used in this project exceeds six thousand for the end of the nineteenth century, and now, this number decreases almost three times. The main contribution to the contemporary global temperature in this analysis follows from satellite measurements. The global temperature change ΔT, obtained through processing in the specified way with averaging for five and fifteen years, is represented in Fig. 5.18. In addition, fluctuations do not exceed 0.2 °C for changes in the global temperature of the Earth, obtained from comparison of results of measurements in winter and summer, during the day and at night, in the Northern and Southern Hemispheres.

According to data of Fig. 5.18, a nonmonotonic evolution of the global temperature from the end of nineteenth century is observed. Indeed, during the years 1880–1910, there was a weak cooling, which was replaced by a more strong cooling, and in 1910–1940, there was a warming, which went on to a cooling in 1940–1950. The subsequent period of time during from 1950 up to 1980 was characterized by fluctuations around a certain average value, and since the eighties, there has been a noticeable warming of our planet.

As it follows from Fig. 5.18, a noticeable warming occurs starting from the eighties of the last century. At the same time, the approximation of data for the change in the global temperature during this period by a linear time dependence gives for the rate of Earth's warming [152]

$$\frac{\mathrm{d}\Delta T}{\mathrm{d}t} = 0.018\,\mathrm{K/year}, \tag{5.3.1}$$

Table 5.6 Change in the global temperature of the Earth expressed in °C, where the average temperature for the twentieth century is taken as zero [154, 155]

	All land May 2018	Northern Hemisphere May 2018	Southern Hemisphere May 2018	All the land All the land May 2019	Northern Hemisphere May 2019	Southern Hemisphere May 2019
Land	1.21	1.27	1.06	1.16	1.25	1.13
Oceans	0.60	0.60	0.54	0.73	0.81	0.69
Land+Oceans	0.77	0.91	0.62	0.85	0.93	0.77

The statistical error of this value is $\Delta T = 0.09K$, as it follows from treatment of Fig. 5.18 data [152]. This means that the linear nature of the global temperature increase in time is realized for time intervals exceeding 5 years. According to a direct measurement made at the Mauna Loa Observatory in 1977–2006, the rate of increase in the average temperature in the area of this observatory is 0.021 °C/year [153]. As is seen, within the limits of accuracy, this value coincides with that given by formula (5.3.1).

Note that the definition of the global temperature is based on the assumption that evolution of the average temperature proceeds in various parts of the globe more or less identically. On the other hand, the data in Table 5.6 show that this assumption is violated. Indeed, according to these data, the strongest warming occurs at the land of the Northern Hemisphere which gives the main contribution to the photosynthesis of our planet .

5.3.2 *Earth's Climate in Past*

Considering the thermal state of the Earth in past, we note that in this case, we are dealing only with a local temperature. The measurement of a local temperature in past is based on the isotopic analysis of sediments and is a subject of study in paleontology and paleoclimatology (for example, [156–161]). The isotopic analysis of sediments allows one to determine the local temperature in past. We below present in Fig. 5.19 the example of determination of the local temperature in Antarctica because in this case the treatment of results was made carefully. In these measurements, air bubbles were extracted from ice pieces which were taken from different depths of the Earth's surface. In treatment, the ice depth corresponded to a time of their formation, so that the concentration of carbon dioxide in the bubbles of some samples one can consider as the concentration of carbon dioxide in the Antarctica atmosphere at an appropriate time.

The air temperature in the bubbles at time of their formation is determined from the analysis of the isotopic composition of oxygen molecules. In the Earth's crust, the oxygen isotopes ^{18}O and ^{16}O are contained in the proportions 0.2% and 99.76%[162].

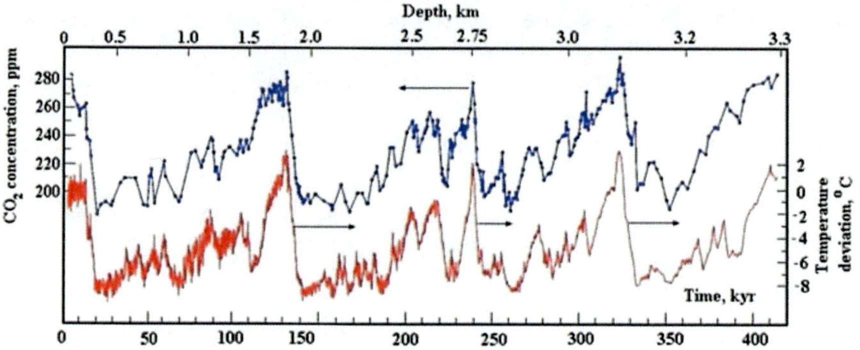

Fig. 5.19 Evolution of atmospheric carbon dioxide concentration and Earth's surface temperature in past in the Antarctic region, as it follows from the analysis of ice deposits in Antarctica near the Vostok meteorological station [163, 164]. Time is directed oppositely

However, since the energy of zero oscillations in the carbon dioxide and oxygen molecules is different for these isotopes of oxygen atoms, then in the presence of thermodynamic equilibrium in a gas-containing oxygen and carbon dioxide, testifies about the temperature at moment of formation of sediments.

Let us take into account that the isotopic equilibrium is established as a result of exchange reactions involving oxygen atoms formed under the action of cosmic rays. Then, the isotopic distributions for carbon dioxide molecules and oxygen molecules are determined by the temperature that is realized in these processes. Although due to the low intensity of cosmic rays, the time for establishing isotopic equilibrium is long, but the temperature analysis of the atmosphere is also associated with long times of the processes occurred.

Figure 5.19 contains evolution of the carbon dioxide concentration and temperature over past hundreds of millennia in Antarctica. They are obtained from the analysis of air bubbles located in the ice deposits, which were extracted near the Vostok meteorological stations. These results, as well as the subsequent analysis of evolution of these parameters of Antarctica up to 400 thousand years ago, indicate a correlation between the atmospheric temperature near the Earth's surface and the concentration of carbon dioxide molecules in the atmosphere in this time, and we below. analyze this information.

As it follows from data of Fig. 5.19, the concentration of carbon dioxide in the atmosphere in past during hundreds of thousands years has varied from 172 to 300 ppm. This result is in accordance with the study of sediments in the Sierra Leone region in the Atlantic Ocean near the west coast of Africa, according to which the concentration of atmospheric carbon dioxide in this area was between 213 ppm and 283 ppm in the time period between 900 thousand years and 2.1 million years ago [165]. But now the concentration of carbon dioxide molecules in the atmosphere is approximately 416 ppm, that is significantly higher than this value in past.

Next, the correlation between the temperature change and the concentration of atmospheric carbon dioxide follows from data of Fig. 5.19. These values oscillate in time according to data of Fig. 5.19 that indicates the alternation of ice ages with periods of warming of the planet, and Fig. 5.19 contains, at least, four periods of ice ages. In addition, from data of Fig. 5.19, it follows that the amplitude of the temperature change in the Antarctic region was approximately 12 K, which contradicts to the main statement of the Paris agreements of 2015 [166], according to which the red line, exceeding which can lead to irreversible Earth's warming, if the Earth's heating exceeds 2 K.

Thus, data of Fig. 5.18 for global temperature are the most reliable, since they are based on a strict average of a local temperature over different geographical points of the Earth. In considering a longer time period, we rely on data relating to certain areas of the Earth where the temperature change may differ from that occurring in other areas of the Earth's surface. Nevertheless, these data provide a qualitative understanding of evolution of the Earth's temperature and confirm the above interpretation of results.

In analyzing the nature of the Earth's climate change in the last millennium, certain information about the temperature of the Earth during this time period is determined by solar activity, expressed in terms of the number of sunspots. Data on the number of sunspots have been available from the seventeenth century, and it is believed that the solar activity correlates with the radiative flux from the Sun. In this regard, the coldest time refers to the period of 1645–1715, the so-called Maunder minimum, when sunspots were completely absent.

It is important the written records of the Earth's temperature and the weather of certain areas of the Earth in recent millennia, although they are usually fragmentary. These data indicate a warming in England in the 10th–eleventh centuries, when grapes ripened in England, and the river Thames was not frozen in winter [167, 168]. At that time, vikings captured Greenland, which surface according to its name was not covered by glaciers in that time. In addition, vikings entered America long before Columbus. However, in the fourteenth century, a cold period occurred in Northern Europe. As a result, the river Thames in London began to freeze annually, and vikings left Greenland [167, 168].

Analyzing the thermal state of our planet and evolution of the temperature of its regions, we do not set out to study these problems in series, but will try to give some facts for this. In all cases of determination of the Earth's temperature in past, the isotopic analysis of sediments is used, which is based on the dependence of the the distribution function over isotopes on a time where a sediment passes from the atmosphere inside the Earth. Determination of this temperature as a clock that gives a time of transition of studied molecules from the atmosphere to the Earth's interior in the form of sediments. Simultaneously, the atmosphere temperature is determined at that time. This is the so-called geochronological method (for example, [169, 170]), which allowed us to look deep into past.

In this way, isotopic methods are used, including both stable and radioactive isotopes. In this case, the concept is used that as long as this element is contained in compounds of the atmosphere, there is a certain distribution of radioactive isotopes

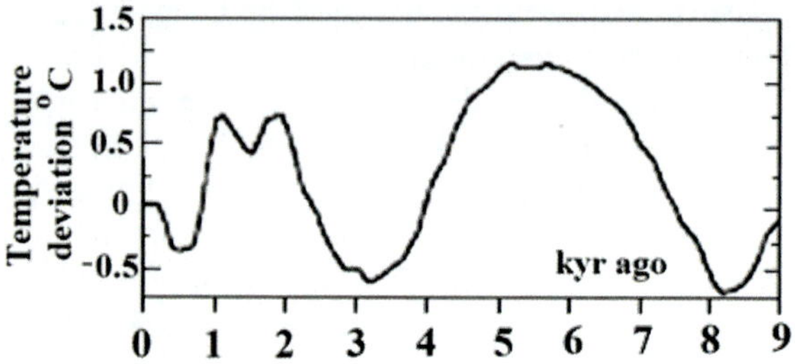

Fig. 5.20 Evolution of the temperature of Greenland from the SO_4^{+2} concentration in its glacier [172]

for this element, which is established by the action of cosmic rays. After these isotopes pass into the sediments, this equilibrium is ceased and the concentration of the radioactive isotopes decreases in time. Then, on the basis of the concentration of this radioactive isotope, it is possible to determine a time when this deposit becomes outside the action of cosmic rays. This approach is the content of a separate branch of science—geochronology (for example, [162, 169–171]). This method allows one to determine the evolution of the Earth's temperature in past (for example, [156–158, 161]).

Along with isotopic methods, a time of formation of this deposit can be determined on the basis of a depth of a studied layer. In reality, this approach is combined with the isotope analysis that increases the reliability and accuracy of determination the transition time of a studied substance from the atmosphere to sediments. Above, we give in Fig. 5.19 evolution of the atmosphere temperature and the concentration of carbon dioxide molecules in the Antarctica region based on the analysis of air bubbles extracted from the ice. Another example of the atmosphere temperature in past is which we give in Fig. 5.20 on the basis of the analysis of the sulfate concentration in ice samples extracted from the Greenland glacier. A time of transition of the deposit in the glacier follows from a depth of sediment location, and the atmosphere temperature results from the ratio of the concentrations of different sulfates. Thus, there is a series of methods to determine the atmospheric temperature in past.

5.3.3 *Accompanying Processes*

Changes in the Earth's energetic balance are accompanied by changes in other parameters of the Earth and the atmosphere. Heating of the Earth causes melting of ice on the Earth's surface, and this, in turn, leads to an increase in the sea level or the ocean level. Below we consider the melting of glaciers on the Earth's surface and the rise in the level of the world's oceans (sea level), which is caused by this. The sea level change is determined now on the basis of radar satellite measurements. The sea level rise is shown in Fig. 5.21 [173] from January 1993. Without dwelling on the details of these results, we note that the sea level rise occurs monotonically. In addition, in 2018, the average rate of the ocean level rise according to NASA data

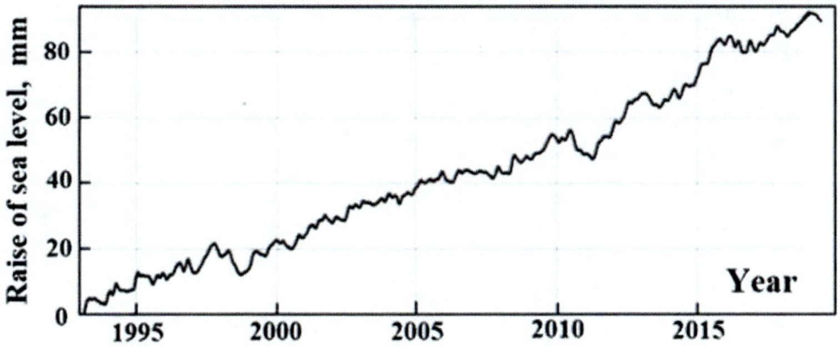

Fig. 5.21 Sea level rise, based on the NASA data [173]

was 3.3 mm/year [174], and the average rate of this this rise starting from 1993 is approximately 3 mm/year according to data of Fig. 5.21.

Let us compare the rate of rise of the sea level with the rate of exchange between open water on the Earth's surface and atmospheric water. According to [29–32], the precipitation rate, and hence, the water evaporation rate, is 3.9×10^{20} g/year. This value corresponds to the rate of water exchange between the atmosphere and the Earth's surface, which is approximately 80 m/year. As is seen, this value exceeds the rate of rise of the sea level in more than 200 times. Thus, sea level rise occurs under conditions of equilibrium between open water on the Earth's surface and atmospheric water. The sea level rise occurs as a result of glacier melting, and the rate of this process corresponds to melting of 2×10^3 km^3 of ice per year.

In considering the water balance in glaciers with taking into account their melting, we note that there are more than 132 thousand glaciers on Earth, which cover an area of 740 thousand km^2 [175]. If the ice from all the glaciers is transformed into water and it evenly covers the surface of the entire planet, then the thickness of the water layer formed by the melting of these glaciers will be 35 cm according to [175], 50 cm according to [176], 60 cm according to [177] and 43 cm according to [178]. As is seen, the amount of water in glaciers, which height is 50 cm on average, is only an order of magnitude higher than the amount of atmospheric water. If the latter is transformed into a liquid state and is distributed over the Earth's surface uniformly, it forms a layer of thickness 2.5 cm.

Figure 5.22 contains the seasonally averaged change in the mass of water contained in glaciers. From this, it follows that a typical time after which half of the water in the glaciers will be melted is approximately 50 years [180]. If we assume a direct transition of melting ice into the ocean, then since the sea level is noticeably higher in January than in July, the main contribution to the melting of glaciers is made by the Southern Hemisphere. When analyzing the amount of water in glaciers, we distribute this ice uniformly over the entire surface of the globe and determine the thickness of the layer, which is melted during the considered period of time. According to data of Fig. 5.22, the thickness of the glacier layer, which is transformed into water annually, is approximately 7 mm. As is seen, this value is about twice compared

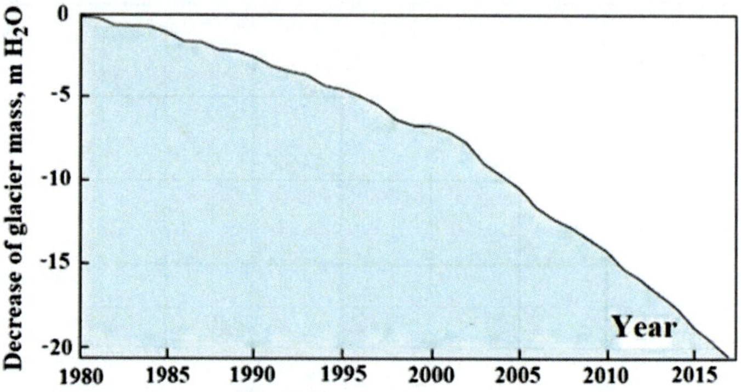

Fig. 5.22 Reducing mass of water in glaciers [179]

Fig. 5.23 Glacier at the base of Lyal'ver peak in the Bezenga region (Caucasus) [181]

to the height to which the sea or ocean level rises. At the same time, the increase in the mass of water in the atmosphere for a year is two orders of magnitude higher than the indicated rate of change in the mass of water on the Earth's surface. From this, one can conclude that the water excess can be associated with the transfer of water to underground storage or with the transformation of water into other compounds as a result of chemical processes.

Glaciers exist in mountains or in cold regions of the Earth, in particular, near the poles of the Earth, at the Arctic and Antarctica. As an example, Fig. 5.23 represents a glacier in the Caucasus Mountains. Glaciers in the mountains exist at the base of the mountains and fill the valleys between the mountain peaks. In the cold season, glaciers are covered with snow, and the thickness of the water layer formed during the season due to falling snow significantly exceeds the amount of melting ice. This means that the change in the mass of melting ice during the year has an oscillatory character, and the amplitude of the oscillations significantly exceeds the annual change in the specific mass of water in glaciers.

Thus, the rise in the ocean level is a long geological process associated with the melting of glaciers. The characteristic time of this geological process is the ratio of the difference in the height of the Earth's topography, which is about 1 km, to the rate of rise of the world ocean, and a typical time of this geological process is about 300 thousand years. In addition, during the ice evolution, the glacier melting alternates with the growth of glaciers, so that the glaciers are a reservoir for storing excess water along with the oceans. This geological process involves the processes of volcanic eruptions and earthquakes, which lead to the displacement of the soil. The process of melting glaciers generates water, which provides a rise in the sea level. The water excess from the glacier melting fills underground voids formed in volcanic eruptions and earthquakes and also participates in chemical processes involving water, which lead to formation of solid ground compounds.

One can connect the glacier melting with the planet warming. In particular, the [182] estimate gives that this warming from the end of the nineteenth century to the end of the twentieth century was 0.7 °C per century. This estimate corresponds to a more precise definition of the thermal state of the planet, shown in Fig. 5.18. We also estimate the power that must be expended for the observed melting of glaciers. Since the energy spent on melting, a unit of mass of ice is 334 J/g [183], and the layer of the annually melting layer of ice is approximately 7 mm. From this, we obtain for the power the value of 4×10^{13} W which is required for the observed ice melting. This value is twice of the total power of contemporary industry, but it is small compared to the power of processes of the energetic balance in the atmosphere.

Thus, the heating of our planet is accompanied by other processes on the Earth's surface and in its atmosphere. Above, we presented a geological process that establishes a balance between water frozen in glaciers, as well as open water on the Earth's surface and in the atmosphere. A typical time for this process, which is about 50 years, is the time when half of glacier ice is melted. This time is long compared to a time of establishing an equilibrium between open water on the Earth's surface and atmospheric water, which is approximately 9 days, i.e., the geological process involving water and glaciers develops more slowly than processes with participation of atmospheric water. On the other hand, slow geological processes are an indicator of changes in the Earth's state and connect various properties of the Earth.

5.3.4 Equilibrium Climate Sensitivity

An increase of the global temperature results from additional fluxes of energy which are absorbed by the Earth's surface. In particular, if the concentration of atmospheric CO_2 molecules increases, an additional radiative flux occurs due to this change, and this additional radiative flux is proportional to the change in the concentration of CO_2 molecules. Let us introduce the parameter ECS (Equilibrium Climate Sensitivity)

[184], which represents the change in the global temperature with a twofold increase in the concentration of atmospheric carbon dioxide molecules, i.e., it is defined as

$$\mathrm{ECS} \equiv \ln 2 \frac{\mathrm{d}\Delta T}{\mathrm{d}\ln c(\mathrm{CO_2})}, \tag{5.3.2}$$

where ΔT is the change of the global temperature, and $c(CO_2)$ is the concentration of carbon dioxide molecules in the atmosphere.

Formula (5.3.2) allows one to define the ECS parameter currently. Taking into account the nature of the change in the concentration of carbon dioxide molecules in the atmosphere and the global temperature in accordance with the formula (5.3.1), one can obtain

$$\mathrm{ECS} = (2.1 \pm 0.4)\,\mathrm{K} \tag{5.3.3}$$

Note that this value is not associated with certain assumptions and follows from the corresponding measurements. In addition, in formula (5.3.3), the concentration of atmospheric carbon dioxide molecules plays the role of an indicator of the atmosphere state and is not the cause of the observed change in the global temperature. Besides that, the contribution of the greenhouse effect to this change when the concentration of carbon dioxide molecules doubles is a small part of the change in the energy flux to the Earth's surface.

On the basis of Fig. 5.13, it is possible to determine the equilibrium sensitivity of the climate in past. Indeed, as this figure shows, the temperature change within a single glacial period is equal approximately 12 K, while the concentration of carbon dioxide molecules lies between 180 and 280 ppm. From this, based on the formula (5.3.2), we get ECS = 27 K in past. This shows that the concentration of carbon dioxide molecules is only an indicator of the thermal state of the planet.

In a general case, the change in the global temperature ΔT results from an additional flux ΔJ absorbed by the Earth's surface. In particular, we below consider the greenhouse effect, where the change of the concentration of some greenhouse components causes an additional energy flux to the Earth's surface. This flux, in turn, creates an increase in the global temperature. To analyze this relationship, it is convenient to introduce the climate sensitivity S [37, 185] as the ratio of the global temperature change ΔT to the additional flux ΔJ absorbed by the Earth's surface, as well as the inverse value, the radiative forcing $F = 1/S$, based on the relations

$$S = \frac{\Delta T}{\Delta J}, \quad F = \frac{1}{S} = \frac{\Delta J}{\Delta T} \tag{5.3.4}$$

One can expect that the climate sensitivity S is a convenient parameter for analyzing changes in the global temperature. However, in reality, this value is sensitive to additional factors. In particular, it depends on the interaction between the atmosphere and ocean, as well as to ocean flows which are not taken into account in global models. The altitude of clouds can change as a result of a shift in the global temperature, and this, in turn, influences on the value of the climate sensitivity. Therefore, the accuracy of determination of the climate sensitivity S is low. In particular, the processing for individual cases of the sediment analysis for past events at various geographical locations gives the climate sensitivity in the range from 0.3 to 1.9 $m^2 \cdot K/W$ for events collected in [185]. Similar processing of past events collected in [186] leads to the climate sensitivity S in the range from 0.25 to 0.79 $m^2 \times K/W$.

In addition, we give the results of some estimations for the climate sensitivity S. Namely, the climate sensitivity is 0.55 $m^2 \cdot K/W$ according to [37], 0.64 $m^2 \cdot K/W$ according to [187], 0.49$m^2 \cdot K/W$ according to [188], and 0.42 $m^2 \cdot K/W$ according to [189]. Statistical averaging of these values gives for climate sensitivity S [190]

$$S \approx 0.5 \frac{m^2 \cdot K}{W} \tag{5.3.5}$$

We estimate the accuracy of the climate sensitivity S obtained in this way as 50%.

References

1. W.H. Dines, Quart. J. Roy. Meteorol. Soc. **43**, 151 (1917)
2. https://www.en.wikipedia.org/wiki/solar-constant
3. https://www.en.wikipedia.org/wiki/Solar-irradiance
4. P.V. Foukal, J. Lean, Science **247**, 556 (1990)
5. R.C. Wilson, H.S. Hudson, Nature **351**, 42 (1991)
6. C. Fröhlich, J. Lean. Geophys. Res. Lett. **25**, 4377 (1998)
7. C. Fröhlich, J. Lean. Astron. Astrophys. Rev. **12**, 273 (2004)
8. S. Mekaoui, S. Dewitte, Solar Phys. **247**, 203 (2008)
9. G. Kopp, J.C. Lean, Geophys. Res. Lett. **38**, L01706 (2011)
10. G.L.Stephens, J.Li, M.Wild et.al. Nat. Geosci. **5**, 691(2012)
11. M. Lockwood, W.T. Ball, Proc. Roy. Soc. **476**, N2238 (2020)
12. R.C. Wilson, Solar Phys. **74**, 217 (1981)
13. https://en.wikipedia.org/wiki/Solar-cycle
14. https://en.wikipedia.org/wiki/Wolf-number
15. R. Wolf. Mitt. Nat. Forsch. Ges. Bern **130**, 169(1848)
16. S.H. Schwabe, Astron. Nachricht. **21**, 233 (1843)
17. https://en.wikipedia.org/wiki/Albedo
18. https://earthobservatory.nasa.gov/images/84499/measuring-earths-albedo
19. H.G. Houghton, J. Meteorol. **11**, 1 (1954)
20. https://en.wikipedia.org/wiki/Electromagnetic-absorption-by-water
21. http://www1.lsbu.ac.uk/water/water-vibrational-spectrum.html
22. C.M.R. Platt, Quart. J. Roy. Meteorolog. Soc. **102**, 553 (1976)
23. S. Twomey, Geofis. Pure Appl. **43**, 227 (1959)
24. S. Twomey, J. Atmos. Sci. **34**, 1149 (1977)
25. M. Wendisch, P. Yang, *Theory of Atmospheric Radiative Transfer* (Wiley, Singapore, 2012)
26. W. Wien, Wied. Ann. Phys. Chem. **58**, 662 (1896)
27. F. Reif, *Statistical and Thermal Physics* (McGrow Hill, Boston, 1965)
28. R.W.Healy, T.C.Winter, J.W.Labaugh, O.L.Franke. *Water budgets : Foundations for effective water-resources and environmental management.* (Reston, Virginia, U.S.Geological Survey Circular 1308, 2007)
29. A. Baumgartner, E. Reichel, *The World Water Balance* (Elsevier, Amsterdam, 1975)
30. J.P. Peixoto, A.H. Oort, *Physics of Climate* (Am er. Inst. Phys, Washington, 1992)

31. K.E.Trenberth, L.Smith, T.Qian et.al. *J.Hydrometeorol.* **8**, 758(2007)
32. https://www.en.wikipedia.org/wiki/water-circle
33. *Understanding Climate Change* (Washington, Nat.Acad.Science, 1975)
34. B.M. Smirnov, *Microphysics of Atmospheric Phenomena* (Switzerland, Springer Atmospheric Series, 2017)
35. B.M. Smirnov, *Introduction to Plasma Physics* (Mir, Moscow, 1977)
36. B.M. Smirnov, *Physics of Weakly Ionized Gases* (Mir, Moscow, 1980)
37. M.L. Salby, *Physics of the Atmosphere and Climate* (Cambr. Univ. Press, Cambridge, 2012)
38. V. Ramanathan, J. Meteorol. Soc. Jpn. **65**, 151 (1987)
39. V. Ramanathan, J. Geophys. Res. **92**, 4075 (1987)
40. V. Ramanathan, Science **240**, 293 (1988)
41. A. Raval, V. Ramanathan, Nature **342**, 758 (1989)
42. V. Ramanathan, R.D. Cess, E.F. Harrison et al., Science **243**, 57 (1989)
43. V. Ramanathan, E.F. Harrison, B.R. Barkstrom, Phys. Today **42**, 22 (1989)
44. J.T. Kiehl, K.E. Trenberth, Bull. Am. Meteorol. Soc. **78**, 197 (1997)
45. K.E. Trenberth, J.T. Fasullo, J.T. Kiehl, Bull. Am. Meteorol. Soc. **90**, 311 (2009)
46. K.E. Trenberth, J.T. Fasullo, Surf. Geophys. **33**, 413 (2012)
47. J.T. Fasullo, K.E. Trenberth, Science **338**, 792 (2012)
48. M. Wild, D. Folini, Ch. Schär et al., Clim. Dyn. **40**, 3107 (2013)
49. F. Vignova, J. Michalsky, T. Stoffel, *Solar and Infrared Radiation Measurements* (CRC Press, Boca Raton, 2012)
50. M.F. Modest, *Radiative Heat Transfer* (Elsevier, Amsterdam, 2013)
51. O. Boucher. *Atmospheric Aerosols. Properties and Climate Impacts.* (Dordrecht, Springer, 2015)
52. D.L. Hartmann, *Global Physical Climatology* (Elesevier, Amsterdam, 2016)
53. G.R. North, K.-Y. Kim, *Energy Balance Climate Models* (Wiley, Weinheim, 2017)
54. T.S.L. Ecuyer et al., J. Climate **28**, 8319 (2015)
55. V. Ramaswamy, A.W. Collins, B.J. Haywood et al. American Meteorological Society. Monograph, Chapter **14**(2019) https://doi.org/10.1175/AMSMONOGRAPHS-D-19-0001.1
56. B.M. Smirnov, EPL **114**, 24005 (2016)
57. *U.S. Standard Atmosphere* (Washington, U.S. Government Printing Office, 1976)
58. C. Le Quere et al., Earth Syst. Sci. Data **10**, 2141 (2018)
59. https://www.carbonbrief.org/analysis-fossil-fuel-emissions-in-2018-increasing-at-fastest-rate-for-seven-years
60. M.A. Ruderman, J.W. Chamberlain, Planet Space Sci. **23**, 247 (1975)
61. B.W. Carroll, D.A. Ostlie, *An Introduction to Modern Astrophysics* (Pearson, London, 2013)
62. B.V. Somov, *Cosmic Plasma Physics* (Kluwer, Dordrecht, 2000)
63. N. Meyer-Vernet, *Basic of Solar Wind* (Cambr. Univ. Press, Cambridge, 2007)
64. R.P. Feynman, R.B. Leighton, M. Sands. *The Feynman Lectures of Physics* (Reading, Addison-Wesley, 1964). vol. 2
65. K. Berger, *The Earth flash.* In *Lightning*. ed. R.H. Golde. (London, Acad. Press, 1977, p. 119)
66. M. Rycroft et al., Space Sci. Rev. **137**, 83 (2008)
67. M.J. Rycroft et al. in *Planetary Atmospheric Electricity* ed. F. Leblanc et al. (Heidelberg, Springer, 2008) p. 83
68. E. Williams, Atmosp. Res. **91**, 140 (2009)
69. B.R. Berrisford, P. Kallberg, S. Kobayashi et al., Quant. J. Roy. Meteorol. Soc. **137**, 1381 (2011)
70. N.B. Vargaftic, *Tables of Thermophysical Properties of Liquids and Gases* (Halsted Press, New York, 1975)
71. B.M. Smirnov, *Reference Data on Atomic Physics and Atomic Processes* (Springer, Heidelberg, 2008)
72. https://www.nationalgeographic.org/encyclopedia/chlorophyll/
73. R. Hill, Nature **139**, 881 (1937)
74. https://en.wikipedia.org/wiki/Chloroplast

75. *The structure and function of plastids* ed. by R.R. Wise, J.K. Hoober (Dordrecht, Springer, 2006)
76. Plastids, *Methods and Protocols*. ed by E. Marechal (New York, Springer, 2018)
77. H.G. Jones, *Plants and Microclimate: A Quantitative Approach to Environmental Plant Physiology* (Camb. Univ. Press, Cambridge, 1992)
78. D.O. Hall, K.K. Rao, *Photosynthesis* (Cambr. Univ. Press, Cambridge, 1999)
79. D.W. Lawlor. *Photosynthesis* (Oxford, BIOS, 2001)
80. J.E. Bidlack, K.R. Stern, S. Jansky *Introductory Plant Biology* (McGraw-Hill, New York, 2003)
81. *Handbook of Photosynthesis* (Boca Raton, CRC Press, 2005)
82. R.E. Blankenship *Molecular Mechanisms of Photosynthesis* (Wiley, London, 2014)
83. https://www.rsc.org
84. R.F. Sage, R.K. Monson, *C4 Plant Biology* (Academic, San Diego, 1999)
85. G.E. Edwards, D.A. Walker, *C3, C4: Mechanisms and Cellular and Environmental Regulation of Photosynthesis* (Blackwell, Oxford, 1983)
86. https://en.wikipedia.org/wiki/RuBisCO
87. I. Andersson, S. Knight, G. Schneider et al. *Nature* **337**, 229(1989) Crystal structure of the active site of ribulose-bisphosphate carboxylase
88. E.M. Salvucci, W.L. Ogren, Photosynth Res. **47**, 1 (1996)
89. T.D. Sharkey, Photosynth Res. **53**, 835 (2019)
90. R.F. Sage, T.L. Sage, F. Kocacinar, Photorespiration and the evolution of C4 photosynthesis. Annu. Rev. Plant Biol. **63**, 19 (2012)
91. https://en.wikipedia.org/wiki/Photorespiration
92. A.S. Raghavendra, R.F. Sage, *C4 Photosynthesis and Related CO_2 Concentrating Mechanisms* (Springer, Dordrecht, 2011)
93. https://en.wikipedia.org/wiki/C4-carbon-fixation
94. http://www.ciesin.org/docs/004-038/004-038a.html
95. https://en.wikipedia.org/wiki/Photosynthetic-efficiency
96. R.E. Blankenship, D.M. Tiede, J. Barber et al., Science **332**(6031), 805 (2011)
97. R.T. Barber, A.K. Hilting, *History of the Study of Plankton Productivity* (Blackwell, Oxford, 2002)
98. P.G.Falkowski. Sci. Am. **287**, N22, 54(2002)
99. J.H. Ryther, Science **166**, 72 (1969)
100. E. Steeman-Nielsen, J. Cons. Int. Explor. Mer. **18**, 117 (1952)
101. J. Marra, Aquat. Microb. Ecol. **56**, 123 (2009)
102. M.J. Behrenfeld, P.G. Falkowski, Limnol. Oceanogr. **42**, 1 (1997)
103. https://en.wikipedia.org/wiki/Calcium-carbonate
104. M. Grosjean, J. Goiot, Z. Yu, Global Planet Change **152**, 19 (2017)
105. R.E. Zeebe, D.A. Wolf-Gladrow, *CO_2 in Seawater: Equilibrium, Kinetics, Isotopes* (Elsevier, Amsterdam, 2001)
106. https://en.wikipedia.org/wiki/ocean-storage-of-carbon-dioxide
107. T. Pedersen, Y. Rosental, S. Seitzinger et al., Science **290**, 291 (2000)
108. *The Carbon Cycle* ed. T.M. Wigley, D.S. Schimmel. (Cambridge, Cambr. Univ. Press, 2000)
109. D. Archer, *The Global Carbon Cycle* (Princeton, Princ. Univ. Press, 2010)
110. https://en.wikipedia.org/wiki/Carbon-cycle
111. https://en.wikipedia.org/wiki/Carbon-dioxide-in-Earth/atmosphere/media/File:Carbon-cycle
112. https://earthobservatory.nasa.gov/features/CarbonCycle
113. L. Pauling. *General Chemistry* (San Francisco, 1970)
114. E.A. Kasatkina, O.I. Shumilov, M. Krapiec, On periodicities term climatic variations near 68° N, 30° E. Adv. Geosci. **13**, 25 (2007)
115. J.M. Kauffman, J. Sci. Explor. **4**, 723 (2007)
116. P. Friedlingstein, M. O'Sullivan1, M.W. Jones et.al. Earth Syst. Sci. Data **11**, 1783(2019)
117. C.B. Field, M.J. Behrenfeld, J.T. Randerson, P. Falkowski, Science **281**, 237 (1998)

118. F.P. Chavez, M. Messie, J.T. Pennington, Annu. Rev. Mar. Sci. **3**, 227 (2011)
119. L.R. Welp, R.F. Keeling, H.A.J.M eijer et al. Nature **477**, 579(2011)
120. https://earthsky.org/earth/scientists-calculate-a-new-rate-for-global-photosynthesis
121. ftp://aftp.cmdl.noaa.gov/products/trends/co2/co2
122. https://www.esrl.noaa.gov/gmd/ccgg/trends
123. https://en.wikipedia.org/wiki/Keeling-curve
124. http://www.fao.org/3/ac836e/AC836E03.htm
125. H. Ritchie (2919) https://ourworldindata.org/life-by-environment
126. https://en.wikipedia.org/wiki/Biomass-(ecology)
127. Y.M. Bar-On, R. Philips, R. Milo, PNAS **115**, 6508 (2018)
128. Ch.Beer, M.Reichstein, E.Tomelleri et al. *Science* **329**, 834(2010)
129. http://www.globalcarbonproject.org
130. C. Le Quere, R. Moriarty, R.M. Andrew et al., Earth Syst. Sci. Data **7**, 47 (2015)
131. C. Le Quere, R. Moriarty, R.M. Andrew et al., Earth Syst. Sci. Data **7**, 349 (2015)
132. R.B. Jackson, J.G. Canadell, C. Le Quere et al., Nat. Clim. Change **6**, 7 (2016)
133. C. Quere, R.M. Andrew, J.G. Canadell et al., Earth Syst. Sci. Data **8**, 605 (2016)
134. G.P. Peters, C. Le Quere, R.M. Andrew et al., Nat. Clim. Change **7**, 848 (2017)
135. M. Roser, H. Ritchie https://ourworldindata.org/food-supply
136. https://www.worldometers.info/world-population/world-population-by-year/
137. https://en.wikipedia.org/wiki/Cereal
138. https://knoema.ru/atlas/
139. https://en.wikipedia.org/wiki/Wheat
140. T.R. Malthus, *An Essay on the Principle of Population* (London, St. Paul's Churchyard, 1798)
141. T.R. Malthus, *Principles of Political Economy Considered with a View of Their Practical Application* (John Murray, London, 1820)
142. https://en.wikipedia.org/wiki/Thomas-Robert-Malthus
143. https://en.wikipedia.org/wiki/Malthusian-catastrophe
144. M. Shermer, Scientific Amer. **314**, 72 (2016)
145. https://ourworldindata.org/land-use
146. J.E. Hansen, D. Johnson, A. Lacis et al., Science **213**, 957 (1981)
147. J. Hansen, M. Sato, R. Ruedy et al. http://www.columbia.edu/~jeh1/mailing/2016/20160120-Temperature2015
148. J. Hansen, M. Sato, R. Ruedy. http://www.columbia.edu/~jeh1/mailing/2014/20140121-Temperature2013
149. J.E. Hansen, R. Ruedy, M. Sato et al., J. Geophys. Res. **106**, 947 (2001)
150. J. E. Hansen, R. Ruedy, M. Sato, K. Lo. Rev.Geophys. **48**, RG4004(2010)
151. J.E. Hansen, R. Ruedy, M. Sato, K. Lo. http://www.data.giss.nasa.gov/gstemp/2011
152. B.M. Smirnov. *Physics of global atmosphere* (Dolgoprudny, Intellect, 2017; in Russian)
153. B.D. Malamud, D.L. Turcotte, C.S.B. Grimmond, Clim. Past **7**, 975 (2011)
154. https://www.ncdc.noaa.gov/sotc/global/201805
155. https://www.ncdc.noaa.gov/sotc/global/201905
156. V.A. Zubakov, I.I. Borzenkova, *Global Paleoclimate of the Late Cenozoic* (Elsevier, Amsterdam, 1990)
157. M. Benton, D.A.T. Harper, *Introduction to Paleobiology and the Fossil Record* (Wiley-Blackwell, Chichester, 2009)
158. T.M. Cronin, *Paleoclimates?: Understanding Climate Change Past and Present* (Columbia Univ. Press, New York, 2010)
159. https://en.wikipedia.org/wiki/Paleontology
160. https://en.wikipedia.org/wiki/Paleoclimatology
161. M.L. Bender, *Paleoclimate* (Princeton, Princ. Univer. Press, 2013)
162. G. Faure, D. Mensing, *Isotopes-Principles and Applications* (Wiley, Hoboken N.J., 2005)
163. J.R. Petit, J. Jouzel, D. Raynaud et al., Nature **399**, 429 (1991)
164. https://www.co2.earth/21-co2-past
165. B. Hönisch, N.G. Hemming, D. Archer et al., Science **324**, 1551 (2009)

166. http://unfccc.int/resource/docs/2015/cop21
167. N. Marsh, H. Svensmark, Space Sci. Rev. **94**, 215 (2000)
168. L.I. Dorman, Adv. Space Rev. **35**, 496 (2005)
169. G. Faure, *Principles of Isotope Geology* (Cambr. Univer. Press, Cambridge, 1986)
170. A.P. Dickin, *Radiogenic Isotope Geology* (Cambr. Univer. Press, Cambridge, 1995)
171. https://en.wikipedia.org/wiki/Geochronology
172. G.A. Zielinsky et al., Science **264**, 948 (1994)
173. https://en.wikipedia.org/wiki/Sea-level-rise
174. http://climate.nasa.gov/system/charts/12-15-seaLevel
175. A. Grinsted, The Cryosphere **7**, 141 (2013)
176. J.A.Church et al. *3rd International Panel Change Climate* (Cambridge, Cambr. Univ. Press, 2001), pp. 639–693
177. V. Radič, R. Hock, J. Geophys. Res. **115**, F01010 (2010)
178. M. Huss, D. Farinotti, J. Geophys. Res. **117**, F04010 (2012)
179. https://www.climate.gov/news-features/understanding-climate/climate-change-glacier-mass-balance
180. http://climate.nasa.gov
181. https://www.liveinternet.ru/users/amayfaar/post116828433
182. J. Oerlemans, Science **264**, 243 (1994)
183. https://en.wikipedia.org/wiki/Latent-heat
184. https://en.wikipedia.org/wiki/Climate-sensitivity
185. Palaeosens Project Members, Nature **491**, 683 (2012)
186. L.B. Stap, P. Köhler, G. Lohmann, Earth Syst. Dynam. **10**, 333 (2019)
187. J. Feichter, E. Roeckner, U. Lohmann, B. Liepert, J. Clim. **17**, 2384 (2004)
188. J. Hansen et al., J. Geophys. Res. **110**, D18104 (2005)
189. B.M. Smirnov, J. Phys. D. Appl. Phys. **51**, 214004 (2018)
190. B.M. Smirnov, *Transport of Infrared Atmospheric Radiation* (de Gruyter, Berlin, 2020)

Chapter 6
Emission of the Atmosphere

Abstract Emission of the atmosphere as a thin air layer located over the Earth's surface results from radiation of water and carbon dioxide molecules, as well as due to water microdroplets which constitute the clouds. Since the emission spectrum of the atmosphere consists mostly of individual broaden spectral lines due to vibration–rotation and rotation states of molecules, the infrared atmospheric spectrum consists of thousands broaden peaks. Therefore, in evaluation of the radiative flux from the atmosphere, we are based on the "line-by-line" method combined with thermodynamic equilibrium between the radiation field and air molecules, as well as the model of standard atmosphere for molecules with accounting for a weak altitude dependence for the atmospheric temperature. Assuming clouds as an opaque matter to be located above a certain altitude h_{cl} and to have a sharp boundary, one can describe the radiative flux created by atmospheric molecules at a given frequency ω by two atmospheric parameters, namely an optical thickness u_ω and an opaque factor $g(u_\omega)$ for atmospheric layer below clouds. It is of principle for these evaluations the HITRAN database that contains parameters radiative transitions in greenhouse molecules located in atmospheric air. The altitude h_{cl} of the cloud boundary (or the radiative temperature T_{cl} for clouds) follows from comparison of the total radiative flux of the atmosphere to the Earth's surface according to the energetic balance of the Earth, as well as the calculated radiative flux created by atmospheric molecules. As a result, the partial radiative fluxes for each component are determined at a given frequency, as well as the total radiative fluxes to the Earth due to each component. In particular, the contribution to the total radiative flux toward the Earth is approximately 64% due to H_2O molecules, 18% due to clouds, 17% due to CO_2 molecules, and about 1% due to CH_4 and N_2O trace molecules. The greenhouse instability is represented that is realized if one water molecule during its residence in the atmosphere emits such radiative flux that causes evaporation more than one water molecule from the Earth's surface. The threshold of this instability occurs if the contemporary global temperature increases by 7K.

B. M. Smirnov, *Global Energetics of the Atmosphere*, Springer Atmospheric Sciences, https://doi.org/10.1007/978-3-030-90008-3_6

6.1 Character of Atmospheric Emission

6.1.1 *Thermodynamics of Emission of Flat Atmospheric Layer*

According to the energetic balance of the Earth and atmosphere, the radiative fluxes from the atmosphere outside and toward the Earth are separated and are created by different regions of the atmosphere. Hence, in the analysis of interaction between the Earth's surface and atmosphere through the radiation field, one can ignore an atmospheric radiation with an environment. In formulating the algorithm of evaluation of atmospheric radiative fluxes, we use general principles of emission of atmospheric air and transport of radiation through it [1–10].

The atmosphere as a source of infrared radiation is a weakly nonuniform gaseous layer located over the Earth surface. Indeed, a thickness of an atmospheric layer which creates the infrared radiation both toward and outside the Earth is (3–6) km, whereas the Earth's radius $R_\oplus = 6370$ km exceeds it significantly. Hence, we consider radiating atmospheric air as a plane layer of a gas. Usually horizontal nonuniformities exceeds remarkably atmospheric altitudes which are responsible for atmosphere radiation. In addition, the air temperature varies weakly in a radiating layer and thermodynamic equilibrium is established between atmospheric molecules and radiation field. This allows one to introduce the radiative temperature T_ω for each radiative frequency ω and is close to the temperature of air located in a radiation regions. Correspondingly, under these conditions, we consider an air layer emitted radiation as a weakly nonuniform atmospheric layer.

As a general principle for a radiation source, we use the Kirchhoff law [11] or the principle of detailed balance between the processes of emission and absorption. According the Kirchhoff law, each emitter is simultaneously the absorber. This principle allows one to use the absorption coefficient k_ω which is the parameter of the absorption process also for the analysis of the emission process.

The absorption coefficient is defined on the basis of the Beer–Lambert law [12, 13] according to which the flux j_ω of photons of a given frequency ω which propagates a direction x through a given matter is depleted as

$$\frac{dj_\omega}{dx} = -k_\omega j_\omega \tag{6.1.1}$$

One more parameter of propagation of the radiative flux through an absorbed gas is the optical thickness u_ω of a layer of a thickness h from the Earth's surface up to a given point which is defined as

$$u_\omega = \int_0^h k_\omega(h)dh \tag{6.1.2}$$

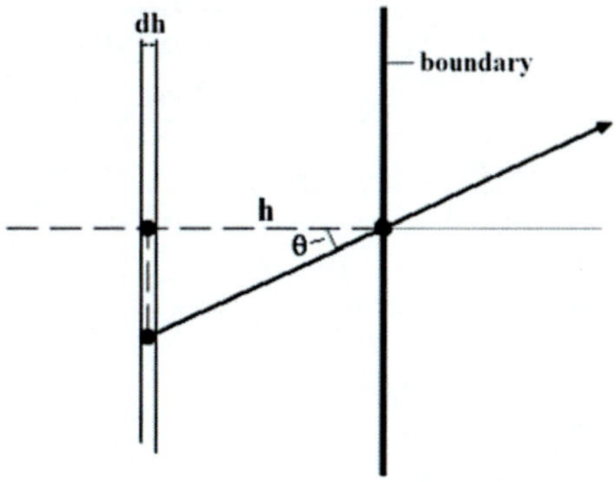

Fig. 6.1 Geometry of photon propagation from a given plane of a thickness dh. The radiative flux from this plane is the sum of elementary fluxes

So, emission of the atmosphere which is created by molecules or microdroplets has a random character, and hence, it is noncoherent one. In addition, the angle emission of forming photons is isotropic. Because of thermodynamic equilibrium, emission from an atmospheric point of coordinate **r** is characterized by the temperature $T(\mathbf{r})$. Therefore, the energy flux of photons at a given frequency is the nucleus of the Biberman–Holstein equation [14, 15]. This flux is the sum of elementary fluxes in accordance with the geometry of their formation according to Fig. 6.1. This energy flux is determined by the expression [16, 17]

$$J_\omega = \int_0^1 dcos\theta \int_0^{u_\omega} du \exp\left(-\frac{u}{\cos\theta}\right) \cdot \frac{\hbar\omega^3}{4\pi^2 c^2}\left[\exp\left(\frac{\hbar\omega}{T(h)}\right) - 1\right]^{-1} \qquad (6.1.3)$$

If the temperature T of atmospheric air is independent of the altitude h, one can represent the total radiative flux J_ω at a given frequency ω as [18, 19]

$$J_\omega = I_\omega(T) g(u_\omega) \qquad (6.1.4)$$

Here, $I_\omega(T)$ is the radiative flux at a frequency ω for a blackbody which temperature is T, and $g(u_\omega)$ is the opaque factor which is determined by the optical thickness u_ω of the layer. For the optically thick layer $u_\omega \gg 1$, we have $g(u_\omega) = 1$, and the radiative flux at a frequency of a blackbody of a temperature T is given by the Planck formula [20, 21]

$$I_\omega(T) = \frac{\hbar\omega^3}{4\pi^2 c^2 \left[\exp\left(\frac{\hbar\omega}{T}\right) - 1\right]} \qquad (6.1.5)$$

Figure 6.2 gives the frequency dependence for the equilibrium radiative flux $I_\omega(T)$ at temperatures which correspond to radiative temperatures of atmospheric emission.

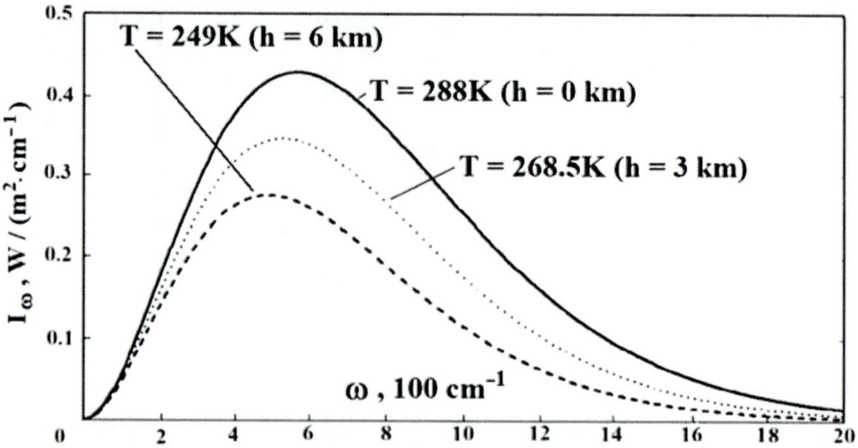

Fig. 6.2 Specific radiative flux from a blackbody at temperatures which are typical for a lower troposphere. The altitude h is indicated at which this temperature is realized

The opaque factor $g(u_\omega)$ of a uniform gaseous layer is given by [18, 22]

$$g(u_\omega) = 2\int_0^1 \cos\theta d\cos\theta \left[1 - \exp\left(-\frac{u_\omega}{cos\theta}\right)\right], \tag{6.1.6}$$

and u_ω is the total optical thickness of the emitting layer. The approximation of the opaque factor in the range $u_\omega \sim 1$ is given by [19, 23]

$$g(u_\omega) = \exp(-1.6u_\omega) \tag{6.1.7}$$

We now transfer from an uniform layer to a weakly nonuniform one. The weakly nonuniform layer implies that the temperature varies weakly in the region of emission. In the framework of the model of standard atmosphere [24], the temperature at a given altitude h is given by

$$T(h) = T_E - h \cdot \frac{dT}{dh}, \tag{6.1.8}$$

where in accordance with the model of standard atmosphere the surface temperature is $T_E = 288K$, and the temperature gradient equals $dT/dh = -6.5$ K/km. Being guided by altitudes below $h = 5km$, we have that the atmospheric temperature differs from the average one within the limits of 20%, and hence, a radiating temperature may be considered as a weakly nonuniform one.

Transiting from a uniform radiating gaseous layer to a nonuniform one, we introduce the effective temperature T_ω as the radiative temperature at a given frequency ω. Correspondingly, formula (6.1.4) takes the form

$$J_\omega = I_\omega(T_\omega)g(u_\omega) \quad (6.1.9)$$

In the case of an optically thick layer $u_\omega \gg 1$, the effective altitude h_ω which is introduced as $T_\omega = T(h_\omega)$ is given by [16, 17]

$$u(h_\omega) = 2/3, \ u_\omega \gg 1 \quad (6.1.10)$$

In the other limiting case $u_\omega \ll 1$, the total optical thickness u_ω of the atmospheric layer is equal [19, 23]

$$u(h_\omega) = u_\omega/2, \ u_\omega \ll 1 \quad (6.1.11)$$

Combining both limiting cases, one can obtain for the optical thickness of an effective layer

$$u(h_\omega) = \frac{u_\omega}{2\exp(-u_\omega) + 1.5u_\omega} \quad (6.1.12)$$

This formula leads to correct limiting expressions for the altitude h_ω of the effective layer.

There are three basic greenhouse components, namely carbon dioxide and water molecules, as well as water microdroplets which form clouds. In addition, three trace greenhouse components, namely methane, nitrogen dioxide and ozone molecules, give the contribution approximately of 1% to the radiative flux which is created by the atmosphere and is absorbed by the Earth's surface. We use below for atmospheric radiation the model where clouds are not located below a certain altitude h_{cl}, and above this altitude the optical thickness of clouds due to water microdroplets of clouds is large. This scheme is presented in Fig. 6.3, and within the framework of this scheme, the radiative flux J_ω is given by [23]

$$J_\omega = I_\omega(T_\omega)g(u_\omega) + I_\omega(T_{cl})[1 - g(u_\omega)], \ T_\omega = T(h_\omega), \ u(h_\omega) = \\ = \frac{u_\omega(h_{cl})}{2\exp(-u_\omega(h_{cl})) + 1.5u_\omega(h_{cl})}, \quad (6.1.13)$$

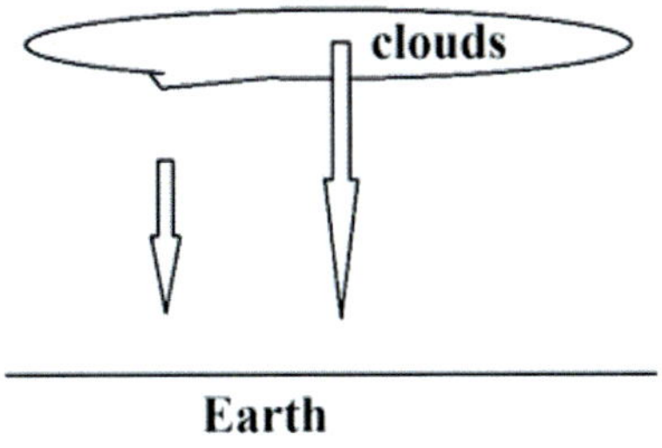

Fig. 6.3 Character of emission from the atmosphere

where T_ω is the radiative temperature for all the molecular atmospheric components, the cloud temperature T_{cl} is independent of a frequency ω and corresponds to cloud location at an altitude h_{cl}, where $T_{cl} = T(h_{cl})$.

6.1.2 Peculiarities of Radiative Transitions in Atmospheric Air

We now analyze methods to determine the rates of radiative transitions in molecules with emission of infrared photons. We are based on general principles of radiative transitions of molecules including selection rules for molecules of a certain symmetry [25–38]. These peculiarities of molecular transitions are added by information for parameters of radiative transitions in molecules [39–43]. These principles and information are used for the analysis of transport of infrared radiation in the atmosphere [6–10]. Taking into account these factors, one can consider the suitable algorithm for evaluation of atmospheric radiative fluxes in the framework of the line-by-line model [1, 5].

Basing on the atmospheric pressure of air and those which are close to it, one can obtain the spectrum of atmospheric emission in the infrared spectrum range to be consisted of individual broaden lines. This means that a typical distance $\Delta\omega$ between neighboring spectral lines is large compared to a width ν of an individual spectral line, i.e.,

$$\Delta\omega \gg \nu \tag{6.1.14}$$

According to this criterion, neighboring spectral lines are not overlapped.

Being guided by pressures of atmospheric air which are close to atmospheric one, we are based on the impact mechanism of broadening of spectral lines according to which [34, 38, 44]

$$a_\omega = \frac{\nu_j}{2\pi[(\omega - \omega_j)^2 + (\nu_j/2)^2]}, \tag{6.1.15}$$

where ν_j is the width of $j-$th spectral line. Since the photon frequency distribution function a_ω is the probability to emit of a photon of a given frequency, its normalization has the form

$$\int a_\omega d\omega = 1 \tag{6.1.16}$$

In conclusion, we note that our goal is to evaluate the radiative fluxes from the atmosphere. The partial radiative flux is determined by formula (6.1.9), and the flux of radiation due to atmospheric molecules is expressed through three quantities, namely the absorption coefficient k_ω at a given frequency, the optical thickness u_ω

at the atmospheric gap between the ground and clouds and the radiative temperature T_ω at a given frequency ω. In turn, these quantities are expressed through radiative parameters of radiative transitions which are taken from the HITRAN data bank. These radiative parameters include the transition frequency ω_j for a given transition j, the width ν_j of this spectral line, the transition intensity $S_{ij}(T_o)$ related to a temperature T_o and the excitation energy ε_{ij} from the ground state for a lower state of this transition.

According to the Kirchhoff law [11], emitted molecules are simultaneously the absorbers. The processes of emission and absorption are detailed inverse ones, and therefore, their rates are connected by the principle of detailed balance. We below derive the connection between the rates of emission and absorption for molecules under simple conditions where radiative transitions result between o and i states only; i.e., we consider the following radiative processes

$$M_o + \hbar\omega \rightleftarrows M_i, \tag{6.1.17}$$

where M is the molecule, and its states are denoted by subscripts. Under thermodynamic equilibrium between the radiation field and molecules, we have the following relation between rates of emission and absorption processes

$$N_i a_\omega A_i = N_o \sigma_\omega i_\omega \tag{6.1.18}$$

Here, N_o and N_i are the number densities of molecules in the lower o and upper i of transition states, a_ω is the photon distribution function [44], so that $a_\omega d\omega$ is the probability that created photons have the frequencies between ω and $\omega + d\omega$, so that this distribution function is normalized according to the relation (6.1.16). Next, A_i is the second Einstein coefficient for this radiative transition or the rate of this transition from an upper state i to the lower state o, and i_ω is the partial photon flux which is equal according to its definition

$$i_\omega = \frac{I_\omega}{\hbar\omega}, \tag{6.1.19}$$

and the energy flux of radiation is given by formula (6.1.5).

We have in the first approximation for the rotation energy of a linear molecule which rotation quantum number is j

$$\varepsilon_j = Bj(j+1), \tag{6.1.20}$$

where B is the rotation constant, and the photon energy at the center of the spectral line for a vibration–rotation transition from the rotation state j is equal

$$\hbar\omega_j = \hbar\omega_o - B \pm B(2j+1) \approx \hbar\omega_o \pm \hbar\Delta\omega j/2, \tag{6.1.21}$$

where $\hbar\omega_o$ is the energy of the vibration transition, $\hbar\Delta\omega = 4B$ is the energy difference for neighboring transitions in the limit of large j with accounting for the symmetry of a linear molecule of the CO_2 type, sign minus relates to P absorption branch, while sign plus corresponds to R-branch, and we are guided by large rotation momenta j.

Restricting by one vibration transition ($o \to i$), we have from the Boltzmann distribution of molecules over rotation states in the limit $T \gg B$

$$N_{vj} = N_v \frac{2B}{T}(2j+1)\exp\left[-\frac{Bj(j+1)}{T}\right], \quad \sum_j N_{vj} = N_v \tag{6.1.22}$$

Here, v is the vibration molecule state and j is its rotation number. Taking the cross section σ_ω from formula (6.1.18), we have the following expression for the absorption cross section

$$\sigma_\omega = \frac{\pi^2 c^2}{\omega^2} A_i \frac{\hbar|\omega - \omega_o|)}{T} \exp\left[-\frac{\hbar^2(\omega-\omega_o)^2}{4BT}\right]\left[1 - \exp\left(-\frac{\hbar\omega}{T}\right)\right]\sum_j a_{\omega-\omega_j}, \tag{6.1.23}$$

where we keep in mind the CO_2 molecule, and therefore, we take $\hbar\Delta\omega = 4B$.

In addition, being guided by atmospheric pressures, we have the impact broadening of spectral lines [34, 38, 44] that results from collisions of a radiating molecule with air ones. In this case, the frequency distribution function of photons a_ω for some vibration– rotation transitions is given by formula (6.1.16). We use below the expression for the absorption cross section (6.1.23) in the subsequent analysis of emission and absorption of CO_2 molecules.

6.1.3 Emission of Atmospheric CO_2 Molecules

Basing on the model (6.1.13) for atmospheric emission and transiting to radiation of the real atmosphere in infrared spectrum range, we take as atmospheric radiators three its components, molecules of water and carbon dioxide, as well as water microdroplets which constitute clouds. In this model, one can separate in a space molecules and clouds, so that radiating molecules are located in a gap between the Earth's surface and clouds. On the basis of the space distribution of CO_2 and H_2O molecules in accordance with formulas (2.1.1) and (2.1.19) with using parameters of molecule emission, one can determine the molecular part of the radiative flux in formula (6.1.13).

Note that along with H_2O and CO_2 molecules, one can account for other molecules, especially in specific cases of the atmosphere with a heightened concentration of some molecular components. But within the framework of the model of standard atmosphere in which average concentrations of molecular components are included, only atmospheric CH_4, N_2O and O_3 molecules create radiative fluxes to the Earth above $1W/m^2$. Therefore, we restrict only these molecules considering them as impurities.

Thus, in the first approach, molecular components of the atmosphere are a water vapor and carbon dioxide. In accordance with formulas (2.1.1), (2.1.19) and (2.1.21), the distribution of the number densities of carbon dioxide $N(CO_2)$ and water $N(H_2O)$ molecules over an altitude h are given by

$$N(H_2O) = N_w \exp\left(-\frac{h}{\lambda}\right), \quad N_w = 3.4 \times 10^{17}\mathrm{cm}^{-3}, \ \lambda = 2\,\mathrm{km}$$

$$N(CO_2) = N_c \exp\left(-\frac{h}{\Lambda}\right), \quad N_c = 1.1 \times 10^{16}\mathrm{cm}^{-3}, \ \Lambda = 10\,\mathrm{km} \qquad (6.1.24)$$

In addition, the atmosphere temperature $T(h)$ as an altitude h function is

$$T(h) = T_E - \frac{\mathrm{d}T}{\mathrm{d}h}h, \ T_E = 288\mathrm{K}, \ \frac{\mathrm{d}T}{\mathrm{d}h} = 6.5\mathrm{K/km} \qquad (6.1.25)$$

It should be noted that a nonuniformity of the atmosphere consists in the difference of temperatures for different atmospheric layers. This leads to a frequency dependence for the radiative temperature in formula (6.1.13). As for the optical thickness u_ω of the atmosphere, in this consideration with using formula (6.1.24), we have for this value according to its definition (6.1.2)

$$u_\omega(h_\omega) = k_\omega(H_2O)\lambda\left[1 - \exp\left(-\frac{h_\omega}{\lambda}\right)\right] + k_\omega(CO_2)\Lambda\left[1 - \exp\left(-\frac{h_\omega}{\Lambda}\right)\right] \qquad (6.1.26)$$

where h_ω is a current altitude, $k_\omega(H_2O)$ and $k_\omega(CO_2)$ are the absorption coefficients by water and carbon dioxide molecules near the Earth's surface.

In considering the absorption coefficient due to atmospheric molecules, accounting for the structure of the expression for this quantity [34, 38, 44], it is convenient to represent it in the form [45]

$$k_\omega = N(H_2O)\sum_j S(H_2O)a(\omega - \omega_j) + N(CO_2)\sum_j S(CO_2)a(\omega - \omega_j), \qquad (6.1.27)$$

where $N(H_2O)$ and $N(CO_2)$ are the total number densities of water and carbon dioxide molecules at a given point, $S(H_2O)$, $S(H_2O)$ are the spectral line intensities for jth transition between molecular states of molecules H_2O and CO_2) correspondingly, ω_j is the frequency of the center of a given spectral line, and $a(\omega - \omega_i)$ is the distribution function of emitted or absorbed photons over frequencies [44] which is given by formula (6.1.16).

We now represent the method of determination of radiative flux in accordance with formula (6.1.13). Taking $k_\omega(CO_2)$ and $k_\omega(H_2O)$ as the absorption coefficients of near-surface air due to CO_2 atmospheric molecules and H_2O atmospheric molecules correspondingly. Introducing the total optical thickness of the atmosphere at a given frequency $u_\omega(h_{cl})$ according to formula (6.1.26), one can determine the effective altitude h_ω at a given frequency as due to molecules with taking into account that the atmosphere is screening by clouds located at an altitude h_{cl}. Taking the effective altitude to be small compared to a typical scale of the altitude $\Lambda \gg h_\omega$, we have

$$u_\omega(h_\omega) \equiv k_\omega(H_2O)\lambda\left[1 - \exp\left(-\frac{h_\omega}{\lambda}\right)\right] + k_\omega(CO_2)\Lambda$$
$$= \frac{u_\omega(h_{cl})}{2\exp[-u_\omega(h_{cl})] + 1.5u_\omega(h_{cl})} \quad (6.1.28)$$

Formula (6.1.28) is the equation for the effective altitude h_ω which is responsible for emission of atmospheric molecules, and the radiative temperature for emission of atmospheric molecules is connected with this quantity by formula (6.1.25) that has the form

$$T_\omega = T_E - h_\omega \frac{dT}{dh}, \quad (6.1.29)$$

where the global temperature is $T_E = 288K$ for the model of standard atmosphere, and the temperature gradient is $dT/dh = 6.5$ K/km. Equation (6.1.28) contains also the effective altitude where radiation from the atmosphere is absorbed. This allows us to determine the radiative flux J_ω from the atmosphere to the Earth on the basis of formula (6.1.13) that has the form

$$J_\downarrow = \int J_\omega d\omega = \int \{J_\omega(T_\omega)g[u_\omega(h_{cl})] + I_\omega(T_{cl})\{1 - g[u_\omega(h_{cl})]\}d\omega, \quad (6.1.30)$$

where the total radiative flux $J_\downarrow$ from the atmosphere is given in Table 7 for various versions of the energetic balance of the Earth and atmosphere.

We now consider the part of atmospheric emission that is created by molecules. In this analysis, we are guided by CO_2 molecules for two reasons. First, a CO_2 molecule as the linear one gives a simple example that allows one to demonstrate general peculiarities of molecular emission of the atmosphere. Second, carbon dioxide molecules are of interest as an important greenhouse component of the atmosphere, and the change of their concentration due to human activity is of interest for the analysis of the real atmosphere.

Figure 6.4 gives the vibration spectrum of the CO_2 molecule and explains the character of the emission spectrum of a gas with CO_2 molecules in the infrared spectrum range. This molecule is characterized by three types of vibration in accordance with its linear structures. In the case of a symmetric oscillation, the oxygen

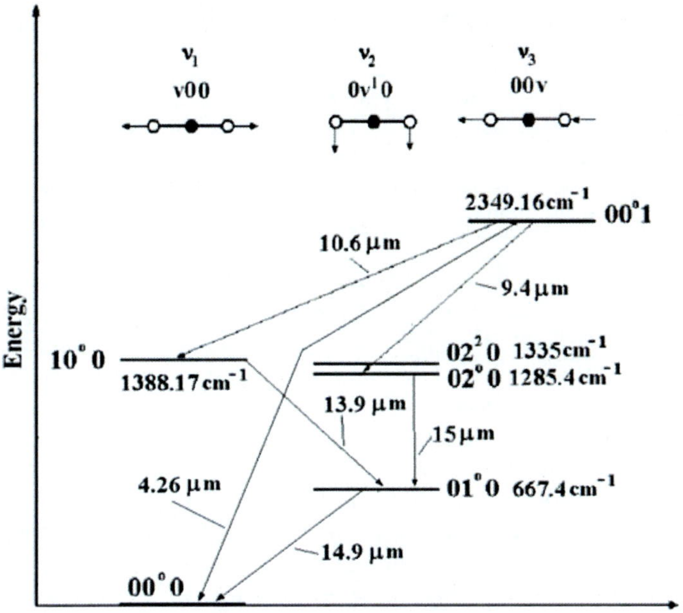

Fig. 6.4 Types of the lowest vibration states of the CO_2 molecules and the wavelengths of radiative transitions between the lowest vibration states in this molecule

atoms move along the axis of the CO_2 molecule, so that during the oscillation the distances between oxygen atoms and the carbon atom remains motionless. In the case of antisymmetric oscillations, oxygen atoms also move along the axis of the molecule, but now the distance between oxygen atoms is conserved. For the torsional oscillation, oxygen atoms move in a perpendicular direction with respect to the molecular axis. Each vibrational transition is accompanied by a change in the specific rotational state of the molecule. The change in the vibrational and rotational states during the radiative transition of a molecule is dictated by the so-called selection rules, which select certain final states of the molecule at a given initial state as a result of radiative transitions [25–29, 39, 40].

For the CO_2 molecules as a linear one, the selection rules with respect to a rotation transition have the form

$$j_f = j \pm 1, \;\; j_f = j, \tag{6.1.31}$$

where j and j_f are the initial and final rotation numbers correspondingly. The sign minus relates to P absorption branch, while sign plus corresponds to R-branch, and in the case of the Q-branch, the rotation number does not vary.

For the CO_2 molecule as a linear one, the rotation constant B is equal according to its definition

$$B = \frac{\hbar^2}{2m_O r_O^2} = 0.39cm^{-1}$$

Here, $r_o = 1.162$ Å[42] is the equilibrium distance between carbon and oxygen atoms, m_O is the mass of an oxygen atom, and $I = mr_O^2$ is the inertia moment of the molecule.

In addition, we note the rotation symmetry of the carbon dioxide molecule. Indeed, the abundance of the isotope ^{12}C in the Earth's crust is 98.9%, and the abundance of the isotope ^{16}O is 99.76% [46]. The nuclear spin of each of these isotopes is zero, so that practically all the CO_2 molecules in the nature have the zero spin of nuclei. Restricting by indicated isotopes, one can obtain the symmetry for molecule reflection with respect to the plane which is perpendicular to the molecular axis and passes through the carbon atom. This operation is analogous to exchange by oxygen molecules which are identical because of zero nuclear spin. In addition, it corresponds to reflection of electrons and nuclei of the carbon atom with respect to the symmetry plane. Since the nuclear spin of the carbon atom is zero, and its electron state is conserved at this operation, we obtain the total conservation of the state of the $^{12}C^{16}O_2$ as a result of the above operations.

Accounting for the separation of electron, vibrational and rotational degrees of freedom, and constructing the total wave function of the CO_2 molecule as the product of the electron, vibrational and rotational wave functions, we take into account the above symmetry of this molecule. Since its electron state is symmetric both for inversion and for reflection with respect to the symmetry plane, the vibrational and rotational states must have a certain symmetry. Indeed, the rotational wave function for the state with the rotation momentum j of the molecule changes as $(-1)^j$ as a result of inversion [47] that selects rotational states which can be realized for a given vibrational state. The wave functions of symmetric and antisymmetric oscillations along the molecular axis for this molecule are not changed as a result of the inversion operation, whereas the vibrational wave function of torsion oscillations is conserved for an even value of the torsion quantum number and changes its sign for odd values of the torsion quantum number. Hence, only states with even or odd rotation numbers of the CO_2 molecule may be stable [47].

One can apply the above results for linear molecules to atmospheric CO_2 molecules. We have for the absorption coefficient $k_\omega = N_v \sigma_\omega$ on the basis of formula (6.1.23)

$$k_\omega = \frac{\pi^2 c^2}{\omega^2} N_v A_i \frac{\hbar|\omega - \omega_o|)}{T} \exp\left[-\frac{\hbar^2(\omega - \omega_o)^2}{4BT}\right]\left[1 - \exp\left(-\frac{\hbar\omega}{T}\right)\right]\sum_j a_{\omega-\omega_j}, \tag{6.1.32}$$

where N_v is the total number density of molecules, the second Einstein coefficient is $A_i = 1/\tau_r$, and the spontaneous lifetime τ_r of vibrationally excited state is independent of the rotation number j in the first approximation.

In particular, this formula gives for the average absorption coefficient $\overline{k_\omega}$ which follows from formula (6.1.32) at large widths ν_j or results from an average over frequencies ω

$$\overline{k_\omega} = \frac{\pi^2 c^2}{\omega^2 \Delta\omega} A_j N_v \frac{\hbar|\omega - \omega_o|}{T} \exp\left[-\frac{\hbar^2(\omega - \omega_o)^2}{4BT}\right]\left[1 - \exp\left(-\frac{\hbar\omega}{T}\right)\right]\sum_j a_{\omega-\omega_j} \tag{6.1.33}$$

where $\Delta\omega = 4B/\hbar$ is the frequency difference for neighboring spectral lines, and we use the normalization condition (6.1.16) for the frequency distribution function. Note that in this case the result does not depend on the broadening character of spectral lines.

We have in the first approximation, where the rotation energy is given by formula (6.1.20), and the energy difference for photons related to centers of neighboring spectral lines does not depend on the rotation number j. In particular, for CO_2 molecules, this difference is $\hbar\Delta\omega = 4B = 1.56cm^{-1}$. If we assume the width ν_j of spectral lines to be independent on j, one can use the regular model or the Elsasser model [48] for the analysis of the frequency dependence for the absorption coefficient. This dependence is almost periodical in a narrow frequency range, and under the above conditions one can find this periodical function.

Indeed, let us use the Mittag–Leffler theorem [49] that has the form

$$\sum_{k=-\infty}^{\infty} \left[(x-k)^2 + y^2\right]^{-1} = \frac{\pi \sinh 2\pi y}{y(\cosh 2\pi y - \cos 2\pi x)} \tag{6.1.34}$$

On the basis of this formula one can obtain for the trunk of the absorption coefficient within the framework of the regular model (e.g., [1, 5, 50–52])

$$k_\omega = \overline{k_\omega}\frac{\sinh(\frac{\pi\nu}{\Delta\omega})}{\cosh(\frac{\pi\nu}{\Delta\omega}) - \cos\frac{2\pi(\omega-\omega_o)}{\Delta\omega}}, \tag{6.1.35}$$

where $\overline{k_\omega}$ is given by formula (6.1.33), and for CO_2 molecules $\Delta\omega = 4\,\mathrm{B}/\hbar = 1.56\,\mathrm{cm}^{-1}$.

As it follows from formula (6.1.35), the absorption coefficient as a frequency function oscillates with maxima at centers of spectral lines and with minima between them. The ratio of the maximal k_{max} and minimal k_{min} values of the absorption coefficient at low values of the parameter $\hbar\nu/B$ is equal

$$\frac{k_{max}}{k_{min}} = \left(\frac{2\Delta\omega}{\pi\nu}\right)^2 = \left(\frac{8B}{\pi\hbar\nu}\right)^2 \tag{6.1.36}$$

Table 6.1 Parameters of radiative transitions in the CO_2 molecule that are responsible for emission of the Earth's atmosphere

Transition	$\hbar\Delta\omega_v$, cm^{-1}	A_i, s^{-1}
$000 \to 010$	667	3.1
$010 \to 100$	721	1.8
$010 \to 02^00$	619	1.2
$010 \to 02^20$	668	2.7

In particular, in the case of CO_2 molecules located in air at room temperature and atmospheric pressure, this ratio is equal approximately 40.

The above formulas allow one to calculate the absorption coefficient due to atmospheric carbon dioxide. Table 6.1 contains parameters of radiative transitions for atmospheric CO_2 molecules. In this table, the energy $\hbar\Delta\omega_v$ of vibration transitions relates to low rotation numbers, and the values of the Einstein coefficients A_i refer to rotation numbers $j \approx 15$, where the distribution function of carbon dioxide molecules over rotation states has the maximum.

6.1.4 *HITRAN Data Bank in Spectroscopy of Atmosphere*

As it follows from the above analysis for the CO_2 molecules, the spectrum of absorption and emission of atmospheric air with these molecules in the infrared spectrum range consists of a large number of broaden and overlapped spectral lines with a large difference between the absorption coefficient at centers of spectral lines and between them. Therefore, for the analysis of emission of the atmosphere under various conditions, the information about radiative parameters for thousands transitions in optically active molecules of the atmosphere is necessary. Now, this information is collected in the HITRAN (HIgh resolution TRANsmission) data bank [53–55]. Data from this bank are of interest for various problems (e.g., [56, 57]), including the atmospheric ones. In order to use the parameters of the HITRAN data bank, it is necessary to express the absorption cross section (6.1.23) and the absorption coefficient k_ω at a given frequency ω (6.1.32) through these parameters. Then, the absorption coefficient due to a certain radiative transition may be represented in the form

$$k_\omega^{ij} = N_v S_{ij} a_\omega, \tag{6.1.37}$$

where i means a sort of optically active molecules, an index j corresponds to a certain radiative transition of these molecules or a certain spectral line, N_i is the number density of molecules of a given sort which are found in the ground state, a_ω is the distribution function of emitted photons over frequencies, S_j is the transition intensity for a given individual spectral line. It is clear that the absorption coefficient due to this radiative transition is proportional to the number density of molecules which takes part in this transition, as well as to the probability a_ω to emit a photon of this frequency ω. Hence, formula (6.1.37) is the definition of the transition intensity S_{ij}.

In evaluating the radiative parameters of the atmosphere, we take the rates of radiative transitions in molecules from the HITRAN data bank [53–57]. We then represent the intensity of radiative transition S_{ij} of a certain vibration–rotation or rotation–rotation transition j of a molecule i on the basis of formula [45, 55]

$$S_{ij} = \frac{\pi^2 c^2}{\omega_j^2} A_j \frac{g_j}{q(T)} \exp\left(-\frac{E_j}{T}\right)\left[1 - \exp\left(-\frac{\hbar\omega_j}{T}\right)\right], \tag{6.1.38}$$

where $g_j = 2j + 1$ is the statistical weight of the lower transition state, ε_j is the excitation energy for the initial rotation state which in the simplest approximation is given by formula $E_j = Bj(j + 1)$, $q(T)$ is the statistical sum which for the above approximation is equal $q(T) = 2T/B$ at low temperatures, $\varepsilon_j = \hbar\omega_o + Bj(j + 1)$ is the transition energy, so that $\hbar\omega_o$ is the energy of vibration excitation. Note that we express the transition energy or transition frequency in spectroscopy units cm^{-1} which can be converted in energy units.

In particular, in the case of vibration–rotation transitions with the ground vibration state for a lower state of transition, the parameters of formula (6.1.38) are equal

$$g_i = 2j + 1,\ \varepsilon_{ij} = \varepsilon_o + Bj(j + 1),\ q(T) = \frac{T}{B},\ T \gg B \tag{6.1.39}$$

where j is the rotational quantum number for the lower state of the molecule, B is the rotational constant for this molecule, ε_{ij} is the excitation energy from the ground rotational state and a given vibrational state v. These formulas are valid for CO_2 molecules. Basic molecular greenhouse components are H_2O and CO_2 molecules. Figures 6.5 and 6.6 contain intensities of these molecules in atmospheric air.

In evaluating the radiative fluxes from the atmosphere, we take radiative parameters from the HITRAN data bank, where the radiative parameters are given for atmospheric air pressure and a certain air temperature T_o. One can transit from the intensity $S_{ij}(T_o)$ related to the temperature T_o to that $S_{ij}(T)$ at a temperature T with accounting for the exponential dependence of the population of a lower transition state. One can represent the intensity for a given vibration–rotation transition is determined by

$$S_{ij}(T) = S_{ij}(T_o)\exp\left(\frac{E_j}{T_o} - \frac{E_j}{T}\right) \tag{6.1.40}$$

where E_{ij} is the energy of excitation of a low transition state.

In this evaluation, we are guided by strong spectral lines such that at the centers of these lines the optical thickness satisfies to the relation $u_\omega > 1$. In accordance with typical parameters of spectral lines due to water and carbon dioxide molecules,

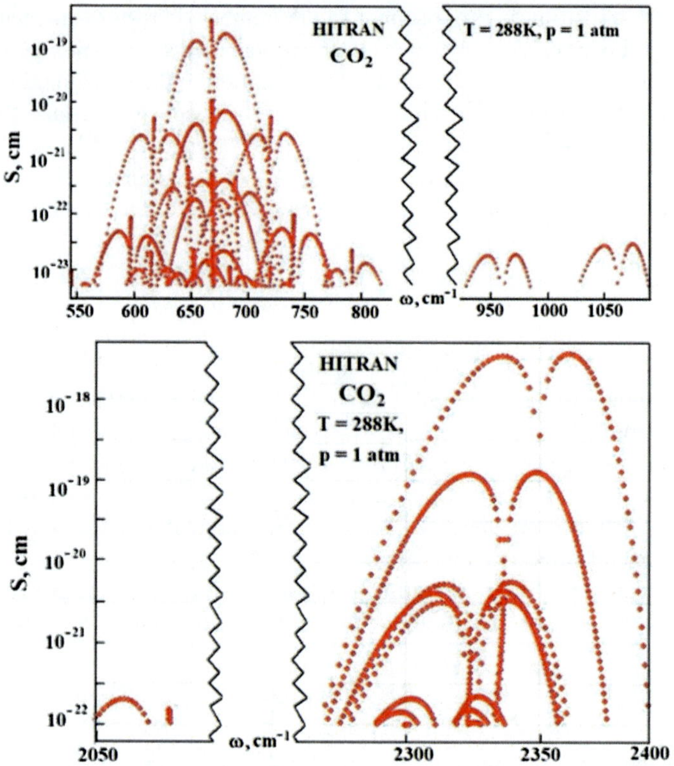

Fig. 6.5 Intensity of absorption of CO_2 molecules in atmospheric air at the temperature $T = 288K$ according to data of the HITRAN bank [53, 54]. The frequency range $(580 - 760)\text{cm}^{-1}$ refers to the basic absorption band of CO_2 molecules at room temperature, and the transitions with the center at $2349\,\text{cm}^{-1}$ correspond to the resonant vibration transition $000 \to 001$ in accordance with Fig. 6.4

one can select from the HITRAN data bank radiative transitions for these molecules whose intensities are satisfied to the relation

$$S_j(H_2O) \gg 1 \times 10^{-24}\,\text{cm}, \;\; S_j(CO_2) \gg 2 \times 10^{-23}\,\text{cm} \tag{6.1.41}$$

Because emission of individual molecules has a random character and leads to the total noncoherent radiation, the absorption coefficient of atmospheric air is the sum of the absorption coefficients for individual molecules and transitions; i.e., this can be represented in the form

$$k_\omega = \sum_{i,j} N_i \frac{S_{ij}\nu_j}{2\pi[(\omega - \omega_j)^2 + (\nu_j/2)^2]} \tag{6.1.42}$$

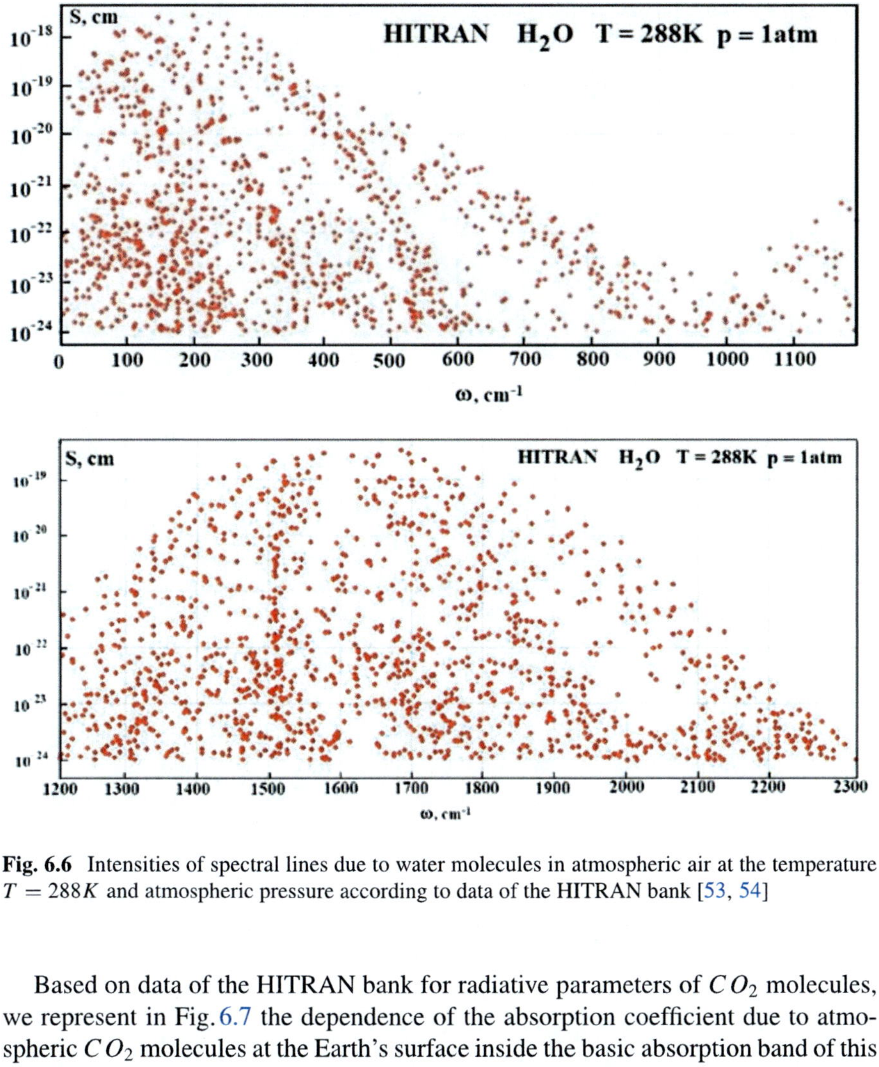

Fig. 6.6 Intensities of spectral lines due to water molecules in atmospheric air at the temperature $T = 288K$ and atmospheric pressure according to data of the HITRAN bank [53, 54]

Based on data of the HITRAN bank for radiative parameters of CO_2 molecules, we represent in Fig. 6.7 the dependence of the absorption coefficient due to atmospheric CO_2 molecules at the Earth's surface inside the basic absorption band of this molecule. The contemporary number density of carbon dioxide molecules is equal $N_v = 1.1 \times 10^{16}\,\text{cm}^{-3}$ according to formula (6.1.24). Within the framework of the model of standard atmosphere, the Earth's surface temperature is taken $T = 288K$ at atmospheric pressure. One can see that the atmospheric absorption spectrum due to CO_2 molecules includes three vibration transitions, and the lowest transition of Table 6.1 is screened by the transition from the ground vibration state of this molecule. These three vibration transitions form the absorption band for the CO_2 molecule at room temperature.

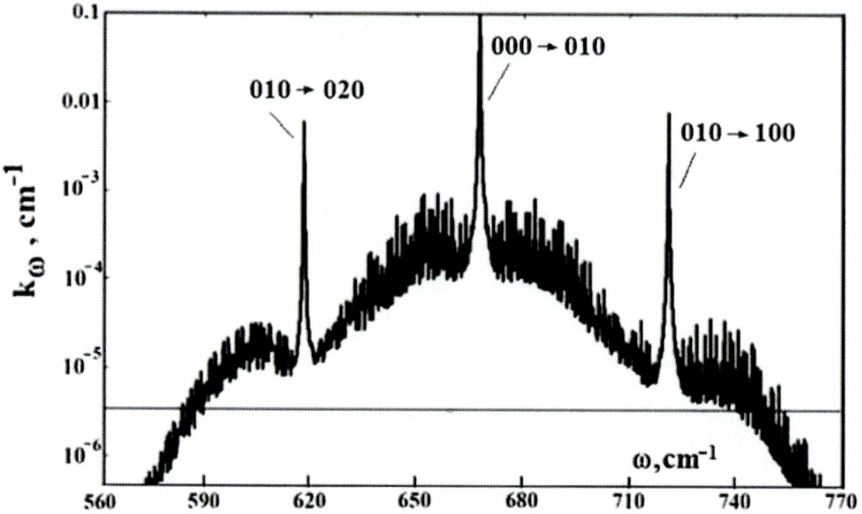

Fig. 6.7 Absorption coefficient in atmospheric air due to CO_2 molecules near the Earth's surface for the basic absorption band of carbon dioxide molecules at room temperature. The solid line corresponds to the atmosphere optical thickness $u_\omega = 2/3$ [58]

In addition, we give in Fig. 6.8 the absorption coefficient of CO_2 molecules in atmospheric air near the violet edge of the absorption band. In this case, the absorption spectrum corresponds to the vibrational transition $01^o0 \to 10^o0$ and corresponds to the R-band of the rotation transition. In this scale, the spectrum consists of broaden overlapped spectral lines for individual rotation transitions.

6.2 Emission of Standard Atmosphere

6.2.1 Model of Atmospheric Emission

Our goal now is to determine radiative fluxes from the atmosphere. In order to understand the character of atmospheric emission, we first consider the simple model [59] with a frequency-independent absorption coefficient k_ω in the case of a large total optical thickness of the atmosphere. We also assume that the absorption coefficient decreases monotonically with an altitude increase and, for definiteness, approximate the altitude dependence for the absorption coefficient as

$$k_\omega(h) = A \exp\left(-\frac{h}{\lambda}\right), \quad u(\infty) = A\lambda \tag{6.2.1}$$

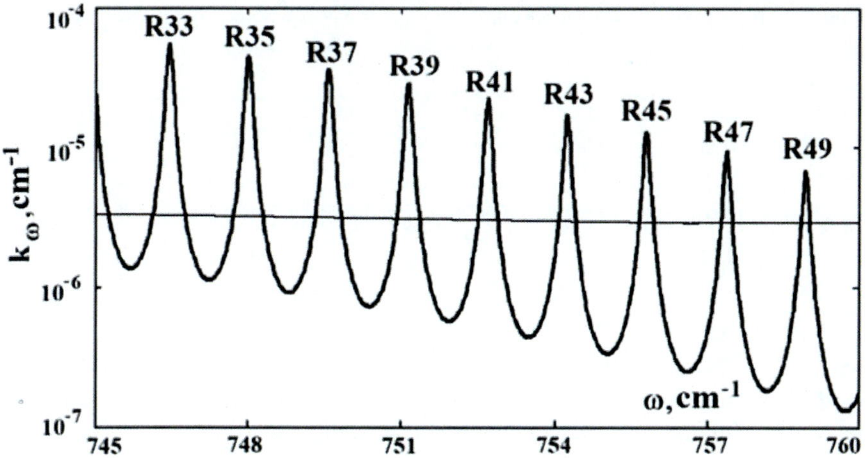

Fig. 6.8 Absorption coefficient in atmospheric air near the Earth's surface as a result of vibration–rotation transitions in atmospheric CO_2 molecules for frequencies near the boundary of the absorption band. Numbers indicate the initial rotational number of the molecule, the solid line corresponds to the atmosphere optical thickness $u_\omega = 2/3$ [23]

In contrast to [2, 59], we now use the total radiative fluxes from the atmosphere from Table 7, namely

$$J_\downarrow = (335 \pm 7)W/m^2, \; J_\uparrow = (215 \pm 8)W/m^2 \tag{6.2.2}$$

This model with data of Table 7 allows us to determine the radiative temperature which is independent now of the frequency and hence may be determined on the basis of the Stephan–Boltzmann law

$$J_\downarrow = \sigma T_\downarrow^4, \; J_\uparrow = \sigma T_\uparrow^4, \tag{6.2.3}$$

One can obtain for radiative temperatures of emission toward the Earth $T_\downarrow$ and outside $T_\uparrow$. We have now for these parameters [2, 59]

$$T_\downarrow = (277 \pm 2)\,\text{K}, \; T_\uparrow = (248 \pm 2)\,\text{K} \tag{6.2.4}$$

Values of the radiative temperature for this model $T_\downarrow$ and the corresponding altitudes $h_\downarrow$ for various versions of the energetic balance for our planet from Table 7 are represented in Table 6.2.

We also obtain for the altitudes which are responsible for atmospheric emission to the Earth $h_\downarrow$ and outside it $h_\uparrow$ on the basis of formula (2.1.2) that give

$$T_\downarrow = T_E - \frac{\mathrm{d}T}{\mathrm{d}h} h_\downarrow, \; T_\uparrow = T_E - \frac{\mathrm{d}T}{\mathrm{d}h} h_\uparrow,$$

Table 6.2 Parameters of the cloud boundary according to different sources for the energetic balance of the Earth and atmosphere. The numbers of sources are taken from Table 7

Number	1	2	3	4	5	average
$J_\downarrow$, W/m^2	327	327	333	346	342	335 ± 7
$T_\downarrow$, K	276	276	277	280	279	277 ± 2
$h_\downarrow$, km	1.9	1.9	1.7	1.3	1.4	1.6 ± 0.2
h_{cl}, km	5.6	5.6	4.8	3.2	3.7	4.6 ± 0.7
T_{cl}, K	252	252	257	267	264	258 ± 6

where $T_E = 288\,\mathrm{K}$ is the temperature of the Earth's surface for the model of standard atmosphere and $\mathrm{d}T/\mathrm{d}h = 6.5\,\mathrm{K/km}$. From this, we find the altitudes which are responsible for atmospheric emission in this direction

$$h_\downarrow = (1.6 \pm 0.2)\,\mathrm{km}, \;\; h_\uparrow = (6.1 \pm 0.3)\,\mathrm{km} \tag{6.2.5}$$

Using the approximation (6.2.1) and requiring according to formula (6.1.13) that the atmospheric optical thickness from the effective layer to the boundary is 2/3, one can obtain the equation for parameters of the approximation (6.2.1)

$$A\lambda[1 - \exp(-h_\downarrow/\lambda)] = 2/3, \;\; A\lambda \exp(-h_\uparrow/\lambda) = 2/3 \tag{6.2.6}$$

These values allow us to determine the parameters of the approximation (6.2.1) on the basis of a general scheme [2, 59] with accounting for parameters (6.2.5)

$$A = (0.50 \pm 0.05)\,\mathrm{km}^{-1}, \;\; \lambda = (4.8 \pm 0.4)\,\mathrm{km} \tag{6.2.7}$$

From this, it follows also for the optical thickness of the atmosphere

$$u = A\lambda = 2.4 \pm 0.2 \tag{6.2.8}$$

Below one can be guided by these parameters in the analysis of emission of the real atmosphere.

6.2.2 *Character of Infrared Radiation from Atmosphere to Earth*

We now construct a more real model for emission of the atmosphere on the basis of the above analysis for the absorption coefficient of atmospheric air with optically active molecules with accounting also for the contemporary understanding of atmospheric physics and processes in the atmosphere [1, 2, 5, 50, 60–71]. As a result, one can formulate the model [23] which allows one to analyze the emission of the real atmosphere. This model includes the peculiarities as follows.

1. Parameters of the model of standard atmosphere [24] are used. Within the framework of this model, the number density of air molecules $N(air)$ depending on the altitude h over the Earth's surface is determined by formula

$$N(air) = N_a \exp\left(-\frac{h}{\Lambda}\right), \quad N_a = 2.55 \times 10^{19}\text{cm}^{-3}, \quad \Lambda = 10\,km$$

The moisture of the standard atmosphere is $\eta = 80\%$ near the Earth. In addition the temperature T_E near the Earth's surface and the temperature gradient $\mathrm{d}T/\mathrm{d}h$ are equal

$$T_E = 288K, \quad \frac{\mathrm{d}T}{\mathrm{d}h} = -6.5\,\text{K/km}$$

2. Under pressures of the order of atmospheric one, the width of separate spectral lines $\Delta\nu$ is small compared the difference of photon frequencies $\Delta\omega$ for neighboring spectral lines according to (6.1.14)

$$\Delta\nu \ll \Delta\omega$$

This leads to the line structures of the spectrum of atmospheric molecules; i.e., the radiative spectrum consists of broaden spectral lines of radiating atmospheric molecules. Hence, the "line-by-line" model [1, 5] is the basis of these evaluations and integral radiative fluxes follow from these partial fluxes.

3. The emitting atmosphere model includes three basic greenhouse components, namely H_2O molecules, CO_2 molecules and liquid water microdroplets as the basic condensed phase in the atmosphere. In addition, traced components as CH_4 molecules and N_2O molecules are trace components in this scheme.

4. Along with the local thermodynamic equilibrium for atmospheric components, this equilibrium takes place between the radiation field and atmospheric air. For the model of standard atmosphere, the atmosphere temperature is independent of a geographic coordinate and depends weakly on the altitude h over the Earth's surface. This allows one to reduce the radiation of optically active molecules of a weakly nonuniform layer to that of a layer of a constant temperature T_ω [16, 17] that is the radiative temperature at a given frequency.

5. Parameters of radiative transitions of greenhouse molecules are taken from the HITRAN data bank [53–55], and therefore, we use the formalism for rates of molecular radiative processes of this data bank [45].

6. The energetic balance of the Earth and its atmosphere is taken into account. According to this balance, radiative fluxes toward the Earth and outside are determined by different atmospheric regions and are separated; i.e., the radiative fluxes to the Earth which are connected with its temperatures do not depend on processes in high layers of the troposphere.

7. Basic greenhouse components are separated so that clouds are located starting from at a certain altitude, and they are characterized by a sharp boundary. Radiation from greenhouse molecules is created in a gap between the Earth's surface and clouds.

The radiative flux from the atmosphere to the Earth J_ω at a given frequency ω, constructed on the above model is based on Fig.5.9 and is given by formula (6.1.13)

$$J_\omega = I_\omega(T_\omega)g(u_\omega) + I_\omega(T_{cl})[1 - g(u_\omega)]$$

Here, T_ω is the radiative temperature for emitted molecules which are located in a gap between the Earth and cloud boundary, T_{cl} is the temperature of the cloud boundary, $I_\omega(T)$ is the radiative flux of a blackbody with a temperature T at this frequency that is given by the Planck formula in accordance with formula (6.1.3), and the opaque factor $g(u_\omega)$ of a uniform gaseous layer which is found in thermodynamic equilibrium with air and is determined by formula (6.1.6) [18, 22]

$$g(u_\omega) = 2\int_0^1 \cos\theta d\cos\theta \left[1 - \exp\left(-\frac{u_\omega}{cos\theta}\right)\right],$$

The opaque factor $g(u_\omega)$ is the probability for radiation isotropically emitted by the Earth's surface to reach the clouds. The frequency dependence of the opaque factor is represented in Fig. 6.9. In addition, the averaged value of the opaque factor is given in Fig. 6.10. The average is made over the frequency range of 20 cm^{-1} that means the frequency range of 10 cm^{-1} below a given frequency and the same value above it.

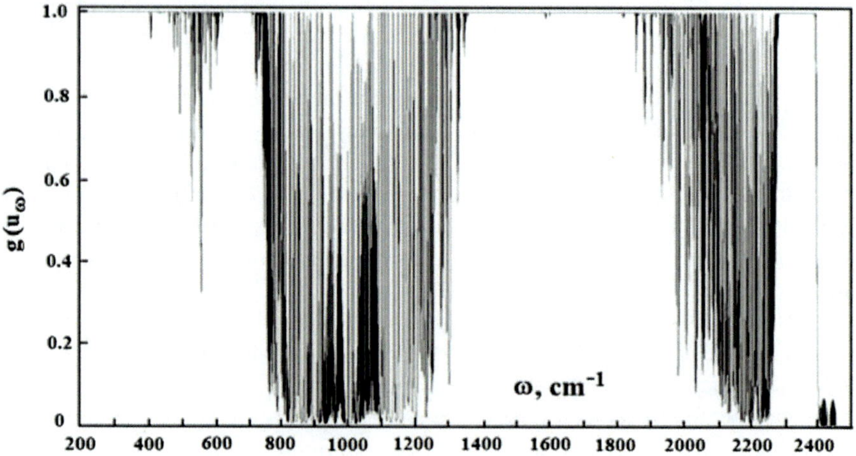

Fig. 6.9 Opaque factor $g(u_\omega)$ due to absorption of atmospheric molecules in the atmospheric gap between the Earth's surface and clouds if the cloud boundary is located at the altitude $h_{cl} = 4.6$ km [72]

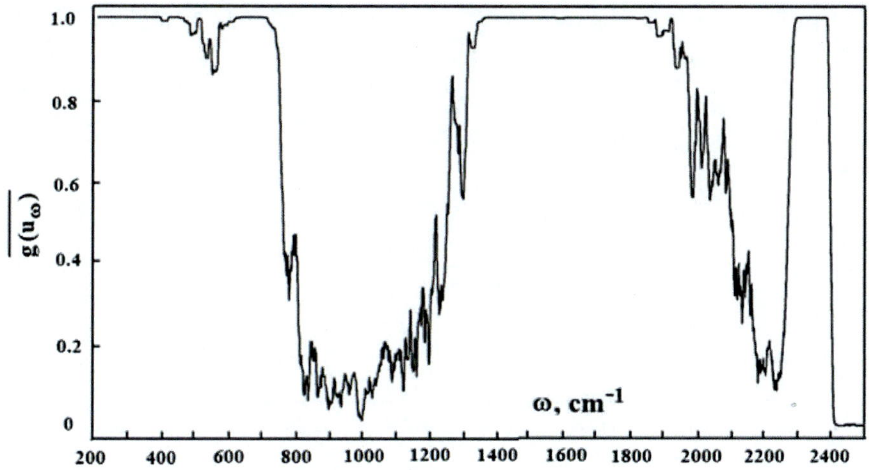

Fig. 6.10 Average opaque factor $\overline{g(u_\omega)}$ over a frequency range of 20 cm^{-1} for absorption of atmospheric molecules located in the gap between the Earth's surface and clouds. The cloud boundary is located at the altitude $h_{cl} = 4.6$ km [72]

The opaque factor characterizes the part of the radiative flux that is emitted by the Earth's surface and attains the clouds. Figure 6.11 contains the radiative flux $I_\omega(T_E)$ that is emitted by the Earth, as well as the radiative flux $\overline{g(u_\omega)}I_\omega(T_E)$ at a given frequency which reaches the clouds for the model of standard atmosphere. From Figs. 6.9 and 6.11, it follows that atmospheric air with optically active molecules, mostly H_2O and CO_2 ones, has the transparency window in the frequency range approximately between 800 cm^{-1} and 1200 cm^{-1}, while in other frequency range, the atmosphere with indicated molecules is opaque.

We have also for the average radiative flux which is emitted by the Earth and reaches the cloud boundary

$$J_\uparrow = \int \overline{g(u_\omega)} I_\omega(T_E) d\omega \tag{6.2.9}$$

Taking $T_E = 288K$, one can obtain on the basis of the model of standard atmosphere

$$J_E = \int I_\omega(T_E) d\omega = 390\,\mathrm{W/m^2}$$

in accordance with the energetic balance of the Earth, and $J_\uparrow = 120W/m^2$. As is seen, a part of the thermal radiative flux that passes through the atmosphere and reaches the clouds is above 30% of the emitted radiative flux.

From formulas (6.1.13) and (6.2.9), one can determine the radiative flux from the atmosphere to the Earth's surface due to atmospheric molecules and clouds, and this flux depends on the cloud temperature T_{cl} or the boundary altitude h_{cl} for

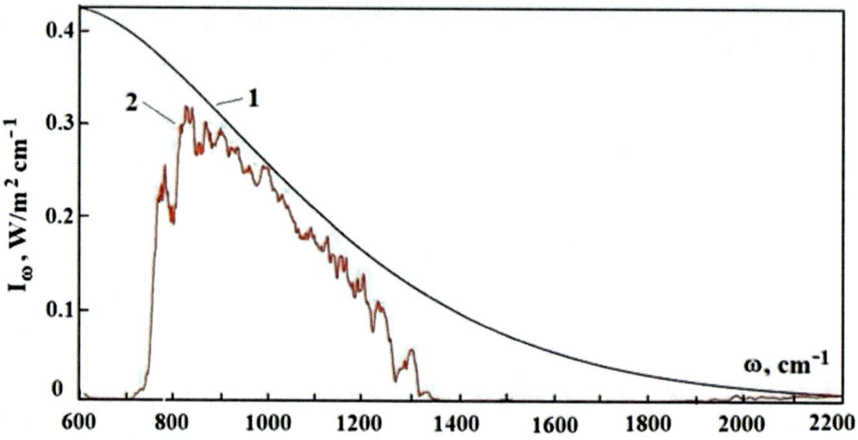

Fig. 6.11 Frequency dependence for the radiative flux $I_\omega(T_E)$ from the Earth's surface (1) and its part $g(u_\omega)I_\omega(T_E)$ which attains the boundary of clouds [72]

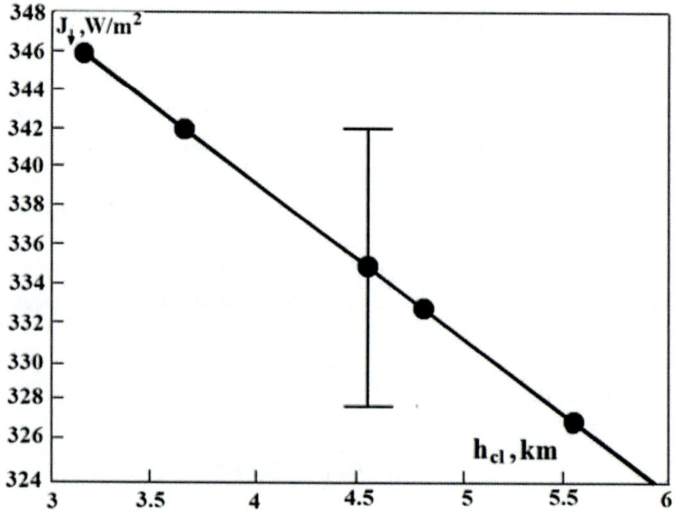

Fig. 6.12 Dependence of the total radiative flux from the atmosphere to the Earth $J_\downarrow$ at the altitude of the cloud boundary h_{cl} [72]

clouds. This dependence is represented in Fig. 6.12. In addition, these parameters are given in Table 6.2 for different versions of the energetic balance of the Earth and its atmosphere.

Within the framework of this model, we characterize the clouds by one parameter h_{cl} that is the altitude of the cloud boundary, though in reality this is the effective average altitude of cloud absorption. Note that clouds exist over a given surface point during a restricted time, and their distribution over altitudes has a random character.

For this model, we find the cloud altitude by comparing the radiative flux J_ω due to radiating molecules located in the gap between the Earth and clouds, and the total radiative flux to the Earth which follows from the energetic balance of the Earth (Table 7) and is given by

$$J_\downarrow \equiv \int J_\omega d\omega \tag{6.2.10}$$

The cloud altitude h_{cl} is connected with the temperature T_{cl} of cloud temperature through the relation of the standard atmosphere model in accordance with formula (6.1.24)

$$T_{cl} = T_E - \frac{dT}{dh} h_{cl},$$

where the global temperature is equal $T_E = 288K$ and the temperature gradient is $\mathrm{d}T/\mathrm{d}h = 6.5\,\mathrm{K/km}$.

6.2.3 *Radiative Fluxes to the Earth*

On the basis of the above model, one can determine the radiative flux emitted by the atmosphere. For this goal, we find first the radiative temperature of molecules located between the Earth's surface and cloud boundary. The radiative temperature T_ω characterizes the radiative flux at a given frequency which is the temperature of atmospheric air at an effective altitude h_ω according to formula (6.1.13)

$$T_\omega = T_E - \frac{\mathrm{d}T}{\mathrm{d}h} h_\omega$$

Figure 6.13 contains the radiative temperature T_ω of the standard atmosphere, and the average radiative temperatures over the range of $20\,\mathrm{cm}^{-1}$ are given in Fig. 6.14.

Some peculiarities of these calculations on the basis of the above model [23] should be noted. One can see that various parameters of atmospheric emission, such as the opaque factor, the radiative temperature, the radiative flux, have a line structure, i.e., are combined from those of individual spectral lines. Averaging of these parameters over a frequency is made for demonstration of typical values of the quantity under consideration. In these evaluations, an averaging is made over the frequency range of the width of $10\,\mathrm{cm}^{-1}$ below and above a frequency under consideration. From this, it follows the importance of information for rates of radiative transitions in molecules, as well as for parameters of broadening of these spectral lines. In reality, we need in such information for thousands of spectral lines, and this marks the importance of the HITRAN database that gives this information.

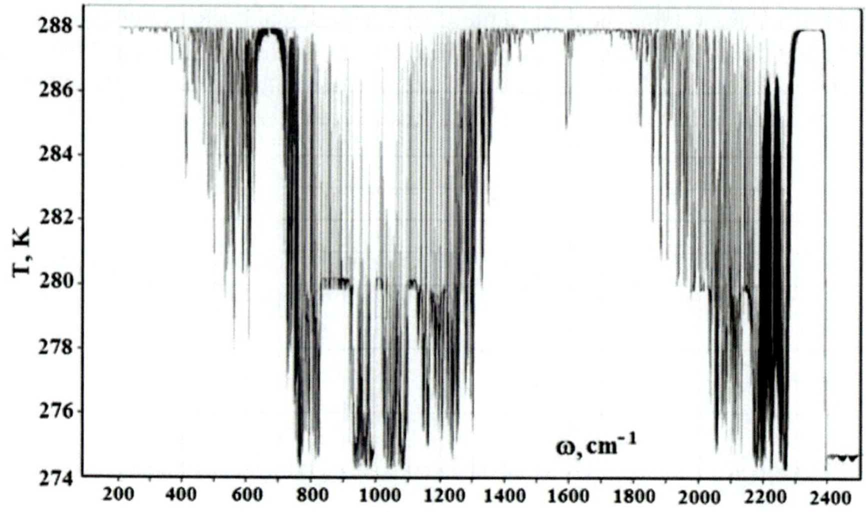

Fig. 6.13 Radiative temperature T_ω due to optically active atmospheric molecules located in the atmospheric gap between the Earth's surface and clouds which boundary is located at the altitude $h_{cl} = 4.6\,\mathrm{km}$ [72]

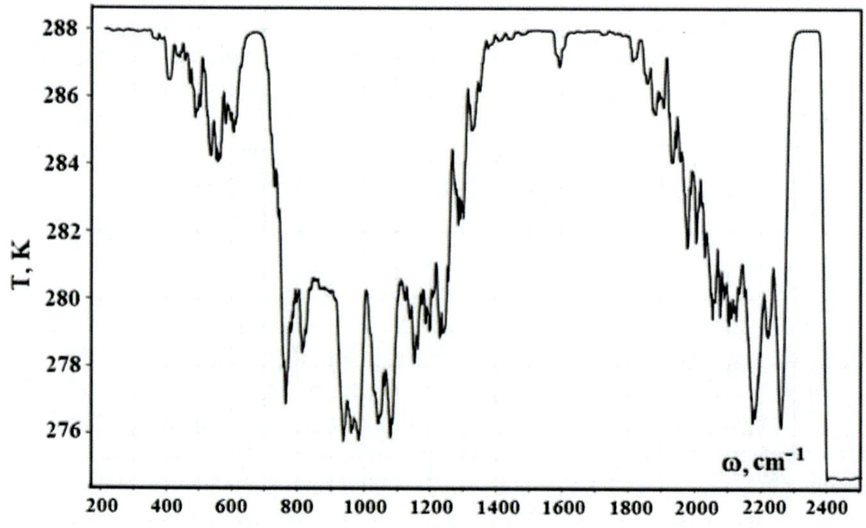

Fig. 6.14 Radiative temperatures T_ω resulted from emission of atmospheric molecules which are located in the atmospheric gap between the Earth's surface and clouds. The radiative temperature is averaged over a range of $20\,\mathrm{cm}^{-1}$, and the cloud boundary is found at the altitude $h_{cl} = 4.6\mathrm{km}$ [72]

Table 6.3 Radiative fluxes in W/m^2 for indicated frequency ranges and the total radiative flux due to a given component [72]. Results of evaluations [23] are given in square parentheses

$\Delta\omega$, cm^{-1}	0–580	580–760	760–1200	1200–2600	Sum
H_2O-molecules	140	18	15	42	(215 ± 2) [166]
H_2O-droplets	2	1	57	3	(63 ± 7) [96]
Water in total	142	19	72	45	(278 ± 10) [262]
CO_2-molecules	0	52	3	1	(56 ± 2) [60]
CH_4-molecules	0	0	2	0	2 [4]
N_2O-molecules	0	0	1	0	1 [3]
$\overline{J_\downarrow}$, W/m^2	142	71	75	49	337
$\overline{J_E(\omega)}$, W/m^2	145	73	117	54	389
$\overline{T_\omega}$, K	285	286	281	282	–
$\overline{g}$	0.99	0.97	0.17	0.96	–

Evaluate values of the cloud boundary h_{cl} given in Fig. 6.12 and Table 6.2 are obtained on the basis of equation (6.1.13). The expression (6.1.13) is used for the partial radiative flux in the case of various versions of the Earth energetic balance represented in Table 5.1. We assume further that the radiative temperature T_ω in formula (6.1.13) to be independent on the cloud altitudes. In reality, this dependence exists, though it is weak. Therefore, we solve this problem in steps.

We give in Table 6.3 and Fig. 6.15, as well in Figs. 6.17, 6.18, and 6.19, the radiative fluxes from the atmosphere to the Earth. These data are based on evaluations [72]. These evaluations are analogous to [23], but they are made in the frequency range up to 2600 cm^{-1} instead of 1200 cm^{-1} in [23]. The frequency range below 1200 cm^{-1} is enough for emission of CO_2 molecules because frequencies above 1200 cm^{-1} do not give the contribution to emission of CO_2 molecules. But this frequency range is of importance for radiation of H_2O molecules.

Note that the radiative flux of the emission of a blackbody at frequencies above 2600 cm^{-1} gives the contribution of 0.1% of the total radiative flux from the Earth, whereas for frequencies above 1200 cm^{-1}, this contribution is equal 16%. In reality, H_2O molecules are the basic absorbers in the frequency range between 1200 cm^{-1} and 2600 cm^{-1}, while in evaluations [23] clouds are presented as the basic absorber in this frequency range. Therefore, a part of the radiative flux in this frequency range is transferred from clouds due to evaluations [23] to water molecules in evaluations [72]. The results for CO_2 molecules are determined by the frequency range below 1200 cm^{-1}. In addition, CH_4, N_2O and O_3 molecules as trace molecules give a small contribution to the total radiative flux from the atmosphere. This total contribution is estimated as 1% [72].

In evaluation of the radiative flux in a certain frequency range for a given atmospheric component, we divide the frequency range in some parts in accordance with the character of the emission process. Indeed, emission of atmosphere to the Earth's

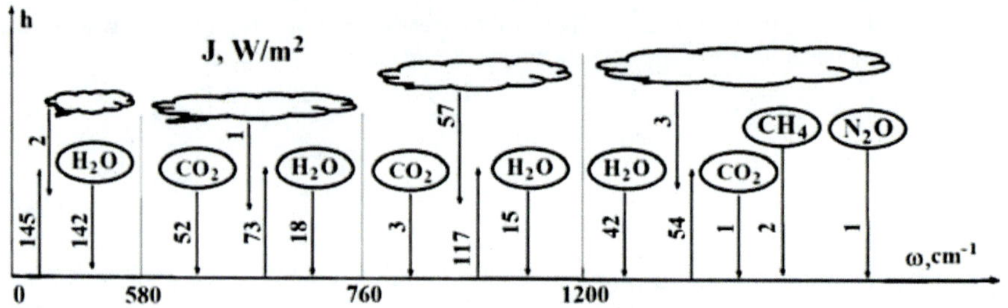

Fig. 6.15 Partial radiative fluxes from the atmosphere to the Earth due to various greenhouse components are given near the arrows for principal frequency ranges and are expressed in W/m^2 [23]

surface is created by H_2O molecules for frequencies below 580 cm^{-1}, whereas emission due to CO_2 molecules dominates in the frequency range between 580 cm^{-1} and 780 cm^{-1}. The frequency range between 780 cm^{-1} and 1200 cm^{-1} may be considered as the window of transparency for atmospheric molecules, and emission of water molecules dominates again in the frequency range above 1200 cm^{-1}.

This statement is confirmed by data of Fig. 6.16 where the optical thickness is given for the atmospheric layer located between the Earth's surface and clouds. The solid line $u_\omega = 2/3$ separates the range of optical thicknesses in large and small ones, so that the opaque range corresponds to $u_\omega \gg 1$, while $u_\omega \ll 1$ relates to the transparent range.

From this it follows also that emission of molecules of trace gases CH_4, N_2O and O_3 is small, and that for other trace gases is negligible. Note that the number density of ozone molecules in the stratosphere is higher by one-two orders of magnitude than that from the troposphere. Infrared emission of stratospheric ozone molecules from the stratosphere does not reach the ground because it is absorbed by clouds, and only emission of ozone from lower troposphere attains the ground.

In this analysis, we divide the frequency range in several parts. In the first part, from zero up to 580 cm^{-1}, water molecules dominate in emission and absorption the infrared radiation of the atmosphere. The second spectrum part from 580 cm^{-1} up to 800 cm^{-1} corresponds to the absorption band of CO_2 molecules. Atmospheric emission due to CO_2 molecules is created mostly in this frequency range. The third spectrum part of Table 6.3, Figs. 6.15 and 6.17, ranged from 800 cm^{-1} up to 1200 cm^{-1}, corresponds to the transparency window of the atmosphere with respect to greenhouse components. This follows from Fig. 6.11 and also from Fig. 6.18, according to which the radiative flux from the Earth's surface attained clouds is close to that emitted by the Earth. In this case, the Earth's surface emits as a blackbody of the temperature $T_E = 288$ K, and we are guided by the model where the atmosphere is transparent in the range 800 cm^{-1} $< \omega <$ 1200 cm^{-1}, whereas it is opaque in other frequency ranges. The fourth spectrum part 1200 cm^{-1} $< \omega$ corresponds to the tail of the frequency distribution function of the emitted radiative flux which is created mostly by water molecules as the greenhouse component.

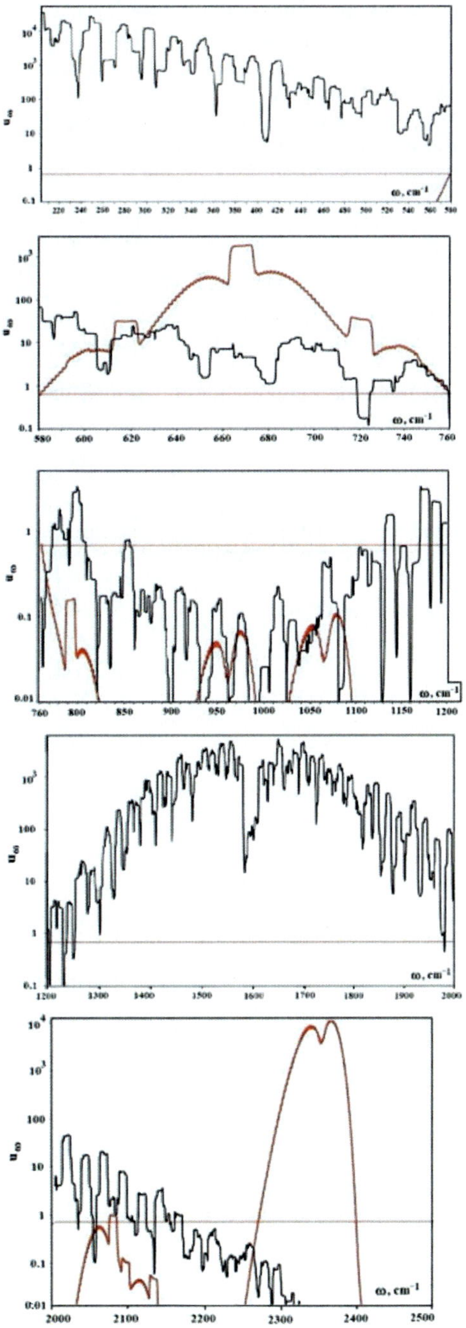

Fig. 6.16 Frequency dependence for the optical thickness (u_ω) of the atmospheric gap located between the Earth's surface and clouds due to H_2O molecules (black curve) and due to CO_2 molecules (red curve). The frequency average is made over the frequency range of $10\,\mathrm{cm}^{-1}$, and the red solid line corresponds to the value of 2/3 for the optical thickness [72]

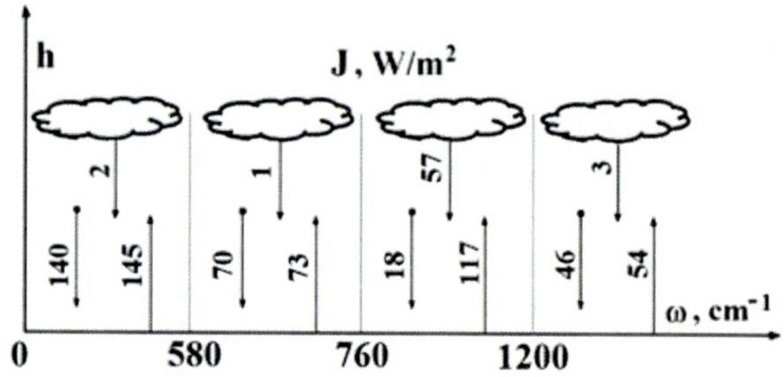

Fig. 6.17 Total radiative fluxes from the atmosphere to the Earth given near the arrows and expressed in W/m^2 for principal frequency ranges due to indicated greenhouse components

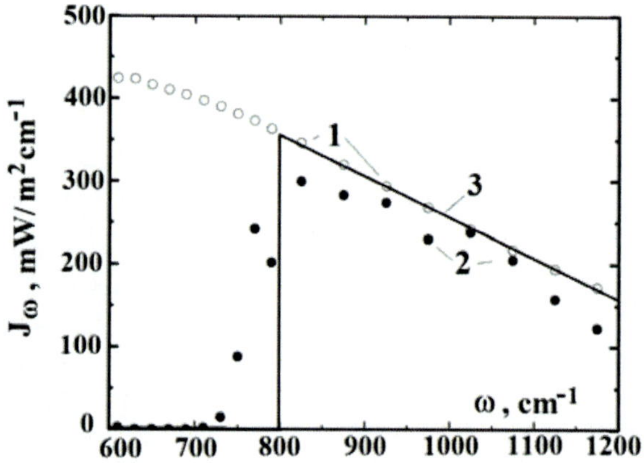

Fig. 6.18 Radiative flux J_E from the Earth's surface (1), and its part J_E' that reaches clouds (2). The fluxes are expressed in W/m^2. The solid line describes the model where the atmospheric gap between the Earth's surface and clouds is opaque in the spectrum range $\omega < 800\,\text{cm}^{-1}$ and is transparent in the range $\omega > 800\,\text{cm}^{-1}$ [72]

It should be noted that evaluations of the radiative fluxes are fulfilled for various versions of the energetic balance for the Earth and its atmosphere that is given in Table 7. In our evaluations, these versions of the Earth's energetic balance lead to different values for the cloud boundary h_{cl} given in Fig. 6.12, as well as to different cloud temperatures T_{cl}. Errors of Table 6.3 follow from data from different versions for the Earth's energetic balance.

It is convenient to treat data of Table 6.3 in a simply way on the basis of average parameters of atmospheric emission for each frequency range. Indeed, on the basis of formula (6.1.13), one can calculate the average radiative temperature $\overline{T_\omega}$ and the average opaque factor $\overline{g}$. An average is made over each range of frequencies of Table 6.3. On the basis of these parameters and formula (6.1.13), one can obtain for the average radiative flux to the Earth from the atmosphere.

$$\overline{J_\omega} = \overline{g} \int I_\omega(\overline{T_\omega}))d\omega + (1 - \overline{g}) \int I_\omega(T_{cl})d\omega, \qquad (6.2.11)$$

The first term in the right-hand side of this formula describes the radiative flux that is created by molecules in the atmospheric gap between the Earth and clouds in this frequency range.

In addition, Fig. 6.19 contains the radiative fluxes from the atmosphere to the Earth due to each greenhouse component and Table 6.4 gives the portion due to

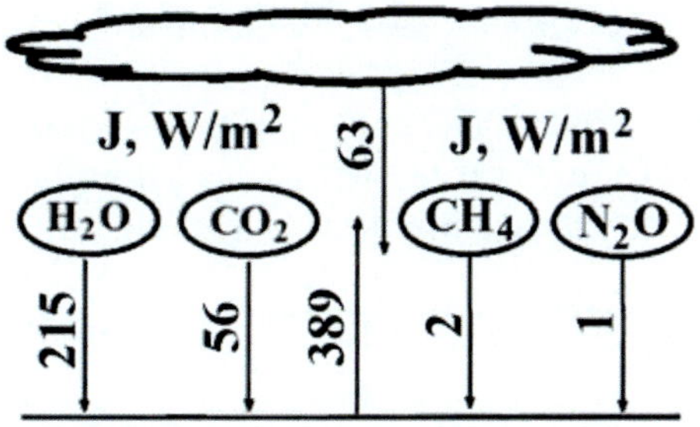

Fig. 6.19 Total radiative fluxes from the atmosphere to the Earth indicated near the arrows and expressed in W/m^2 due to various greenhouse components

Table 6.4 The contribution of greenhouse components to the radiative flux from the atmosphere to the Earth according to [72]. Radiative fluxes in W/m^2 for indicated frequency ranges and the total radiative flux due to each greenhouse component. Results of evaluations [23] are given in square parentheses

Component	Contribution, %
H_2O-molecules	63 ± 1 [51]
H_2O-droplets	19 ± 2 [29]
CO_2-molecules	17 ± 1 [18]
Trace molecules	1 [2]

each greenhouse. One can compare the data of Table 6.4 for evaluations [72] in the frequency range to $2600\,\mathrm{cm}^{-1}$ and those [23] for frequencies up to $1200\,\mathrm{cm}^{-1}$. According to spectra of greenhouse components, Table 6.4 gives nearby values of contribution from CO_2 molecules since its absorption takes place at frequencies below $1200\,\mathrm{cm}^{-1}$. In the same manner, the total radiative flux from atmospheric water (water molecules and microdroplets) is nearby for both cases. But the contribution from water molecules or water microdroplets is different for these cases because water molecules absorb for frequencies above $1200\,\mathrm{cm}^{-1}$.

6.2.4 Emission of Molecules at Microscopic Level

We now analyze the effectivity of emission of atmospheric molecules on the basis of the above data for radiation of the atmosphere. We use the concept of standard atmosphere with parameters averaged over time and globe. Then, in considering CO_2 molecules, we take their number density near the Earth's surface $N(CO_2) = 1 \times 10^{16}\,\mathrm{cm}^{-3}$, and using formula (2.1.1) for their altitude distribution, we obtain for the number $n(CO_2$ of CO_2 molecules in a vertical atmospheric column at altitudes below clouds, whose effective altitude is $h = 4.6\,\mathrm{km}$, as

$$n(CO_2) = 3.7 \times 10^{25}\,\mathrm{m}^{-2}$$

Since the specific power of atmospheric emission due to CO_2 molecules is $J_\downarrow(CO_2) = 56\,\mathrm{W/m}^2$ according to data of Fig. 6.18, one can obtain from this for the specific power of emission $p(CO_2)$ for radiation per one CO_2 molecule

$$p(CO_2) = \frac{J_\downarrow(CO_2)}{n(CO_2)} = 1.5 \times 10^{-24}\,\mathrm{W} = 1 \times 10^{-5}\,\mathrm{eV/s} \tag{6.2.12}$$

One can compare this with the radiative power air due to CO_2 molecules if the thermodynamic equilibrium is supported between molecules in the ground vibration state and the excited one. In addition, photons are emitted isotropically and leave the atmosphere without absorption. Taking for estimation that the photon energy is equal to the excitation energy of the first vibrationally excited state $\hbar\omega = 667\,\mathrm{cm}^{-1}$ which corresponds to the maximum of the absorption coefficient as a frequency function. We below take also the value of the second Einstein coefficient which is a reciprocal value with respect to the radiative lifetime τ_r and is equal $1/\tau_r = 1.6\,\mathrm{s}^{-1}$; one can find for the following estimation for the value of the radiative power per one molecule $p'(CO_2)$ for the transparent atmosphere

$$p'(CO_2) = \frac{\hbar\omega}{4\tau_r}\exp\left(-\frac{\hbar\omega}{T}\right) = 1 \times 10^{-3}\mathrm{eV/s},$$

where the temperature T is taken to be equal to the global temperature $T_E = 288\,\mathrm{K}$. As is seen, only 1% of photons emitted by CO_2 molecules and directed to the ground, reach it.

We now determine the total energy $\varepsilon(CO_2)$ that releases as a result of emission of one atmospheric CO_2 molecule. Taking the residence time $\tau = 4\,year$ [73] for atmospheric CO_2 molecules, one can obtain the total photon energy $\epsilon(CO_2)$ emitted by one CO_2 molecule during its residence in the atmosphere

$$\epsilon(CO_2) = \tau p(CO_2) = 12\,\mathrm{keV} \tag{6.2.13}$$

Let us repeat the same operation for H_2O molecules. Taking the number density of water molecules near the Earth's surface as $N(H_2O) = 3.4 \times 10^{16}\mathrm{cm}^{-3}$, and using formula (2.1.19) for their altitude distribution, one can obtain for the number $n(H_2O)$ of H_2O molecules in a vertical atmospheric column at altitudes below clouds, whose effective altitude is $h = 4.6km$. This value is equal for water molecules

$$n(H_2O) = 6 \times 10^{22}\,\mathrm{cm}^{-2}$$

According to Fig. 6.18, the total specific power emitted by water molecules and water microdroplets which constitute clouds is equal $J_\downarrow(H_2O) = 278\,\mathrm{W/m}^2$. From this, one can find the average power created by one water molecule in the atmosphere or the specific power of emission $p(CO_2)$ for radiation per one H_2O molecule

$$p(H_2O) = \frac{J_\downarrow(H_2O)}{n(H_2O)} = 5 \times 10^{-25}\,\mathrm{W} = 3 \times 10^{-6}\mathrm{eV/s} \tag{6.2.14}$$

From this, one can determine the total energy $\epsilon(H_2O)$ released as a result of emission of one atmospheric H_2O molecule. Taking the residence time $\tau = 9days$ for water molecules located in the atmosphere, one can find the energy $\epsilon(H_2O)$ of photons due to

$$\epsilon(H_2O) = \tau \cdot p(H_2O) = 2.2\,\mathrm{eV} \tag{6.2.15}$$

This exhibits that CO_2 molecules located in the Earth's atmosphere are more effective radiators than atmospheric water molecules.

On the basis of the above value (6.2.15) for the energy consumed per one water molecule, we analyze the possibility of the climate instability. This means that evaporation of one water molecule creates the radiative flux to the Earth that leads to formation more than one water molecule. As a result, the number of water molecules in the atmosphere grows exponentially, and a typical growth time is the residence time of water molecules in the atmosphere. The criterion of development of this instability has the form

$$\xi = \frac{\epsilon(H_2O)}{\varepsilon(H_2O)} > 1, \tag{6.2.16}$$

where $\epsilon(H_2O)$ is the average energy which releases in the form of photons during the residence time of a water molecule, and $\varepsilon(H_2O)$ is the energy cost for one water molecule, that is the energy consumed on formation of one water molecules. Value of these parameters are given by formulas (6.2.15) and (5.1.16). They are equal $\epsilon(H_2O) = 2.2\,\mathrm{eV}$ and $\varepsilon(H_2O) = 3.6\,\mathrm{eV}$. From this, it follows under contemporary conditions for the above parameter.

$$\xi = 0.64 \tag{6.2.17}$$

From this, one can conclude that the standard atmosphere under contemporary conditions is stable with respect to the greenhouse effect that is caused by water molecules.

We note the estimating character of formula (6.2.17). Keeping in mind the lack of this estimation that during its development atmospheric parameters vary significantly, one can see that this instability exists. Indeed, its nature is that each water molecule during its residence in the atmosphere creates more than one new molecule through the greenhouse effect. Though under conditions of the standard atmosphere, this instability is not realized according to formula (6.2.17), but it occurs at larger global temperatures. The strongest temperature dependence for the parameter $\xi(T)$ according to formula (6.2.17) belongs to the energy flux J_c due to evaporation of water molecules from the Earth's surface that has the form

$$J_c(T_E) \sim \exp(-\varepsilon_o/T_E)$$

according to formula (2.1.4). From this, one can obtain that the instability under consideration starts from the global temperature $T_E + \Delta T$, so that $\xi(T_E + \Delta T) = 1$. The change of the global temperature that leads to the instability threshold is given by

$$\Delta T = \frac{T_E^2}{\varepsilon_o} \ln\left(\frac{1}{\xi(T_E)}\right) \tag{6.2.18}$$

For the standard atmosphere, we have

$$\Delta T = 7\,\mathrm{K} \tag{6.2.19}$$

The accuracy of the value (6.2.17) is estimated as 20%. This leads to a low accuracy of formula (6.2.18), and on its basis, one can conclude that the greenhouse instability may be realized at an increase of the global temperature $\Delta T \sim 10\,\mathrm{K}$. Possible, the greenhose instability is taken place duing the Eocene epoch.

References

1. R.M. Goody, Y.L. Yung, *Principles ofAtmospheric Physics and Chemistry* (Oxford University Press, New York, 1995)
2. B.M. Smirnov, *Microphysics of Atmospheric Phenomena* (Springer Atmospheric Series, Switzerland, 2017)
3. M. Wendisch, P. Yang, *Theory of Atmospheric Radiative Transfer* (Wiley, Singapore, 2012)
4. M.F. Modest, *Radiative Heat Transfer* (Elsevier, Amsterdam, 2013)
5. R.M. Goody, *Atmospheric Radiation?: Theoretical Basis* (Oxford University Press, London, 1964)
6. K.Ya. Kondratyev, *Radiation in the Atmosphere*. (Academic Press, New York, 1969)
7. E.J. McCartney, *Absorption and Emission by Atmospheric Gases* (Wiley, New York, 1983)
8. K.N. Liou, *An Introduction to Atmospheric Radiation* (Academic Press, Amsterdam, 2002)
9. G.W. Petry, *A First Course in Atmospheric Radiation* (Sunlog Publ, Madison, 2006)
10. W. Zdunkowski, T. Trautmann, A. Bott, *Radiation in the Atmosphere* (Cambridge University Press, Cambridge, 2007)
11. G. Kirchhoff, R. Bunsen, Annalen der Physik und. Chemie **109**, 275 (1860)
12. A. Beer, Annalen der Physik und. Chemie **86**, 78 (1852)
13. J.H. Lambert, *Photometry, or, on the measure and gradations of light, colors, and shade.* (Eberhardt Klett, Augsburg, 1760)
14. L.M. Biberman, JETP **17**, 416 (1947)
15. T. Holstein, Phys. Rev. **72**, 1212 (1947)
16. B.M. Smirnov, *Physics of Weakly Ionized Gases* (Mir, Moscow, 1980)
17. B.M. Smirnov, *Physics of Ionized Gases* (Wiley, New York, 2001)
18. Ya..B.. Zel'dovich, Yu.P. Raizer, *Physics of shock waves and high-temperature hydrodynamic phenomena* (Academic Press, New York, 1966)
19. B.M. Smirnov, JETP **126**, 446 (2018)
20. L.D. Landau, E.M. Lifshitz, *Statistical Physics*, vol. 1 (Pergamon Press, Oxford, 1980)
21. F. Reif, *Statistical and Thermal Physics* (McGrow Hill, Boston, 1965)

22. B.M. Smirnov, *Physics of Weakly Ionized Gas*. (Nauka, Moscow, 1972; in Russian)
23. B.M. Smirnov, *Transport of Infrared Atmospheric Radiation* (de Gruyter, Berlin, 2020)
24. *U.S. Standard Atmosphere*. (U.S. Government Printing Office, Washington, 1976)
25. G. Herzberg, *Molecular Spectra and Molecular Structure* (Van Nostrand Reinhold, Princeton, 1945)
26. G.M. Barrow, *Introduction to Molecular Spectroscopy* (McGraw-Hill, New York, 1962)
27. H.C. Allen, P.C. Cross, *Molecular Vibrators: The Theory and Interpretation of High Resolution Infrared Spectra* (Wiley, New York, 1963)
28. M.A. El'yashevich, *Molecular Spectroscopy*. (Fizmatgiz, Moscow, 1963; in Russian)
29. G. Herzberg, *Molecular Spectra and Molecular Structure: Electronic Spectra and Electronic Structure of Polyatomic Molecules* (Van Nostrand, New York, 1966)
30. J.I. Steinfeld, *Molecules and Radiation* (Dover, New York, 1985)
31. S. Svanberg, *Atomic and Molecular Spectroscopy* (Springer, Berlin, 1991)
32. G. Herzberg, *Molecular Spectra and Molecular Structure: Infrared and Raman Spectra of Polyatomic Molecules* (Malabar, Florida, Krieger, 1991)
33. C. Banwell, E. McCash, *Fundamentals for Molecular Spectroscopy* (McGrow Hill, London, 1994)
34. V.P. Krainov, H.R. Reiss, B.M. Smirnov, *Radiative Processes in Atomic Physics* (Wiley, New York, 1997)
35. P.S. Sindhu, *Fundamentals of Molecular Spectroscopy* (New Age International, Dehli, 2006)
36. S. Chandra, *Molecular Spectroscopy* (Alpha Science International, Dehli, 2009)
37. J.L. McHale, *Molecular Spectroscopy* (CRC Press, Boca Raton, 2017)
38. V.P. Krainov, B.M. Smirnov, *Atomic and Molecular Radiative Processes* (Springer Nature, Switzerland, 2019)
39. L.M. Sverdlov, M.A. Kovner, E.P. Krainov, *Vibrational Spectra of Polyatomic Molecules* (Wiley, New York, 1974)
40. *Vibrational Intensities*. ed. by W.B. Person, G. Zerbi. (Elsevier, Amsterdam, 1980)
41. A.A. Radzig, B.M. Smirnov, *Reference Data on Atoms, Molecules, and Ions* (Springer, Berlin, 1985)
42. S.V. Khristenko, A.I. Maslov, V.P. Shevelko, *Molecules and Their Spectroscopic Properties* (Springer, Berlin, 1998)
43. http://www1.lsbu.ac.uk/water/water-vibrational-spectrum
44. I.I. Sobelman, *Atomic Spectra and Radiative Transitions* (Springer, Berlin, 1979)
45. M. Simeckova, D. Jacquemart, L.S. Rothman, et al. JQSRT **98**, 130 (2006)
46. B.M. Smirnov, *Reference Data on Atomic Physics and Atomic Processes* (Springer, Heidelberg, 2008)
47. L.D. Landau, E.M. Lifshitz, *Quantum Mechanics* (Pergamon Press, Oxford, 1965)
48. W.M. Elsasser, Phys. Rev. **54**, 126 (1938)
49. E.T. Whittaker, G.N. Watson, *Modern Analysis* (Cambridge University Press, London, 1940)
50. M.L. Salby, *Physics of the Atmosphere and Climate* (Cambridge University Press, Cambridge, 2012)
51. B.M. Smirnov, G.V. Shlyapnikov, Sov. Phys. Uspekhi **23**, 139 (1980)
52. B.M. Smirnov, *Plasma Processes and Plasma Kinetics* (Wiley, Berlin, 2007)
53. https://www.cfa.harvard.edu/
54. http://www.hitran.iao.ru/home
55. http://www.hitran.org/links/docs/definitions-and-units/
56. L.S. Rothman, I.E. Gordon, Y. Babikov, et al., JQSRT. **130**, 4 (2013)
57. I.E. Gordon, L.S. Rothman, C. Hill, et al., JQSRT. **203**, 3 (2017)
58. B.M. Smirnov, *Global Atmospheric Phenomena Involving Water* (Springer Atmospheric Series, Switzerland, 2020)
59. B.M. Smirnov, EPL **114**, 24005 (2016)
60. J.T. Houghton, *The Physics of Atmospheres* (Cambridge University Press, Cambridge, 1977)
61. J.V. Iribarne, H.P. Cho, *Atmospheric Physics* (Reidel Publ, Dordrecht, 1980)

62. R.G. Fleagle, J.A. Businger, *Introduction to Atmospheric Physics* (Academic Press, San Diego, 1980)
63. M.L. Salby, *Fundamentals of Atmospheric Physics* (Academic Press, San Diego, 1996)
64. J.H. Seinfeld, S.N. Pandis, *Atmospheric Chemistry and Physics* (Wiley, New York, 1998)
65. D.G. Andrews, *An Introduction to Atmospheric Physics* (Cambridge University Press, Cambridge, 2000)
66. J.H. Seinfeld, S.N. Pandis, *Atmospheric Chemistry and Physics* (Wiley, Hoboken, 2006)
67. J.M. Walace, R. Hobbs, *Atmospheric Science. An Introductory Survey*. (Elsevier, Amsterdam, 2006)
68. M.H.P. Ambaum, *Thermal Physics of the Atmosphere* (Wiley-Blackwell, Oxford, 2010)
69. I. Lagzi, et al., *Atmospheric Chemistry*. (Institute of Geography and Earth Science, Budapest, 2013)
70. R. Caballero, *Physics of the Atmosphere* (IOP Publish, Bristol, 2014)
71. G. Visconti, *Fundamentals of Physics and Chemistry of the Atmosphere* (Springer Nature, Switzerland, 2017)
72. D.A. Zhilyaev, B.M. Smirnov, JETP. **133**, 807 (2021)
73. M. Grosjean, J. Goiot, Z. Yu, Global Planet Change **152**, 19 (2017)

Chapter 7
Greenhouse Effect in Varying Atmosphere

Abstract Change of the radiative flux from the atmosphere to the Earth's surface as a result of a change of the atmosphere composition is evaluated within the framework of the above model for standard atmosphere which uses the line-by-line method and parameters for radiative transitions of atmospheric molecules from the HITRAN data bank. Interaction between optically active molecules of the atmosphere through the radiation field is governed by the Kirchhoff law according to which atmospheric radiators are simultaneously absorbers. As a result, the change Δ in the radiative flux from the atmosphere to the Earth due to change in the concentration of molecules of a certain types differs from the change Δ' of the radiative flux created by molecules of this type. The ratio Δ'/Δ of the above radiative fluxes is equal, three for water molecules, five for CO_2 molecules, two for CH_4 molecules, 1.5 for N_2O molecules and one for O_3 molecules. The simplification in universal climatological models, where the interaction between optically active components is neglected, the Kirchhoff law is ignored, i.e., the values Δ and Δ' assume to be identical in these models. Applying the above analysis to the real atmosphere, we find the change of the global temperature is (0.6 ± 0.3) as a result of doubling of the concentration of CO_2 molecules. On the basis of this analysis and observed evolution of the real atmosphere, one can obtain that an observed change in the concentration of CO_2 molecules leads to the change in the global temperature that is one third of the observed change. Assuming the other part of the global temperature change results from the change of the concentration c(H_2O) of water molecules, one can obtain for the rate of its change $d \ln c(H_2O)/dt = 0.003\,\text{year}^{-1}$. This corresponds to the rate of the change of the atmosphere moisture η as $d \ln \eta/dt = 0.002$, and this change conserves the stability of the real atmosphere more 100 year.

B. M. Smirnov, *Global Energetics of the Atmosphere*, Springer Atmospheric Sciences, https://doi.org/10.1007/978-3-030-90008-3_7

7.1 Radiation of Trace Gases

7.1.1 *Character of Atmosphere Radiation Depending on Its Composition*

Various aspects of radiation from the standard atmosphere toward the Earth were analyzed above. We now consider the influence of the atmospheric composition on its emission. In other words, we analyze below the change of the atmospheric radiative flux depending on the change on its composition. Because the Earth's atmosphere contains several greenhouse components, it is of importance the interaction of greenhouse components through the radiation field due to the Kirchhoff law [1]. Indeed, let us evaluate the radiative flux from a gaseous layer which contains several optically active components, and its temperature is constant over the layer. If its optical thickness is large $u_\omega \gg 1$, the layer emits as a blackbody with the layer temperature.

Let us consider an increase of the amount of one greenhouse component that causes an increase of the radiative flux due to this component from the gaseous layer at a given frequency ω, say, by the value ΔJ_ω if the optical thickness of this layer due to this component is restricted, and this frequency is found inside the absorption band of a given component. If the total optical thickness of this layer at a given frequency is large and the temperature of a gas T is constant over the layer, the total radiative flux from the layer I_ω does not vary at a given operation and is determined by the Planck formula (6.1.5) because of the thermodynamic equilibrium between molecules of the layer and the radiation field. Thus, on one side, this change leads to an increase of the radiative flux due to a given component, and on the other side, the total radiative flux from the layer does not vary. Hence, the radiative flux due to other greenhouse components decreases as a result of an increase of the concentration of a given component for an optically thick layer.

This character of the change of radiative fluxes is operated by the Kirchhoff law [1] according to which radiators of the layer are simultaneously the absorbers. Hence, an additional amount of the first optically active component works as an absorber for radiation of other components. From this it follows that the change of the radiative flux from the atmosphere as a result of change of the amount of a given optically active component is determined by a temperature gradient at frequencies where the optical thickness is not small. Evidently, an interaction between optically active components in the atmosphere is determined by overlapping of spectra for these components. Hence, the precise information is required for spectra of optically active components for determination changes of radiative fluxes as a result of change of the atmosphere composition. HITRAN database contains this information for atmospheric molecules and is an important instrument for the problem under consideration. On the basis of this information, one can analyze the connection between changes in radiative fluxes due to various components and overlapping of spectra of these optically active components. We start this analysis of the greenhouse effect in the atmosphere from trace gases.

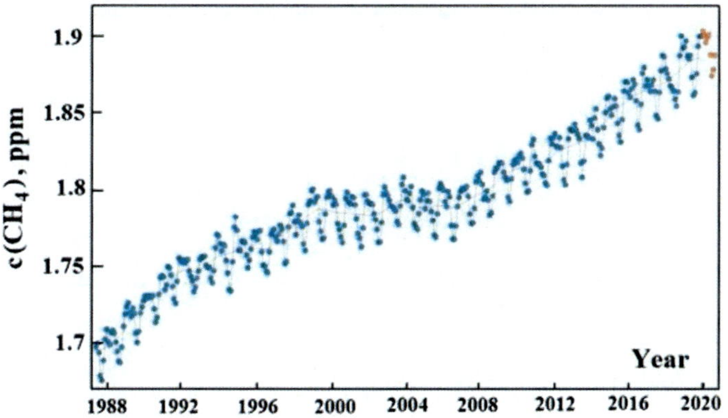

Fig. 7.1 Evolution of the global concentration of methane molecules in the atmosphere last time [4, 5]

7.1.2 *Methane in the Earth's Atmosphere*

Basic greenhouse components of the atmosphere are water and carbon dioxide molecules, as well as atmospheric water in the condensed phase which constitutes clouds. As the latter component, we restrict only by water microdroplets because they contain the basic part of clouds. Trace gases of the atmosphere include methane, nitrogen dioxide and ozone, which give a small, but remarkable contribution to the radiative flux from the atmosphere to the Earth. The peculiarity of emission of trace gases is a restricted frequency range of the absorption band that simplifies the analysis.

The most important trace gas of the atmosphere is methane, and below we consider participation of methane in the greenhouse effect of the atmosphere. The concentration of atmospheric methane molecules increases by 2.5 times from 1750 up to now [2–5]. The contemporary heating of the Earth may cause the thawing of permafrost that leads to extraction of methane from swamps. This gives various speculations and discussions about the methane role in the climate change. Therefore, below we evaluate the greenhouse effect due to emission of atmospheric methane for various scenaria of its evolution in the atmosphere.

We first give the amount of methane in the Earth's atmosphere. The concentration of atmospheric methane molecules varies from 0.72 ppm in the pre-industrial epoch to 1.9 ppm now [2], and we give in Fig. 7.1 evolution of the concentration of atmospheric methane [2–5]. Because the residence time of methane molecules in the Earth's atmosphere is 8 years [6], these molecules are mixed with air molecules uniformly as a result of air convection. It should be noted the contribution of the human activity to the contemporary amount of atmospheric methane, that is (50–65)% [7]. The number density of atmospheric methane molecules is less than that for carbon dioxide molecules which is the greenhouse component by 200 times. Therefore, methane is a trace greenhouse component which contribution to the total radiative flux of the atmosphere is small.

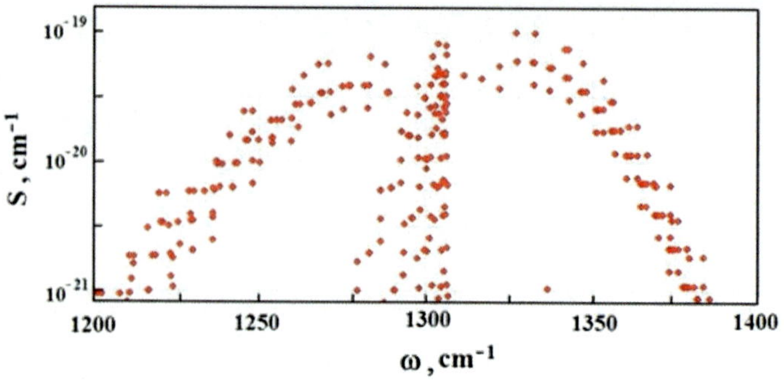

Fig. 7.2 Intensities of spectral lines of the methane molecule located in atmospheric air [12]

The contemporary concentration of methane molecules in atmospheric air that is approximately 1.9 ppm according to data of Fig. 7.1 corresponds to the number density of methane molecules $N_m = 5 \times 10^{13}\,\mathrm{cm}^{-3}$ near the Earth's surface. The absorption band of the methane molecule is centered at the frequency $\omega_o = 1306\,\mathrm{cm}^{-1}$. On the basis of data of Fig. 7.1, one can account for the number density of methane molecules per unit area of the vertical atmospheric column that is responsible for emission of methane molecules and is equal approximately $2 \times 10^{19}\,\mathrm{cm}^{-2}$. In addition, the radiative lifetime of a triply degenerate deformation vibration states is $\tau_v = 0.4\,\mathrm{s}$, and the rotation constant for the methane molecule is $B = 5.2\,\mathrm{cm}^{-1}$ [8–11].

In consideration the emission of methane molecules in the atmosphere, we use the model of §6.2.1 for the radiative fluxes from the atmosphere to the Earth's surface and take parameters of radiative transitions for methane molecules from the HITRAN data bank [12–14]. Figure 7.2 contains values of the intensities of spectral lines for methane molecules located in atmospheric air according to the HITRAN data bank. The absorption band for methane molecules lies in the frequency range approximately from $1240\,\mathrm{cm}^{-1}$ up to $1320\,\mathrm{cm}^{-1}$, and this frequency band is the object of the subsequent analysis. We below evaluate radiative parameters of the Earth's atmosphere due to atmospheric methane being guided by this frequency range.

Figure 7.3 gives the optical thickness of the atmosphere defined by formula (6.1.2) inside the absorption band for methane molecules. Next, within the framework of the standard atmospheric model [15] and the model of atmosphere emission [16], the radiative flux J_ω at a given frequency ω which is emitted by the atmosphere and is absorbed by the Earth, according to formula (6.1.13), is given by

$$J_\omega = I_\omega(T_\omega)g(u_\omega) + I_\omega(T_{\mathrm{cl}})[1 - g(u_\omega)]$$

Here, T_ω is the radiative temperature for atmospheric molecules, T_{cl} is the temperature of the cloud boundary, and $I_\omega(T)$ is the radiative flux of a blackbody with a temperature T at this frequency that is given by the Planck formula (6.1.5). The opaque factor $g(u_\omega)$ of a uniform gaseous layer which is defined by formula (6.1.6) is

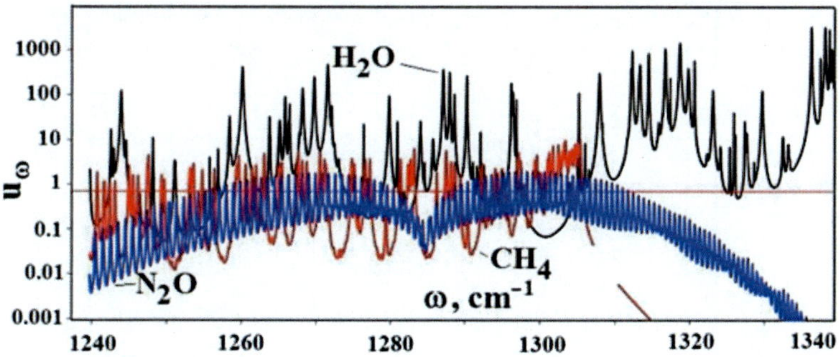

Fig. 7.3 Optical thickness u_ω of the atmospheric gap between the Earth and clouds due to optically active atmospheric molecules CH_4, N_2O and H_2O in the absorption band of the CH_4 molecule. The cloud altitude is $h_{cl} = 4.6$ km

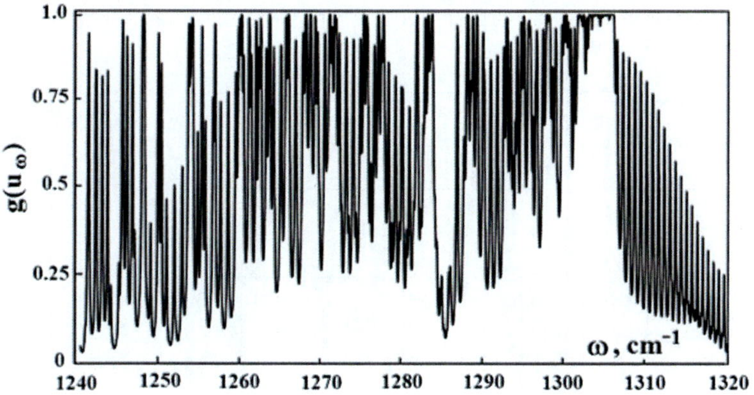

Fig. 7.4 Opaque factor $g(u_\omega)$ for the standard atmosphere in the range of absorption of methane molecules

represented in Fig. 7.4, and the opaque factor of the atmosphere inside the absorption band of methane molecules given in Fig. 7.5 is averaged over a range of frequencies which exceeds the width of an individual spectral line. This dependence corresponds to parameters of the model of standard atmosphere and to an average moisture of the atmosphere $\eta = 80\%$, as well as to a dry atmosphere.

According to the definition, the partial radiative flux $J_\omega(CH_4)$ created by methane molecules is determined by formula

$$J_\omega(CH_4) = \frac{k_\omega(CH_4)}{\kappa_\omega} I_\omega(T_\omega) g(u_\omega), \tag{7.1.1}$$

where $k_\omega(CH_4)$ is the absorption coefficient at an altitude h_ω due to methane molecules, κ_ω is the total absorption coefficient at this altitude, and the radiative temperature T_ω at this altitude h_ω for the model of standard atmosphere is given by

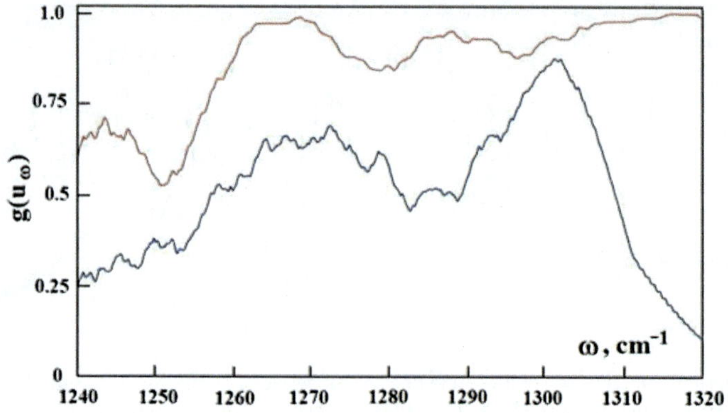

Fig. 7.5 Opaque factor $g(u_\omega)$ of the atmosphere in the frequency range of absorption of methane molecules averaged over a frequency range of $5\,\mathrm{cm}^{-1}$ ($2.5\,\mathrm{cm}^{-1}$ below a given frequency and above it) for the standard atmosphere of a moisture of $\eta = 80\%$ (red) and for a dry atmosphere, where the moisture is $\eta = 0\%$ (dark)

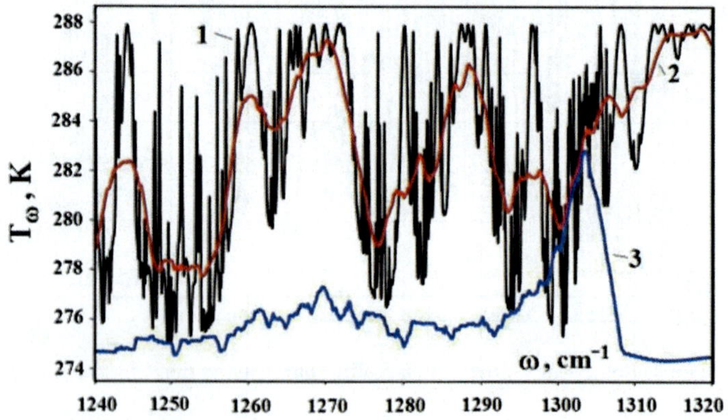

Fig. 7.6 Radiative temperature T_ω of the standard atmosphere (black) in the range of absorption of methane molecules (1). The radiative temperature is averaged over a frequency range of $5\,\mathrm{cm}^{-1}$ ($2.5\,\mathrm{cm}^{-1}$ below a given frequency and above it). These data relate to the standard atmosphere of a moisture of $\eta = 80\%$ (2), as well as to a dry atmosphere (3) given in blue (the moisture is $\eta = 0\%$)

$$T_\omega = T_\mathrm{E} - h_\omega \frac{\mathrm{d}T}{\mathrm{d}h}, \qquad (7.1.2)$$

The radiative temperature T_ω of the standard atmosphere inside the absorption band of methane molecules is represented in Fig. 7.6. Figures 7.7 and 7.8 contain the frequency dependencies for the partial radiative fluxes J_ω created by methane molecules under various conditions.

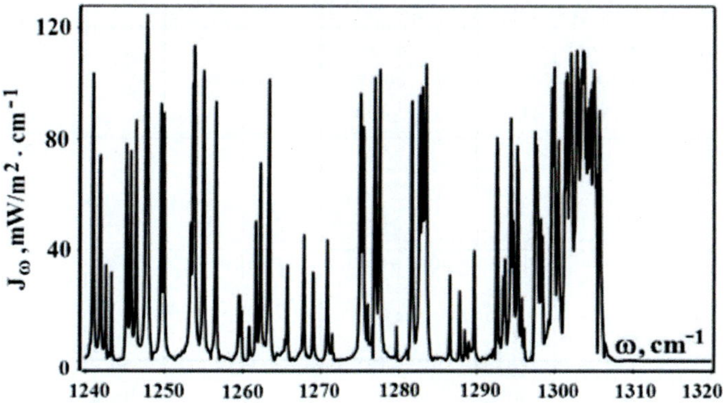

Fig. 7.7 Partial radiative flux $J_\omega(CH_4)$ created by methane molecules of the standard atmosphere

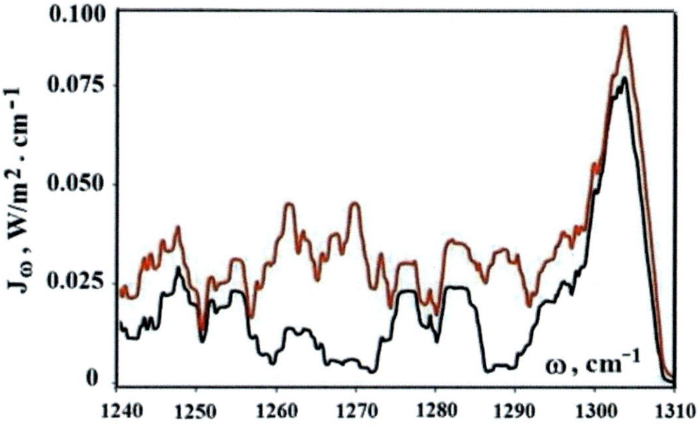

Fig. 7.8 Partial radiative flux $J_\omega(CH_4)$ created by methane molecules is averaged over a frequency range of $5\,\text{cm}^{-1}$ ($2.5\,cm^{-1}$ below a given frequency and above it) for the standard atmosphere (black) and for a dry atmosphere (red)

Let us analyze now the above information for emission of methane molecules located in the atmosphere inside the absorption band of methane molecules. As it follows from this, these parameters as a frequency function have a line structure. This means that spectral lines are not overlapped, so that radiative parameters inside the absorption band at each frequency are determined by a certain radiative transition for the methane molecule. Next, from this, one can demonstrate the character of interaction between various optically active components. Indeed, from Fig. 7.3, it follows that three components, namely H_2O, CH_4 and N_2O molecules partake in emission in the frequency range under consideration, and their interaction may be analyzed.

A certain component does not give contribution to the radiative flux in a frequency range, where the total optical thickness is large $u_\omega \gg 1$, because the emission of a

Table 7.1 Expressed in W/m^2 the values of radiative fluxes due to an indicated atmospheric component inside the frequency range from 1240 to $1320\,cm^{-1}$, i.e., at frequencies of the absorption band of methane molecules, for different atmosphere compositions

Atmosphere composition	CH_4	H_2O	N_2O	total
Dry atmosphere	2.3 (2.8)	0	1.9	8.2
Standard atmosphere	1.3 (1.6)	6.2	0.8	4.2
Atm. without methane	0	6.5	0.9	7.4

The radiative fluxes for all frequencies are given in parentheses

given component is screened by other optically active components. But due to the line structure of the spectrum, in the case of a large optical thickness at centers of spectral lines, the radiative flux is determined by frequency ranges between individual spectral lines where the optical thickness is $u_\omega \sim 1$. Table 7.1 contains the radiative fluxes integrated over the absorption band of the CH_4 molecule. The radiative flux for a component X is given by

$$J(X) = \int J_\omega(X) d\omega \qquad (7.1.3)$$

where the integral is taken over a given frequency range.

As it follows from data of this table, molecules H_2O, CH_4 and N_2O compete through the radiation field due to the Kirchhoff law [1]. According to this law, these molecules are simultaneously radiators and absorbers. For this reason, removal of H_2O molecules or CH_4 molecules from the atmosphere leads to an increase of the radiative flux from the atmosphere due to other greenhouse components. It is taken into account in Table 7.1 that though the frequency range between 1240 and $1320\,cm^{-1}$ is determined the total radiative flux of the atmosphere due to methane molecules, the frequency range outside this band gives an additional small contribution to the total radiative flux created by methane molecules.

7.1.3 Emission of N_2O Atmospheric Molecules

One more greenhouse component of atmospheric air as a trace gas is nitrous dioxide (N_2O). Though the concentration of N_2O molecules in the atmosphere is less by an order of magnitude compared to that for methane molecules, more favorable spectroscopic parameters may provide the existence of an absorption band even at such concentration. We take the concentration of atmospheric N_2O molecules to be 0.3 ppm according to measurements [17, 18], and its concentration grows slightly in time being 0.27 ppm in 18th century, 0.28 ppm in 19th century and 0.29 ppm in 20th century [19]. We also use below spectroscopic parameters of N_2O molecule [8–11, 20, 21]. The rotation constant of this linear molecule is equal $B_o = 0.419\,cm^{-1}$.

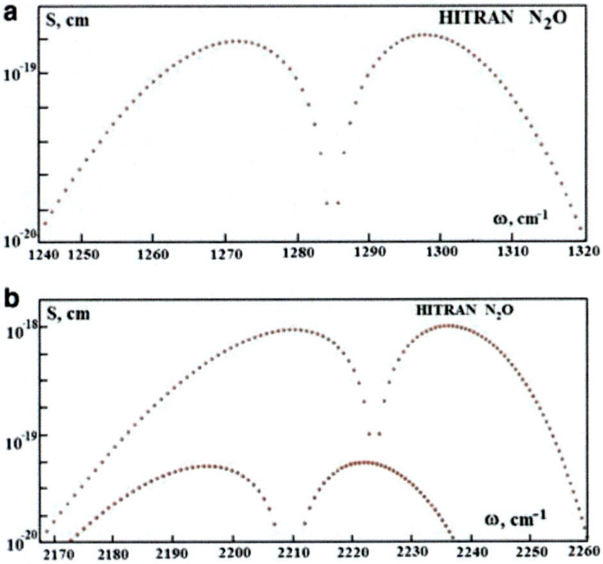

Fig. 7.9 Intensities of spectral lines due to N_2O molecules in atmospheric air at the temperature $T = 288$ K and atmospheric pressure according to data of the HITRAN bank [13, 14] in the absorption band of $1285\,cm^{-1}$ (**a**) and of $2224\,cm^{-1}$ (**b**) [12]

Table 7.2 Radiative fluxes due to an indicated atmospheric component inside the absorption band of the N_2O molecule in the frequency range between 2170 and $2280\,cm^{-1}$. The radiative fluxes are expressed in W/m^2

Atm. composition	N_2O	H_2O	CO_2	Clouds	Total flow
Standard atmosphere	0.24	0.056	0.053	0.053	0.41
Dry atmosphere	0.26	0	0.054	0.061	0.38
CO_2 doubling	0.24	0.055	0.068	0.049	0.42
N_2O doubling	0.32	0.048	0.051	0.035	0.46

The radiative time $\tau_r = 5$ ms corresponds to the antisymmetric vibration state, and its frequency is equal $\nu_3 = 2224\,cm^{-1}$. The radiative time $\tau_r = 80$ ms relates to the frequency $\nu_2 = 1285\,cm^{-1}$ of the deformation vibration state.

Thus, there are two absorption bands for N_2O molecules. The strongest one due to torsion molecule vibrations is overlapped with the absorption band of the CH_4 molecule. Emission of atmospheric N_2O molecules due to radiative transitions between torsion vibrations is represented in Table 7.1 and Fig. 7.3. Intensities of spectral lines for these two bands due to dioxide molecules are given in Fig. 7.9. The second band of radiative transitions due to antisymmetric vibrations is analyzed in Table 7.2.

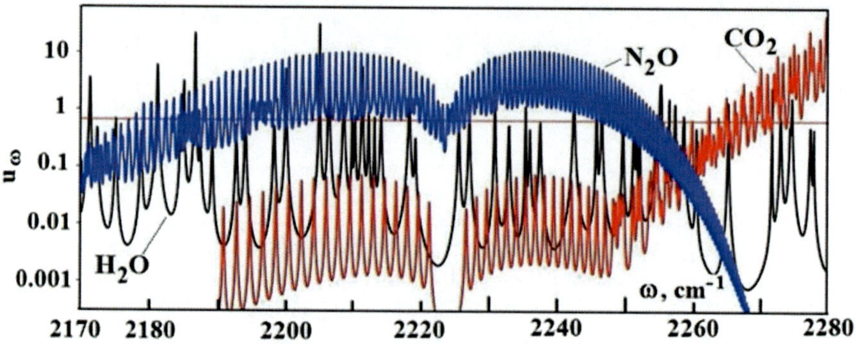

Fig. 7.10 Optical thickness u_ω of the atmospheric gap between the Earth and clouds due to optically active atmospheric N_2O, H_2O and CO_2 molecules in the absorption band of the N_2O molecule. The cloud altitude is $h_{cl} = 4.6$ km, the red line corresponds to the optical thickness of 2/3

Figure 7.10 contains the optical thickness of the standard atmosphere inside the absorption band of the N_2O molecules. As is seen, three types of molecules are absorbers in this frequency range, but absorption ranges due to each molecule are separated, i.e., this frequency range may be divided in ranges of domination of each molecule.

In the analysis of the data of Table 7.2, it should be noted that this frequency range gives a small contribution to the radiative flux of the atmosphere. Indeed, the total radiative flux of the Earth's surface in the frequency range under consideration is equal

$$\int J_E d\omega = 0.68\,\text{W/m}^2,$$

where J_E is the radiative flux from the Earth's surface over the absorption band of the N_2O molecule for the standard atmosphere at the surface temperature $T_E = 288$ K. Because the average radiative flux of the standard atmosphere is approximately 335 W/m^2, one can conclude that emission of the atmosphere inside the absorption band does not exceed 2%. From data of Table 7.2, it follows that the total radiative flux in this frequency range due to molecular components increases as a result of addition of some molecules in the atmosphere and decreases if a part of molecules is removed. But both the contribution of these molecules to infrared atmospheric radiation and their change as a result of composition of atmospheric gases is small.

7.1.4 Tropospheric Ozone as Greenhouse Component

One more trace gas which may be considered as the trace greenhouse component is ozone. Stratospheric ozone is of importance for atmospheric processes. Ozone molecules absorb approximately 3% of the solar radiative flux in its ultraviolet tail

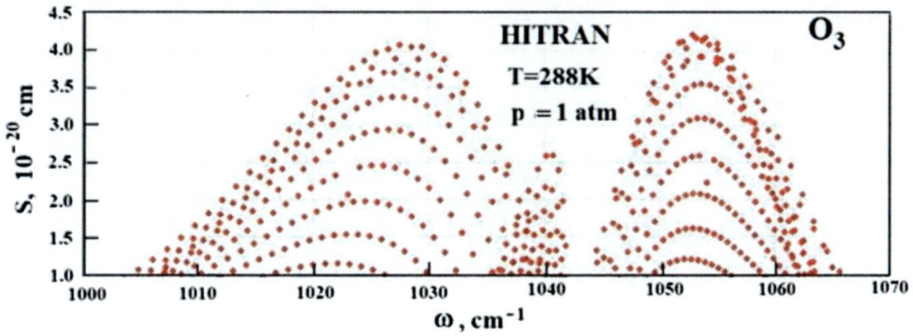

Fig. 7.11 Intensities of spectral lines for ozone molecules located in atmospheric air [12–14] inside of the absorption band for ozone molecules

and in this manner protect living organisms on the Earth surface from the action of ultraviolet radiation. 97% of the amount of stratospheric ozone is located in the altitude range from 15 to 50 km [22–24]. Infrared radiation emitted by stratospheric ozone cannot reach the Earth's surface because of absorption by clouds. Hence, infrared radiation which goes to the Earth's surface is created by tropospheric ozone that is located in the atmosphere below clouds. We below consider tropospheric ozone as a greenhouse component.

The concentration of tropospheric ozone in the atmosphere is an order of magnitude lower than for nitrogen dioxide and is (20–30) ppb [25, 26]. However, the absorption band of an ozone molecule with a center of about $1042\,\text{cm}^{-1}$ is found both in the range of maximum emission for thermal radiation and in the range of atmospheric transparency. In addition, because the concentration of tropospheric ozone depends on an environment, it may differ from its average value significantly. For definiteness, below we take the concentration of tropospheric ozone as 25 ppb. This corresponds to the density of ozone molecules per unit column of the gap between the ground and clouds as

$$n(O_3) = 2 \times 10^{17}\,\text{cm}^{-2}$$

We give in Fig. 7.11 the intensities of spectral lines of ozone molecules located in atmospheric air according to data of HITRAN bank. Radiative vibration transition which is responsible for strongest spectral lines is analogous to torsion transitions in CO_2 and N_2O molecules. Taking the intensity of the strongest spectral lines as $S = 4 \times 10^{-20}\,\text{cm}$ and the width of spectral lines for ozone molecules located in atmospheric air as $\nu \sim 0.1\,\text{cm}^{-1}$, one can estimate the maximal optical thickness of spectral lines of tropospheric ozone

$$u_\omega \sim \frac{\text{Sn}(O_3)}{\nu} \sim 0.1$$

As is seen, the troposphere is transparent with respect to ozone which is located there.

Table 7.3 Contribution of individual optically active components of the atmosphere to the total radiation flux from the atmosphere to the Earth's surface in the region of the absorption band of ozone molecules at frequencies between 990 and 1070 cm^{-1} and with the indicated atmospheric composition. The radiation fluxes are expressed in units of W/m^2

State	O_3	H_2O	CO_2	clouds	Total flow
Standard atmosphere	0.84	1.4	0.48	9.1	11.8
Dry atmosphere	3.4	0	0.62	9.8	13.8
O_3 doubling	1.9	1.4	0.52	8.7	12.5

Table 7.3 for the optical properties of the atmosphere in the range of the absorption band of ozone molecules is an analog of Table 7.2 for the absorption band of nitrogen dioxide molecules. The data in this table refer to the concentration of ozone molecules equal to 25 ppb for a standard atmosphere and 50 ppb for an atmosphere with the doubled concentration of ozone molecules. Since the optical thickness of the atmosphere due to ozone molecules is significantly less than one, the contribution of ozone molecules to the total radiation flux of the atmosphere changes proportionally to the change in the density of ozone molecules, if ozone molecules play the main role in this range of the spectrum.

In spite of the relatively low concentration of ozone molecules in the atmosphere, the contribution of ozone to atmospheric emission is comparable to that due to other trace optically active molecular components of the atmosphere. In this case, the low concentration is compensated by the region of the spectrum favorable for emission, as well as by the transparency of the atmosphere in this region of the spectrum. Figure 7.12 demonstrates the contribution to the radiative flux of the atmosphere inside the spectrum range according to the absorption band of ozone molecules. In this spectrum range, the opaque factor is small, or the total optical thickness of the satmosphere is low, i.e., the atmosphere is transparent at such frequencies. This allows one to consider emission of each greenhouse component independently from other components.

7.2 Change of Greenhouse Effect

7.2.1 Atmospheric Emission at Doubled Concentration of Carbon Dioxide

The greenhouse effect in atmospheric emission results from radiation of optically active gases located in the atmosphere. A change of the concentration of greenhouse gases leads to change of radiative fluxes created by the atmosphere, both partial fluxes at a certain frequency, and the total fluxes from each component. We now consider the character of change in atmospheric emission due to change of basic greenhouse gases, namely due to variation of the concentration of CO_2 and H_2O molecules.

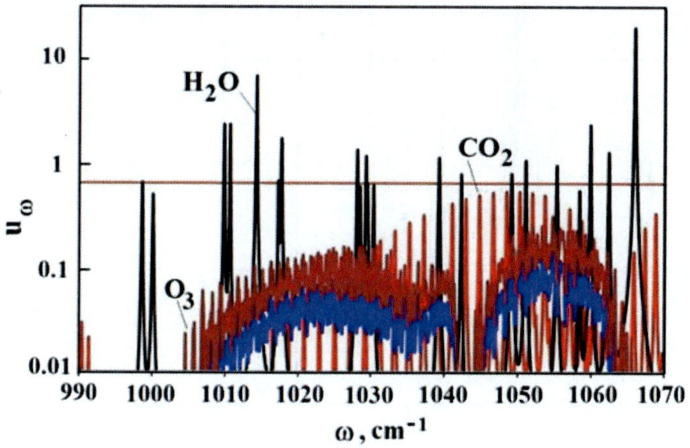

Fig. 7.12 Optical thickness u_ω of the atmospheric layer located between the Earth's surface and clouds for optically active molecules in the region of the absorption band of ozone molecules. The altitude of the clouds is $h_{cl} = 4.6\,\mathrm{km}$

In analyzing a change of the greenhouse effect due to carbon dioxide emission, we consider below doubling of the concentration CO_2 molecules, as it is made often. In this analysis, we are based on the above model of atmospheric emission with three basic greenhouse components, namely H_2O and CO_2 molecules, as well as clouds consisted of water microdroplets. In this model, clouds are located at a certain altitude and has a sharp boundary. Clouds separate the atmosphere in regions which are responsible for atmospheric emission to the Earth and outside. We below are restricted ourselves by lower layers which create the radiation directed to the Earth's surface. These layers are located in a gap between the ground and clouds. For simplicity, we assume the position of the cloud boundary to be unvaried in changes of the atmosphere composition.

Under these conditions, doubling of the concentration of CO_2 atmospheric molecules leads to an increase of the radiative temperature at each frequency due to increase of emission of carbon dioxide molecules. Figure 7.13 gives the difference of radiative temperatures for the doubled and contemporary concentrations of carbon dioxide molecules within the framework of the standard atmosphere model for frequencies where these molecules absorb.

From data of Fig. 7.13, it follows that the difference of radiative temperatures for indicated cases is small at the center of the absorption band, where the atmospheric optical thickness is large. On the contrary, this difference is relatively large at frequencies which correspond to the boundary of the absorption band. Especially large difference corresponds to laser wavelengths centered at 9.4 and 10.6 μm wavelengths where absorption due to water molecules, as well as for carbon dioxide molecules, is weak, and the atmosphere is relatively transparent.

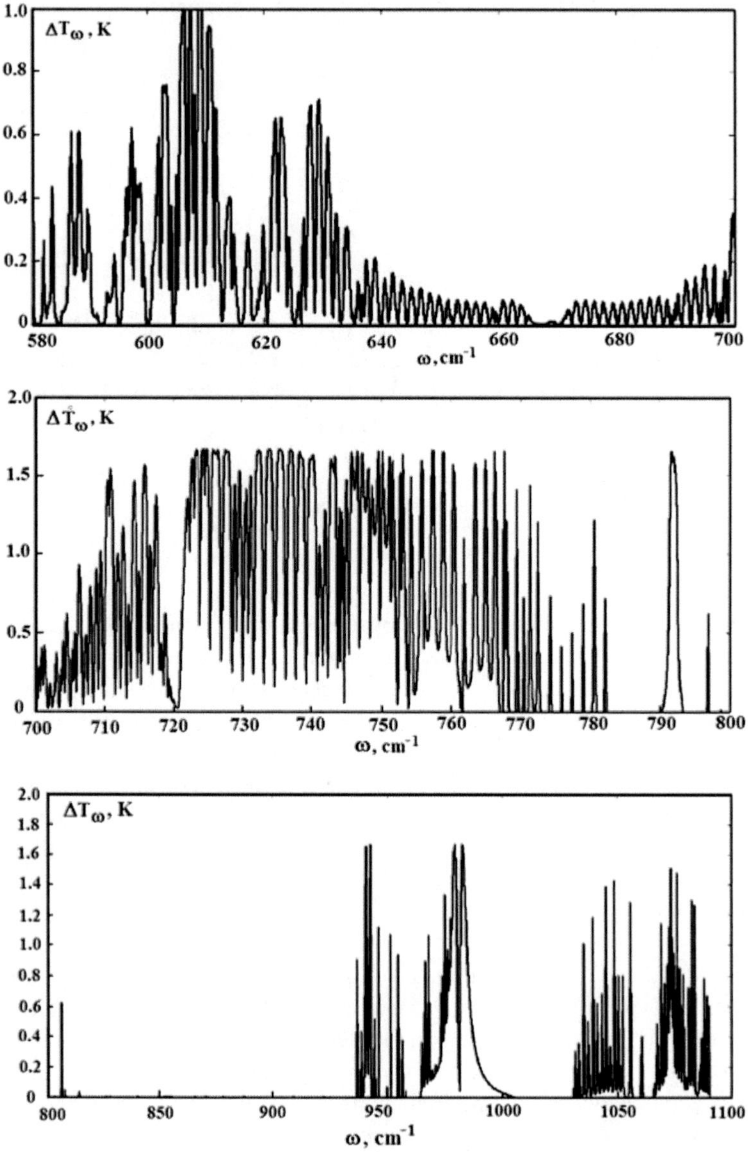

Fig. 7.13 Difference of the radiative temperatures of atmospheric air at the doubled and contemporary concentrations of CO_2 molecules for three frequency ranges with absorption of these molecules

We now consider step-by-step changes in radiative fluxes with variation of the frequency. The results are represented in Table 7.4 for basic greenhouse components. The values Δ_c, Δ_w, Δ_d of Table 7.4 refer to radiative fluxes from the atmosphere to the Earth due to CO_2 molecules, H_2O molecules and water microdroplets of clouds correspondingly, according to the relations

$$\Delta_c = \int [J'_\omega(CO_2) - J_\omega(CO_2)]d\omega,$$
$$\Delta_w = \int [J'_\omega(H_2O) - J_\omega(H_2O)]d\omega, \ \Delta_d = \int [J'_\omega(drop) - J_\omega(drop)]d\omega, \tag{7.2.1}$$

where $J_\omega(CO_2)$, $J_\omega(H_2O)$ and $J_\omega(drop)$ are the radiative fluxes due to indicated components at the contemporary concentration of CO_2 molecules, and $J'_\omega(CO_2)$, $J'_\omega(H_2O)$, and $J'_\omega(drop)$ are the radiative fluxes due to indicated components at the doubling concentration of CO_2 molecules. The change of the total radiative flux Δ is defined as

$$\Delta = \Delta_c + \Delta_w + \Delta_d \tag{7.2.2}$$

Data of Table 7.4 confirm the statement that the variation of partial radiative fluxes for greenhouse components, as well as the change of the total radiative flux from the atmosphere to the Earth, takes place only in spectrum ranges where CO_2 molecules absorb. Besides it, the basic change of the radiative flux as a result of the concentration change for some optically active components takes place in the frequency range where the optical thickness is of the order of one. But the main conclusion of data of Table 7.4 is the validity of the Kirchhoff law [1] according to which radiators located in a gas are simultaneously absorbers. This means that an increase of the concentration of a certain greenhouse component leads to an increase in the radiative flux of this component, as well as to a decrease in the radiative flux due to other components.

From data of Table 7.4, it follows for the change of radiative fluxes as a result of doubling of the concentration for CO_2 molecules

$$\Delta = (1.4 \pm 0.1)\mathrm{W/m^2}, \ \Delta_c = (7.2 \pm 0.1)\,\mathrm{W/m^2}, \tag{7.2.3}$$

where Δ is the change of the total radiative flux from the atmosphere to the Earth as a result of doubling of the concentration of CO_2 molecules, Δ_c is the change created by CO_2 molecules for this changing. From this and the subsequent analysis we have for the change of the average radiative flux $J_\downarrow$ from the atmosphere to the Earth as a result of any change of the concentration $c(CO_2)$ of atmospheric CO_2 molecules is given by formula

Table 7.4 Variations of radiative fluxes for the standard atmosphere as a result of doubling of the CO_2 concentration in an indicated spectrum range of CO_2 molecules [27]. The frequency ranges $\Delta\omega$ are given in cm^{-1}, and radiative fluxes are expressed in W/m^2. Results of calculations [16] are given in parentheses

Frequency range, cm^{-1}	Δ_c	Δ_w	Δ_d	Δ
580–600	0.70	−0.60	−0.05	0.05 (0.03)
600–620	0.91	−0.75	−0.05	0.11 (0.04)
620–640	0.75	−0.71	0	0.04 (0.02)
640–660	0.23	−0.22	0	0.01 (0.01)
660–680	0.08	−0.08	0	0 (0)
680–700	0.27	−0.26	0	0.01 (0.01)
700–720	0.38	−0.32	0	0.06 (0.06)
720–740	0.56	−0.20	−0.12	0.24 (0.20)
740–760	1.03	−0.42	−0.37	0.24 (0.29)
760–780	0.65	−0.02	−0.45	0.18 (0.14)
780–800	0.25	0	−0.17	0.08 (0.06)
800–850	0.18	0.01	−0.15	0.04 (0.04)
900–950	0.20	0	−0.15	0.05 (0.04)
950–1000	0.35	0	−0.25	0.10 (0.22)
1000–1050	0.18	0	−0.12	0.06 (0.05)
1050–1100	0.45	0	−0.31	0.14
2250–2400	0.02	0	−0.01	0.01
Total, ΔJ	7.19 (7.24)	−3.57 (−3.02)	−2.20 (−2.90)	1.42 (1.32)

$$\frac{dJ_{\downarrow}}{d\ln c(CO_2)} = 2\frac{W}{m^2} \tag{7.2.4}$$

We now analyze the nature of the change in the radiation fluxes created by different greenhouse components from the standpoint of the Kirchhoff law. According to this law, radiators of the atmosphere are simultaneously absorbers, and hence injection in the atmosphere an additional amount of carbon dioxide molecules leads both to an increase of the radiative flux created by CO_2 molecules and to a decrease of the radiative flux due to other greenhouse components because their emission is screened by an additional carbon dioxide. This compensation is stronger, the smaller is the temperature variation depending on the altitude. Since for the standard atmosphere the temperature gradient is 6.5 K/km, the air temperature varies weakly in the atmospheric layer which is responsible for atmospheric emission. Therefore, the change of the radiative flux Δ_c created by additional CO_2 molecules exceeds remarkably the change Δ of the total radiative flux to the Earth. The ratio of these values is

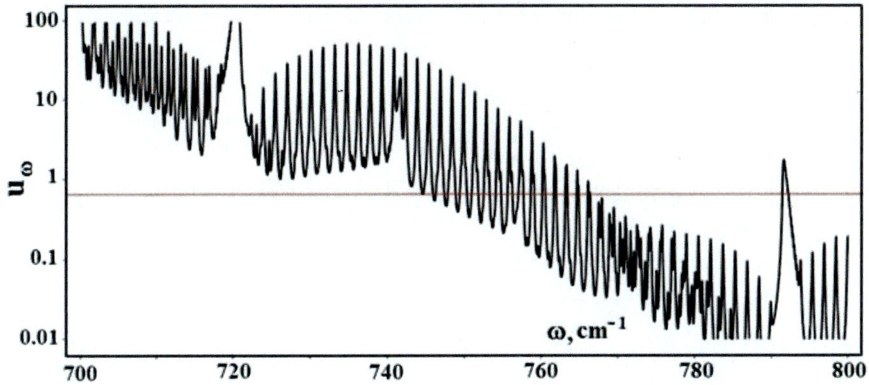

Fig. 7.14 Optical thickness u_ω of the atmospheric gap between the Earth and clouds due to CO_2 and H_2O molecules in the frequency range where the optical thickness of the atmosphere is of the order of one. The solid line corresponds to $u_\omega = 2/3$ and separates roughly the frequency ranges of large and low optical densities

$$\Delta_c/\Delta = 5.2 \pm 0.2 \tag{7.2.5}$$

As it was shown, the main contribution to an increasing radiative flux follows from a frequency range where the optical thickness is of the order of one. Figure 6.16 represents the frequency dependence of the optical thicknesses u_ω of the standard atmosphere at the violet boundary of the absorption band for CO_2 molecules. This absorption band lies approximately in the range from 580 up to 760 cm^{-1}, and the position of the violet boundary of the absorption band follows from this Figure. Comparison of data of Fig. 7.11 and Table 7.4 with those of Fig. 7.14 shows that the maximum differences of radiative temperatures and radiative fluxes for doubled and contemporary concentrations of atmospheric CO_2 molecules correspond to the boundary of the absorption band for carbon dioxide molecules.

We also draw attention to the following fact. Within the framework of the line-by-line method, we obtain that the spectrum of the atmosphere consists of separate broaden spectral lines. This means that the optical thickness varies from the center of a spectral line to a middle between this and neighboring spectral lines strongly, at least, by tens. The difference of radiative fluxes as a result of a change of the concentration of an optically active molecules is created by a frequency range where $u_\omega \sim 1$. Hence, the frequency range which gives the contribution to a change of radiative fluxes is divided in some ranges, where the above relation is fulfilled. There is another character in climatological models with interpolation of the atmospheric spectrum by smooth dependencies.

We also analyze the character of the change of radiative fluxes from the atmosphere to the Earth as a result of doubling in the concentration of atmospheric CO_2 molecules depending on the frequency of emission. Current changes of radiative fluxes are introduced by

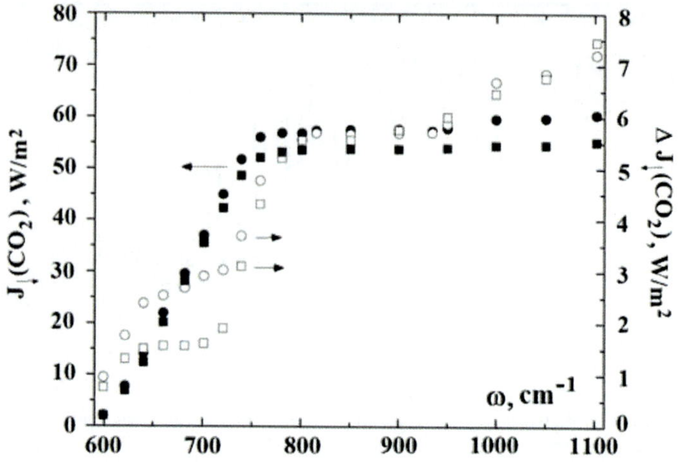

Fig. 7.15 Radiative flux from the atmosphere to the Earth $J_{\downarrow}(CO_2)$ created by atmospheric CO_2 molecules at the contemporary concentration of CO_2 molecules (closed signs) and its change $\Delta J_{\downarrow}(CO_2)$ as a result of doubling of carbon dioxide concentration (open signs), as they are defined by formula (7.2.5). Circles are data of [16], and squares are the results of [27]

$$J_{\downarrow}(\omega) = \int_0^{\omega} J_{\omega}(CO_2)d\omega, \quad \Delta J_{\downarrow}(CO_2) = \int_0^{\omega} [J'_{\downarrow}(CO_2) - J_{\omega}(CO_2)]d\omega, \qquad (7.2.6)$$

where $J_{\omega}(CO_2)$ is the radiative flux at a given frequency from the atmosphere to the Earth created by CO_2 molecules, and $\Delta J_{\downarrow}(CO_2)$ is the change of the partial radiative flux at a given frequency due to emission of atmospheric carbon dioxide molecules. Figure 7.15 contains current values of the radiative flux $J_{\downarrow}(\omega)$ due to CO_2 molecules and the difference of radiative fluxes $\Delta J_{\downarrow}(CO_2)$ for the doubled and current concentrations of atmospheric CO_2 molecules in accordance with formulas (7.2.6). Data given in Fig. 7.15 are evaluations of [16] for the frequency range up to 1200 cm^{-1} which includes absorption bands of CO_2 molecules for thermal emission of the atmosphere, and data [27] account for the frequency range up to 2600 cm^{-1} which includes frequencies which are responsible for thermal emission of the atmosphere.

In the same manner, one can introduce the change of the total radiative flux $\Delta J(\omega)$ from the atmosphere to the Earth defined as

$$\Delta J(\omega) = \int_0^{\omega} [J'_{\omega} - J_{\omega}]d\omega, \qquad (7.2.7)$$

and the quantity (7.2.2) is equal

$$\Delta = \Delta J(\infty), \quad \Delta_c = \Delta J_c(\infty)$$

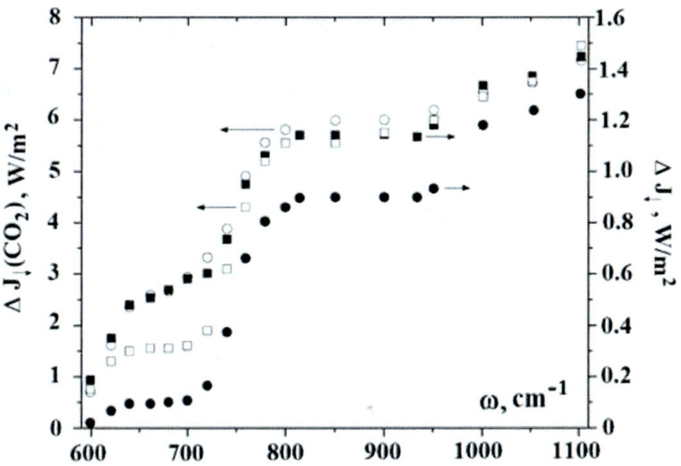

Fig. 7.16 Changes of the radiative flux $\Delta J_{\downarrow}(CO_2)$ from the atmosphere to the Earth as a result of doubling of the concentration of atmospheric CO_2 molecules for the radiative fluxes created by CO_2 molecules (open signs), as well as the change $\Delta J_{\downarrow}$ of the total radiative flux in this variation, (close signs), as they are defined by formulas (7.2.6) and (7.2.7). Circles are data of [16], and squares are the results of [27]

Fig. 7.16 contains the change of the total radiative flux $\Delta J_{\downarrow}(\omega)$ as a result of doubling the concentration of atmospheric CO_2 molecules, as well as the change $\Delta J_{\downarrow}(\omega)(CO_2)$ of the radiative flux created by CO_2 molecules in accordance with formulas (7.2.6) and (7.2.7).

One can expect that under real conditions with a weakly nonuniform atmosphere this physical picture is almost conserved. In particular, an increasing concentration of carbon dioxide molecules causes an increase in the radiative flux due to CO_2 molecules, and this increase is almost compensated by a decrease of the radiative flux created by water molecules and water microdroplets. The difference of changes for these fluxes is small compared to each of them. Just this situation corresponds to Fig. 7.15, so that the change in the total radiative flux toward the Earth is small compared to that resulted from emission of CO_2 molecules.

As it follows from Figs. 7.15 and 7.16, the basic contribution to the radiative flux to the Earth's surface created by CO_2 molecules J_c follows from the absorption band of CO_2 molecules which lies between approximately 580 and 760 cm^{-1}. A change of the radiative flux due to radiation of CO_2 molecules Δ_c, as well as a change of the total radiative flux from the atmosphere to the Earth Δ, is developed mostly near the violet boundary of the absorption band of CO_2 molecules. In addition, laser transitions in CO_2 molecules with centers about 9.4μm and 10.6μm give an additional contribution approximately 30% to the total change Δ of the radiative flux. Note that the frequency range of laser transitions gives approximately 2% to the radiative flux J_c created by CO_2 molecules.

In order to simplify the problem, the wish occurs to replace a change of the total radiative flux Δ toward the Earth by a change of the flux Δ_c due to CO_2 molecules, as it takes place in the absence of other greenhouse components or if the absorption bands for these components are not overlapped. This takes place in most calculations. In addition, only the frequency range of the absorption band between 580 and 760 cm^{-1} is used for this goal. In this case, the change of the radiative flux defined by formula (7.2.6) is equal approximately 4 W/m^2 according to [24, 28]. This evaluation on the basis of the HITRAN spectroscopic data leads to the value that is approximately 20% higher than the indicated one (see Table 7.4). The range of higher frequencies increases the indicated value by approximately 30%. This shows the connection between these results and the previous estimations.

According to this analysis, a change of the radiative flux due to CO_2 molecules as a result of doubling of the concentration of CO_2 molecules exceeds by five times a change of the total radiative flux toward the Earth. In the case of averaged infrared spectrum of carbon dioxide molecules [28], the change of the total radiative flux toward the Earth is approximately 1.1 W/m^2 instead of 1.3 W/m^2 of Table 7.4 with using the real spectrum of water molecules. The change of the radiative flux due to CO_2 molecules as a result of this doubling is 7.2 W/m^2, as it follows from Table 7.4, instead of this value of 4 W/m^2 for averaged spectrum of carbon dioxide molecules. The later coincides with that of papers [24, 29], where this change is given without explanation. As is seen, changes in the total radiative flux and due to CO_2 molecules differ significantly. Note that we are based on parameters of radiative transitions taken from the HITRAN data bank that makes them more reliable and simple.

A change of the radiative flux due to CO_2 molecules as a result of an increase of its atmospheric concentration exceeds significantly a change of the total radiative flux at this operation. One can explain the reason of this fact. The atmosphere under consideration is a weakly nonuniform gas, i.e., its temperature depends weakly on the layer altitude. Let us consider nearby conditions where the atmospheric temperature does not depend on the altitude, whereas optically active components of air provide its large optical thickness. Then, the atmosphere emits as a blackbody, and the radiative flux from it does not vary, if the concentration of an optically active component changes. But the radiative flux grows due to a component with an increasing concentration, and hence the radiative flux due to other optically active components decreases. The reason of this is the screening of radiation of these components by an additional radiators of an increasing component.

Table 7.5 contains changes in radiative fluxes as a result of an indicated change in the concentration of atmospheric CO_2 molecules. We use the following notations for parameters of Table 7.4 which are similar to those of formula (7.2.1)

$$\Delta_c = j_c' - j_c, \quad \Delta_w = j_w' - j_w, \quad \Delta_d = j_d' - j_d, \quad \Delta = j' - j \tag{7.2.8}$$

Here, j_c, j_w, j_d, j are the radiative fluxes due to CO_2 molecules, H_2O molecules, water microdroplets of clouds and the total radiative flux for the standard atmosphere. Quantities j_c', j_w', j_d', j' are the same parameters for a new atmosphere content.

Table 7.5 Changes of radiative fluxes as a result of variation of the concentration of atmospheric carbon dioxide molecules from that $c(CO_2)$ of the standard atmosphere up to $c'(CO_2)$. Changes of radiative fluxes are integrated over all frequencies and are expressed in W/m^2. Values Δ_c, Δ_w and Δ_d are the changes of radiative fluxes integrated over all frequencies due to CO_2 molecules, H_2O molecules and microdroplets of clouds correspondingly, as well as Δ is the change of the total radiative flux from the atmosphere to the Earth integrated over frequencies

$c'(CO_2)/c(CO_2)$	Δ_c	Δ_w	Δ_d	Δ
0.5	−7.23	3.74	2.15	−1.38
0.7	−3.78	1.93	1.11	−0.72
0.9	−1,12	0.57	0.33	−0.21
1	0	0	0	0
1.1	1.11	−0.51	−0.30	0.20
1.5	4.36	−2.12	−1.30	0.85
2	7.19	−3.57	−2.20	1.42

From this analysis, it follows that according to data of Tables 7.4 and 7.5, the radiative flux change Δ_c due to CO_2 molecules as a result of doubling of the concentration of CO_2 molecules is larger by five times than the change of the total radiative flux Δ. In climatological models, the Kirchhoff law [1], according to which radiators are simultaneously absorbers, is ignored. Therefore, the change of the radiative flux Δ_c due to CO_2 molecules is taken in climatological models instead of the change of the total radiative flux Δ. This conclusion results from neglecting the absorption of emitted radiative fluxes by water molecules and water microdroplets of clouds by additive CO_2 molecules.

The analysis [30, 31] of this effect with using the Kirchhoff law shows that it is strong because of a weak temperature dependence on the altitude. Indeed, if the atmosphere temperature is unvaried with an altitude, an increase of the radiative flux due to an increase of the concentration of a given component is compensated by a decrease of the radiative flux due to other components if the optical thickness is large. Correspondingly, if the temperature varies weakly with an altitude, the change of the radiative flux Δ_c as a result of an increase of the concentration of CO_2 molecules exceeds remarkably the total change Δ of the radiative flux, i.e., the ratio Δ_c/Δ is large [30, 31].

Evidently, this ratio is determined by overlapping of spectra for atmospheric CO_2 molecules and other greenhouse components. Hence, the information is required about spectra of greenhouse components and their radiative parameters for evaluations of these radiative fluxes. Therefore it is useful to consider the history of understanding of this problem. Some key papers related to development of this problem are collected in the book [32]. Swedish scientist S.Arrenius was first who sets this problem in the end of 19th century. In the paper of 1896 [33] he asked "Is the temperature of the ground in any way influenced by the presence of heat-absorbing gases in the atmosphere ?". The answer for this question was awkward because of the science state in that time. In addition, the goal of the Arrenius paper was to treat the Langley experiments where the night absorption over the spectrum was measured for solar radiation which was scattered from the moon. For CO_2 molecules the strong absorption was observed near the strongest spectral line of $4.3\mu m$ for this molecule which contribution to thermal emission of the atmosphere is negligible. The absorption spectra are not overlapped for strongest radiative transitions of CO_2 and H_2O molecules which are of importance for the Arrenius problem. In addition, thermal emission of CO_2 molecules in the atmosphere for the wavelengths below 4.3 μm is about 0.5 W/m^2 [16] compared with the radiative flux of 335 W/m^2 from the atmosphere to the Earth.

The subsequent analysis of the CO_2 problem was made more strictly. In particular, the title of the Calender paper of 1949 [34] was "Can carbon dioxide influence climate?". It should be noted that the reliability of numerical calculations for the radiative flux of atmospheric emission is determined by information about radiative transitions in atmospheric molecules. The important contribution in this study was made by Plass and collaborators [35, 36] in fifties. These evaluations were restricted by the spectrum range (12–18)μm which include basic emission of CO_2 molecules in the real atmosphere, and they are based on the regular (Elsasser) model [37] with an identical difference of frequencies for neighboring distances between neighboring spectral lines. In addition, new information was used for parameters of radiative transitions of CO_2 and H_2O molecules in these evaluations. Along with evaluation of the ECS which is defined by formula (5.3.2), it was estimated the influence of H_2O molecules on the global temperature change as a result of doubling of the concentration of atmospheric CO_2 molecules. This effect was estimated as 20% for the change of the radiative flux due to overlapping of spectral lines due to information of that time. This shows the importance of information which follows from the HITRAN database [12–14].

7.2.2 *Change of Atmospheric Emission Due to Water Molecules*

Along with CO_2 molecules, water molecules are of importance for atmospheric emission, and below we consider the influence of a change of their concentration in the atmosphere on the radiative flux from it. By analogy with the CO_2 case, one can obtain that an increase of the concentration of atmospheric H_2O molecules causes an increase the radiative flux from the atmosphere to the Earth. On contrary, removal of H_2O molecules from the atmosphere leads to a decrease the radiative temperature and radiative flux. Figure 7.17 gives a decrease in the radiative temperature of the standard atmosphere if one half of the amount of atmospheric water is removed.

According to data of Fig. 7.17, small frequencies below 500 cm^{-1} do not contribution to the change of the radiative temperature because of a large optical thickness of the atmosphere due to atmospheric water molecules. Therefore changes in radiative fluxes start from frequencies above 500 cm^{-1}. The absorption spectrum of CO_2

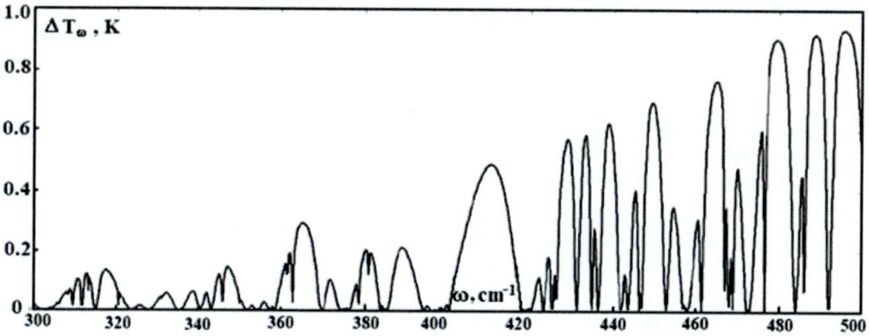

Fig. 7.17 Decrease of the radiative temperature as a result of removal of one half amount of water from the standard atmosphere

Table 7.6 Variations of radiative fluxes from the standard atmosphere to the Earth as a result of change in the concentration of H_2O molecules in atmospheric air. Here, $c(H_2O)$ is the contemporary concentration of H_2O molecules, $c'(H_2O)$ is that at a new concentration of water molecules. Notations Δ_c, Δ_w, Δ_d and Δ refer to the changes of radiative fluxes due to CO_2 molecules, H_2O molecules, water microdroplets of clouds and to the change of the total radiative flux from the atmosphere to the Earth correspondingly. These flux changes are integrated over all frequencies and are expressed in W/m^2

$c'(H_2O)/c(H_2O)$	Δ_c	Δ_w	Δ_d	Δ
0.5	3.84	−15.64	6.52	−4.98
0.7	2.05	−7.94	3.20	−2.53
0.9	0.62	−2.38	0.91	−0.74
1	0	0	0	0
1.1	−0.57	2.10	−0.81	0.66
1.5	−2.52	8.86	−3.36	2.78
2	−4.42	15.15	−5.65	4.71

molecules in the atmosphere starts also at frequencies above 580 cm^{-1}, and there absorption by CO_2 molecules can influence on the change of radiation fluxes due to CO_2 molecules.

We give in Table 7.6 the changes of radiative fluxes in the course of variation of the concentration of atmospheric H_2O molecules. As is seen, an increase in the concentration of water molecules leads to an increase Δ_w of the radiative flux due to these molecules, that is screened partially by other greenhouse components. As a result, the ratio of the radiative flux change Δ_w due to water molecules to the change of the total radiative flux Δ from the atmosphere to the Earth is approximately three.

In addition to Table 7.6, changes of atmospheric radiative fluxes to the Earth are represented in Fig. 7.18 as a result of change in the concentration of water molecules. One can see that this dependence is linear for real changes of this concentration. From this, one can obtain for the change of the total radiative flux $J_\downarrow$ from the atmosphere to the Earth as a result of a change in the concentration $c(H_2O)$ of atmospheric H_2O molecules

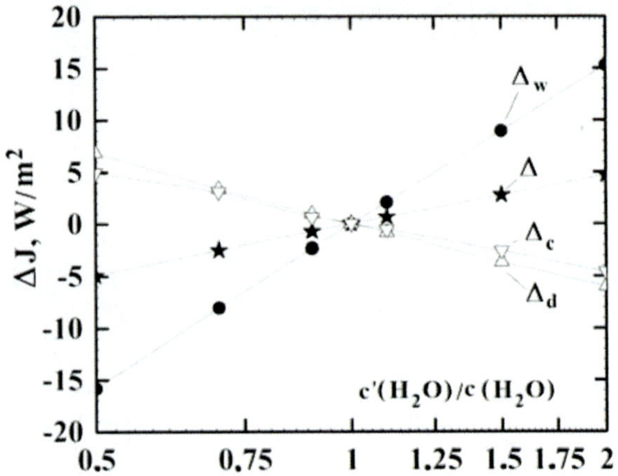

Fig. 7.18 Changes of radiative fluxes from the atmosphere to the Earth due to basic greenhouse atmospheric components. Notations and units are the same, as in Table 7.6

$$\frac{\partial J_{\downarrow}}{\partial \ln c(H_2O)} = \frac{\Delta}{\ln[c'(H_2O)/c(H_2O)]} = 7\,W/m^2, \qquad (7.2.9)$$

As it follows from data of Table 7.6, this derivative is less in 3.5 times than that in the case of CO_2 molecules. In addition, CO_2 molecules are more effective in this change compared with H_2O molecules. Indeed, on average the number density of water molecules near the Earth's surface is higher by 30 times compared with that for CO_2 molecules, and the amount of water molecules in the atmosphere is larger by 7 times compared with that for CO_2 molecules.

It is more a precisely according to data of Table 7.6 that the ratio of radiative fluxes is equal

$$\Delta_w/\Delta = 3.2 \pm 0.1, \qquad (7.2.10)$$

where Δ_w is the change in the radiation flux from the atmosphere to the Earth's surface created by water molecules, Δ is the change in the total infrared radiation flux. Comparing with the case of CO_2 molecules, one can conclude that the effective optical thickness of the spectrum range which is responsible for formation of the above difference in the water case is less than that in the carbon dioxide case.

We also consider the character of a change of the above radiative fluxes as the frequency increases. Table 7.7 contains changes of the radiative fluxes at an increase of the concentration of atmospheric water molecules by 1.25 times in the standard atmosphere. This change leads to an increase the average moisture of the standard atmosphere near the Earth surface from its contemporary average value $\eta = 80\%$ up to the saturated state of $\eta = 100\%$ if the global temperature is not varied. One can see that the frequency range of 500–1500 cm^{-1} which is responsible for creation of thermal radiation of the atmosphere determines also the change of the radiative flux due to H_2O molecules.

Table 7.7 Variations of radiative fluxes from the standard atmosphere to the Earth as a result of an increase of the concentration of H_2O molecules in atmospheric air by 1.25 times. Values Δ_c, Δ_w, Δ_d and Δ are the changes of radiative fluxes which notations and units are the same as in Table 7.6

Frequency range, cm^{-1}	Δ_w	Δ_c	Δ_d	Δ
250–500	0.30	0	−0.16	0.14
500–750	2.02	−1.24	−0.44	0.34
750–1000	1.21	−0.11	−0.71	0.39
1000–1250	0.82	−0.01	−0.44	0.37
1250–1500	0.37	0	−0.08	0.17
1500–2000	0.08	0	0.02	0.07
2000–2500	0.07	0	−0.02	0.01
Total	4.89	−1.36	−1.88	1.48

In conclusion of this analysis, we note also a weak dependence of the radiative flux $J_\downarrow$ from the atmosphere to the ground on the concentration of greenhouse components. Indeed, on the basis of formulas (7.2.4) and (7.2.9) and Table 7.7, we have

$$\frac{\partial \ln J_\downarrow}{\partial \ln c(CO_2)} = 0.036, \quad \frac{\partial J_\downarrow}{\partial \ln c(H_2O)} = 0.025 \tag{7.2.11}$$

This weak dependence justifies comparison of various versions of the Earth's energetic balance given in Table 5.1 for different times where the atmosphere composition varied weakly.

7.3 Change of the Earth's Thermal State

7.3.1 Climate Sensitivity for the Earth's Atmosphere

Above (§5.3.4) we define the climate sensitivity S according to formula (5.3.4) as the conversional coefficient between the change ΔJ of the radiative flux from the atmosphere to the Earth and the global temperature change ΔT. We also introduce the radiative forcing $F = 1/S$ as a reciprocal value with respect to the climate sensitivity. In addition, we use evaluations for this value that gives one after their averaging according to formula (5.3.5)

$$S \approx 0.5 \frac{m^2\,K}{W}$$

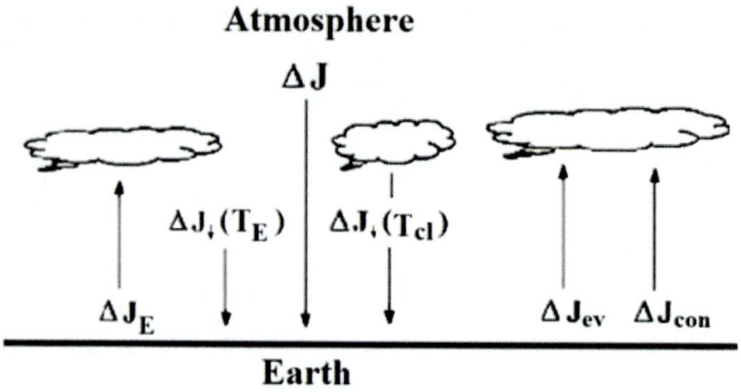

Fig. 7.19 Energy fluxes which compensate an additional energy flux ΔJ to the Earth's surface that leads to an increase of the global temperature. These fluxes are as follows. $\Delta J_{\rm E}$ is an additional radiative flux from the Earth's surface, $\Delta J_{\rm ev}$ is the change of the energy flux due to water evaporation from the Earth's surface, $\Delta J_{\rm c}$ is that due to atmospheric convection, $\Delta J_{\downarrow}(T_{\rm E})$ is the change of the radiation flux from the atmosphere as a result of variation of the global temperature, and $\Delta J_{\downarrow}(T_{\rm cl})$ is this flux change from clouds

Being guided by the scheme of processes given in Fig. 7.19, we determine the climate sensitivity S on the basis of the above evaluations for atmospheric radiative fluxes within the framework of the model of standard atmosphere. If the Earth gets an additional energy flux, its temperature (global temperature) increases. As a result, additional energy fluxes arise which compensate an additional energy flux ΔJ to the Earth's surface. Evaluating these fluxes, one can determine the radiative flux F as

$$F = F_{\rm E} - F_{\downarrow} + F_{\rm cl} + F_{\rm ev} + F_{\rm c} \qquad (7.3.1)$$

The quantities of the right side of this equation are defined by Fig. 7.19. Below we determine each forcing of formula (7.3.1) for the model of standard atmosphere.

Starting from the first term of formula (7.3.1), we assume that the Earth emits as a blackbody. Then, the connection between the change of the radiative flux $\Delta J_{\rm E}$ and the global temperature ΔT in the form of the radiative forcing $F_{\rm E}$ has the following form

$$\Delta J_{\rm E} = 4\sigma T_{\rm E}^3 \Delta T = \frac{4\Delta T}{T_{\rm E}} J_{\rm E}, \ F_{\rm E} = \frac{4J_{\rm E}}{T_{\rm E}} = 5.4\frac{\rm W}{\rm m^2\,K}, \qquad (7.3.2)$$

where $T_{\rm E} = 288$ K is the temperature of the Earth, and $J_{\rm E} = 390\,{\rm W/m^2}$ is an average radiative flux from the Earth's surface, as it follows from the energetic balance of the Earth represented in Table 5.1 and Fig. 5.4.

Figure 7.20 contains the dependence on the Earth's temperature for the radiative flux from the atmosphere to the Earth under the condition that the cloud temperature is not varied under the action of an additional flux. From Fig. 7.20, it follows for the radiative forcing due to infrared emission of the atmosphere

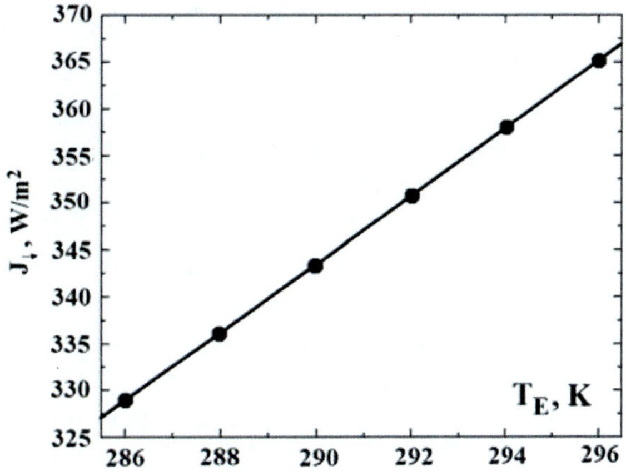

Fig. 7.20 Dependence on the Earth's temperature for the radiative flux from the atmosphere to the Earth due to optically active molecules of the atmosphere for the model of standard atmosphere. The cloud temperature T_{cl} is unvaried. Sold curve corresponds to formula (7.3.3) $\partial J_{\downarrow}/\partial T_E = 3.6\,W/(m^2\,K)$

$$F_{\downarrow} = \frac{\partial J_{\downarrow}}{\partial T_E} = 3.6\frac{W}{m^2\,K}, \tag{7.3.3}$$

From this, one can obtain the radiative forcing due to infrared radiation F_{rad} of the Earth and atmosphere

$$F_{rad} = F_E - F_{\downarrow} = 1.8\frac{W}{m^2\,K} \tag{7.3.4}$$

Let us consider another version of the cloud radiation where the altitude h_{cl} does not vary in the course of change of the global temperature. Then, the cloud temperature T_{cl} varies that leads to an additional radiating forcing F_{cl} due to the change of the cloud temperature that is equal

$$F_{cl} = \frac{4J_{cl}}{T_{cl}} \tag{7.3.5}$$

Taking the cloud temperature $T_{cl} = (258 \pm 6)$ K and the radiative flux from clouds to the Earth $J_{cl} = (61 \pm 7)\mathrm{W/m^2}$ according to the above evaluations, we have for this part of the forcing

$$F_{cl} = (0.95 \pm 0.15)\,\mathrm{W/m^2K} \tag{7.3.6}$$

Note that in this case we assume the temperature derivative over the altitude to be unvaried as a result of changing of the global temperature.

In addition, one can consider the model of constant absorption coefficient over the spectrum, where the effective radiative temperature according to formula (6.2.4) is equal $T_{\downarrow} = (277 \pm 2)$ K. Under this assumption, we have

$$F_{\downarrow} = \frac{4J_{\downarrow}}{T_{\downarrow}} = 4\sigma \cdot T_{\downarrow}^3 = (4.8 \pm 0.1)\frac{\mathrm{W}}{\mathrm{m}^2\,\mathrm{K}}, \tag{7.3.7}$$

where $\sigma = 5.67 \times 10^{-8}\,\mathrm{W/(m^2\,K^4)}$ is the Stephan–Boltzmann constant. Averaging over the above three models, one can obtain for the radiative forcing due to infrared radiation F_{rad} of the Earth and atmosphere

$$F_{rad} = F_E - F_{\downarrow} = (1.2 \pm 0.6)\frac{\mathrm{W}}{\mathrm{m}^2\,\mathrm{K}} \tag{7.3.8}$$

One can see a high uncertainty of the radiative forcing due to infrared radiation of the Earth and atmosphere.

We now consider the radiative forcing due to air convection. One can represent two versions. In the first case, in the course of the change of the global temperature, the tropopause temperature assumes to be unvaried. In this case, according to the model of standard atmosphere [15], the atmospheric temperature has the minimum $T_{min} = 217$ K at the altitude of $h_{min} = 11$ km. In this case, the temperature gradient is $(T_E - T_{min})/h_{min}$ varies with the change of the global temperature T_E. According to formula (5.1.14), the convection flux is $J_{con} = 41\,\mathrm{W/m^2}$ that leads to the radiative forcing F_{con} due to convection

$$F_{con} = \frac{J_{con}}{T_E - T_{min}} = 0.6\frac{\mathrm{W}}{\mathrm{m}^2\,\mathrm{K}}, \tag{7.3.9}$$

In the other version, where the temperature gradient in the atmosphere does not vary, the radiative forcing due to convection is equal $F_{con} = 0$.

In considering the radiative forcing F_{ev} due to water evaporation from the Earth's surface and its condensation in the atmosphere, we assume first the rate of falling of rain or snow to the Earth surface is independent of the global temperature. In this case, we have $F_{ev} = 0$. In other case, where the rate of processes of condensation in the atmosphere and the rate of falling of formed droplets or particles are independent of the temperature, the change of energy processes as a result of the water circulation follows from the change of the number density of water molecules. In this case, the equilibrium of atmospheric water and that located at the Earth's surface leads to the concentration of atmospheric water molecules c near the Earth's surface

$$c \sim \exp(-\varepsilon_o/T),$$

where $\varepsilon_o = 0.44$ eV is the binding of the water molecule for liquid water according to formula (2.1.13). As a result, we obtain for the change of the energy flux ΔJ_{ev} due to water evaporation

$$\Delta J_{ev} = J_{ev}\frac{\varepsilon_o \Delta T}{T_E^2} \tag{7.3.10}$$

Introducing the evaporation forcing F_{ev}, one can obtain

$$\Delta J_{ev} = J_{ev}\frac{\varepsilon_o \Delta T}{T_E^2}, \quad F_{ev} = \frac{J_{ev}\varepsilon_o}{T_E^2} = 3.4\frac{\mathrm{W}}{\mathrm{m^2\,K}} \tag{7.3.11}$$

Summarizing various versions of energy transfer, one can determine the total radiative forcing F according to formula

$$F = F_{rad} + F_{con} + F_{ev} \tag{7.3.12}$$

Using these values for various versions, one can find finally

$$F = 2 \times 10^{\pm 0.5}\frac{\mathrm{W}}{\mathrm{m^2\,K}} \tag{7.3.13}$$

From this, we have for the climate sensitivity

$$S = 0.5 \times 10^{\pm 0.5}\frac{\mathrm{m^2\,K}}{\mathrm{W}} \tag{7.3.14}$$

Though this value coincides with that (5.3.5) averaged over various evaluations, one can note its large uncertainty. A large uncertainty follows also from data in past. In particular, palaeontological study [38] gives for various geographical points of the globe the climate sensitivity which lies in the range from 0.3 up to $1.9(\mathrm{m^2\,K/W})$, whereas according to another analysis of the climate sensitivity in past [39], this value lies within the limits between 0.25 and $0.79(\mathrm{m^2\,K/W})$.

7.3.2 Change of Global Temperature

The climate sensitivity depends on processes which are included in consideration. For this reason, the corridor where the value of the climate sensitivity can vary in one order of magnitude according to formula (7.3.14). We now analyze from the standpoint of the accuracy of the climate sensitivity the parameter which is close to it, namely the equilibrium climate sensitivity (ECS) that is defined by formula (5.3.2). This value characterizes the change of the global temperature as a result of concentration doubling for atmospheric CO_2 molecules.

We first present the values of ECS which we have now under different assumptions. If we use measurements of the change of the concentration of CO_2 molecules and the global temperature in the real atmosphere, one can obtain the value ECS according to formula (5.3.3)

$$\mathrm{ECS} = (2.1 \pm 0.4)\,^\circ\mathrm{C}$$

Table 7.8 Values of ECS or the global temperature change at doubling of the concentration of CO_2 molecules under various model assumptions

Model	ΔT, °C	References
1	0.5	[48, 49]
2	0.4	[28]
3	0.5	[50, 51]
4	0.6	[16]

In this formula, the concentration of CO_2 molecules is only the indicator of the change of the global temperature. Hence, in reality, the value (5.3.3) is the upper limit of the ECS.

One can add to this data obtained on the basis of climatological models. These data are the basis of the intergovernmental panel on climate change [40], according to which ECS ranges from 1.5 °C up to 4.5 °C. This corresponds to statistical averaging of some calculations [41–47] on the basis of climatological models according to which

$$\mathrm{ECS} = (3.0 \pm 1.5)\,^{\circ}\mathrm{C} \tag{7.3.15}$$

As is seen, this value exceeds partially the upper limit of the ECS according to (5.3.3). We above show that climatological models lead to a wrong result because they ignore the Kirchhoff law in replacing the total change of the radiative flux from the atmosphere due to the change the concentration of CO_2 molecules by that due to CO_2 molecules. For a real atmosphere, this leads to increase the resulted change of the global temperature by five times according to formula (7.2.5). For this reason in the subsequent analysis, we do not consider results of climatological models and will be guided by the value [16]

$$\mathrm{ECS} = (0.6 \pm 0.3)\,^{\circ}\mathrm{C} \tag{7.3.16}$$

which are based on the climate sensitivity (5.3.5) and evaluations of greenhouse fluxes of infrared radiation from the atmosphere to the Earth. This value accounts for variation of the concentration of CO_2 molecules only, whereas other atmospheric parameters are unvaried. In spite on a large error of this value, one can conclude from this that evolution of the concentration of atmospheric carbon dioxide cannot explain the observational rate of an increase of the global temperature.

Our statement is that the difference in the temperature changes (7.3.15) and (7.3.16) is explained by ignoring the Kirchhoff law which accounts for indirectly optical interaction between various optically active components of the atmosphere, and this interaction results from overlapping of spectral lines for these components. One can exhibit this by comparison of different models of the same nature which we used for evaluation of the ECS represented in Table 7.8.

The absorption coefficient k_ω of the model 1 [48, 49] of Table 7.8 is a sum of those for molecules of carbon dioxide and that of water molecules and water microdroplets. For simplification, the total absorption coefficient is averaged over frequencies of water molecules and water microdroplets. The absorption coefficient due to CO_2 molecules is determined on the basis of the Elsasser model [37] and is averaged over frequencies for neighboring spectral lines. It is taken for the basic absorption band of CO_2 molecules for thermal emission which is located between 589 and 760 cm^{-1}.

The model 2 [28] of Table 7.8 uses the same assumptions, but averaging of the absorption coefficient over neighboring spectral lines for CO_2 molecules is absent. The model 3 [50, 51] is based on data of the HITRAN bank for spectral parameters of CO_2 and H_2O molecules, as well as the space distributions of CO_2 and H_2O molecules according to the model of standard atmosphere. Water microdroplets and water molecules are distributed over the space similar to atmospheric air molecules. The density of water microdrops is found on the basis of the energetic balance of the Earth. The fourth model [16] of Table 7.8 is analogous to the third one, but water microdroplets of clouds are located starting from a certain altitude. In addition, it accounts for "laser" transitions around the wavelengths of 9.4μm and 10.6μm. These transitions give the contribution about 2% to the total radiative flux created by CO_2 molecules, whereas their contribution to a change of the global temperature by approximately 30%.

From data of Table 7.8, it follows that model assumptions do not influence significantly for changes of the global temperature as a result of an increase of the concentration of atmospheric CO_2 molecules. This leads to the conclusion that the contradiction between the above results and those of climatological models follows from ignoring of the Kirchhoff law in simplification of the character of atmospheric emission. This leads to a large error according to formula (7.2.4).

We now compare the values of ECS which are obtained in various versions. One can consider the value (5.3.3) which follows from measurements as an upper limit for the global temperature increase owing to the greenhouse effect due to emission of CO_2 molecules. The value (7.3.15) from climatological models is larger than the upper limit (5.3.3) that proves that climatological models with ignoring the Kirchhoff law are incorrect. Next, the values according to formula (7.3.16) and Table 7.8 account for directly the contribution of emission of CO_2 molecules. As is seen, in spite of a large error in evaluation of ECS, one can conclude that CO_2 molecules are not the basic radiators in the change of the atmospheric emission. One can assume that the main contribution to a change of the global temperature follows from a change of the concentration of atmospheric H_2O molecules.

We above show the importance of optical interaction for radiative components of the real atmosphere through the overlapping of their spectra. In the case of basic greenhouse components, i.e., H_2O and CO_2 molecules, this leads to the difference of radiative fluxes changes as a result of change in the concentration of molecules for the flux due to varying components and the total flux by several times. We now give one more example in Fig. 7.21 where the global temperature of the Earth varies as a result of a change in the concentration of methane molecules.

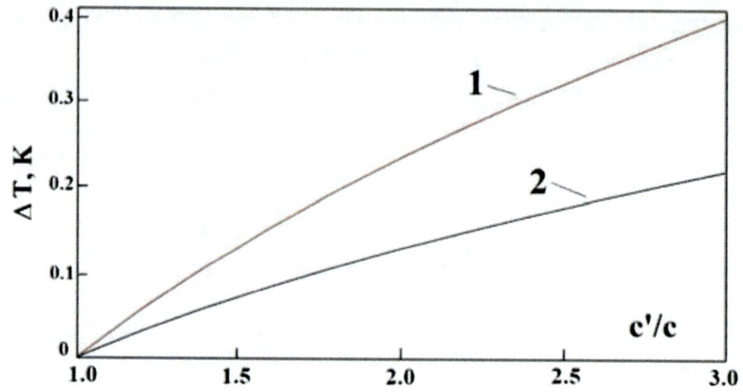

Fig. 7.21 Change of the global temperature as a result of variation for the concentration of methane molecules from c that corresponds to the real atmosphere to c'. 1 relates to the dry atmosphere, and 2 relates to the real atmosphere [27]

Indeed, according to data of Fig. 7.2, the spectrum of methane molecules is overlapped with the spectrum of H_2O molecules. Hence, a change of the radiative flux from the atmosphere to the Earth which follows from a change in the concentration of methane molecules is different for a dry atmosphere and that under standard conditions [15] where the atmospheric moisture near the Earth's surface of 80%. According to data of Fig. 7.19, the ratio of indicated radiative fluxes is approximately two.

7.3.3 *Moisture Evolution in the Greenhouse Effect*

Assuming on the basis of the above analysis that an observed increase of the global temperature during last decades is determined by an increase of the atmospheric water amount, we below analyze the character of processes in this case. For definiteness, we assume that an increase in the concentration of atmospheric CO_2 molecules provides $1/3$ part of the growth rate of the global temperature, whereas $2/3$ of this growth is created by an increase of the concentration of atmospheric water molecules. But simultaneously with an increase of the concentration of water molecules in the atmosphere, the global temperature increases. We now analyze the character of these changes under real conditions.

Based on the assumption that an increase of the amount of atmospheric water is the reason of observed growth of the global temperature during last decades, we determine the rate of an increase in the concentration of atmospheric water molecules that provides observed changes. According to formulas (5.3.3) for the change of the global temperature as a result of doubling of the concentration of CO_2 molecules in the real atmosphere and formula (5.2.7) for the rate of an increase in the concentration of atmospheric CO_2 molecules, one can obtain for the rate of an increase ΔT of the global temperature according to measurements

$$\frac{\mathrm{d}\Delta T}{\mathrm{d}t} = 18\frac{\mathrm{mK}}{\mathrm{year}} \tag{7.3.17}$$

One can transfer this in the change of the radiative flux $J_{\downarrow}$ from the atmosphere to the Earth on the basis of the climate sensitivity $S = 0.5\,\mathrm{m^2\,K/W}$ according to formula (5.3.5)

$$\frac{\mathrm{d}J_{\downarrow}}{\mathrm{d}t} = 36\frac{\mathrm{mW}}{\mathrm{m^2\,year}}, \tag{7.3.18}$$

Along with this, on the basis of atmospheric measurements, we have from calculations for a change of radiative fluxes as a result of a change of the concentrations $c(CO_2)$ and $c(H_2O)$ for CO_2 and H_2O molecules on the basis of formulas (7.2.4) and (7.2.9)

$$\frac{\partial J_{\downarrow}}{\partial \ln c(\mathrm{CO_2})} = 2\frac{\mathrm{W}}{\mathrm{m^2}}, \quad \frac{\partial J_{\downarrow}}{\partial \ln c(\mathrm{H_2O})} = 7\,\mathrm{W/m^2}$$

The rate of accumulation of carbon dioxide in the atmosphere is given by formula (5.2.7)

$$\frac{d \ln c(\mathrm{CO_2})}{\mathrm{d}t} = 6 \times 10^{-3}\mathrm{year}^{-1}$$

From this, one can obtain for the part of the radiative flux $J_{\downarrow}(CO_2)$ to the Earth which is created by CO_2 molecules

$$\frac{\mathrm{d}\Delta J_{\downarrow}(\mathrm{CO_2})}{\mathrm{d}t} = 12\,\mathrm{mW/year} \tag{7.3.19}$$

We assume that a change in the radiative flux to the Earth is determined by CO_2 and H_2O molecules, i.e.,

$$\frac{\mathrm{d}J_{\downarrow}}{\mathrm{d}t} = \frac{\mathrm{d}J_{\downarrow}(\mathrm{CO_2})}{\mathrm{d}t} + \frac{\mathrm{d}J_{\downarrow}(\mathrm{H_2O})}{\mathrm{d}t}$$

From formulas (7.3.18) and (7.3.19), we have for the rate of the radiative flux $\mathrm{d}J_{\downarrow}(H_2O)/dt$ that is created by H_2O molecules

$$\frac{\mathrm{d}J_{\downarrow}(\mathrm{H_2O})}{\mathrm{d}t} = 24\frac{\mathrm{W}}{\mathrm{m^2\,year}} \tag{7.3.20}$$

It should be noted that we assume that in this consideration the change in the radiative flux $J_{\downarrow}$ is determined by a change in the concentration of CO_2 and H_2O molecules, rather than a change of the global temperature. Formula (7.3.20) gives on the basis of formula (7.2.9) for the rate of a change in the concentration of atmospheric water molecules which provides the above rate for the rate of the concentration of atmospheric water molecules

$$\frac{d \ln c(H_2O)}{dt} = 3 \times 10^{-3} \text{year}^{-1} \quad (7.3.21)$$

Let us analyze now evolution of the moisture of atmospheric air which is defined according to formula (2.1.11)

$$\eta = \frac{N_w}{N_{sat}(H_2O)} = \frac{c(H_2O)}{c_{sat}(H_2O)},$$

where N_w is the current number density of atmospheric water molecules, $c(H_2O)$ is the current concentration of H_2O molecules in the atmosphere, $N_{sat}(H_2O)$ is the number density of water molecules at the saturated vapor pressure for a given temperature, $c_{sat}(H_2O)$ is the concentration of water molecules at the saturated vapor pressure. According to formula (2.1.12), the temperature dependence for $N_{sat}(H_2O)$ as well as for $c_{sat}(H_2O)$ has the form

$$N_{sat} \sim \exp\left(-\frac{\varepsilon_o}{T}\right),$$

where $\varepsilon_o \equiv E_{sat}$ is the binding energy of a water molecule with a water macroscopic surface. We assume it to be liquid, and then $\varepsilon_o = 0.44$ eV. From this, we have

$$\frac{d \ln c_{sat}(H_2O)}{dt} = \frac{\varepsilon_o}{T^2} \cdot \frac{d\Delta T}{dt} \approx 0.001 \text{ year}^{-1}, \quad (7.3.22)$$

where we use formula (7.3.17) for the rate of change of the global temperature. As a result, one can obtain

$$\frac{d \ln \eta}{dt} = \frac{d \ln c(H_2O)}{dt} - \frac{\varepsilon_o}{T^2}\frac{dT}{dt} = 0.002 \text{ year}^{-1}, \quad (7.3.23)$$

Evidently, this character of moisture evolution continues until η is below one. From this, one can estimate a typical time $\tau(H_2O)$ of this character of an increase of the concentration of water molecules in the atmosphere

$$\tau(H_2O) \sim 100 \text{ year} \quad (7.3.24)$$

Alongside with the change of the concentrations of H_2O and CO_2 molecules, the global temperature varies in the processes under consideration. We now evaluate the change in the radiative flux from the atmosphere to the Earth under real conditions. Let us evaluate the change in the relative flux $J_\downarrow$ from the atmosphere to the Earth with accounting for both factors. We have

$$\frac{dJ_\downarrow}{dt} = \frac{\partial J_\downarrow}{\partial T_E}\frac{d\Delta T}{dt} + \frac{\partial J_\downarrow}{\partial \ln c(H_2O)}\frac{d \ln c(H_2O)}{dt} + \frac{\partial J_\downarrow}{\partial \ln c(CO_2)}\frac{d \ln c(CO_2)}{dt} \tag{7.3.25}$$

On the basis of formula (7.3.3) and the above evaluations, we have

$$\frac{dJ_\downarrow}{dt} = 0.1\frac{W}{m^2\,year} \tag{7.3.26}$$

The first channel corresponds to the change of the global temperature that causes a redistribution of molecules over excited states. One can see that the contribution of this channel of an increase in the radiative flux to the Earth is twice compared to another one which relates to a change of the concentration of H_2O and CO_2 molecules.

References

1. G. Kirchhoff, R. Bunsen, Annalen der Physik und. Chemie **109**, 275 (1860)
2. https://www.esrl.noaa.gov/gmd/ccgg/trends-ch4/
3. http://www.physicalgeography.net/fundamentals/7h.html
4. https://en.wikipedia.org/wiki/Atmospheric-methane
5. https://www.nationalgeographic.com/environment/article/methane
6. http://www.soest.hawaii.edu/mguidry/Unnamed-Site-2/Chapter residence time of 8 years
7. T.F. Stocker, D. Qin, G.-K. Plattner et al., IPCC. (New York, Cambridge University Press, 2013)
8. L.M. Sverdlov, M.A. Kovner, E.P. Krainov, *Vibrational Spectra of Polyatomic Molecules* (Wiley, New York, 1974)
9. W.B. Person, G. Zerbi (ed.), *Vibrational Intensities* (Amsterdam, Elsevier, 1980)
10. A.A. Radzig, B.M. Smirnov, *Reference Data on Atoms, Molecules, and Ions* (Springer, Berlin, 1985)
11. S.V. Khristenko, A.I. Maslov, V.P. Shevelko, *Molecules and Their Spectroscopic Properties* (Springer, Berlin, 1998)
12. http://www.hitran.org/links/docs/definitions-and-units/
13. https://www.cfa.harvard.edu/
14. http://www.hitran.iao.ru/home
15. *U.S. Standard Atmosphere.* (Washington, U.S. Government Printing Office, 1976)
16. B.M. Smirnov, *Transport of Infrared Atmospheric Radiation* (de Gruyter, Berlin, 2020)
17. D. Pierotti, A. Rasmussen, J. Geophys. Res. **82**, 5823 (1977)
18. B.D. Hall, G.S. Dutton, J.W. Elkins, J. Geophys. Res. **112**, D09305 (2007)
19. T. Machida et al., Geophys. Res. Lett. **22**, 2921 (1995)
20. G. Herzberg, *Molecular Spectra and Molecular Structure: Electronic Spectra and Electronic Structure of Polyatomic Molecules* (Van Nostrand, New York, 1966)
21. G. Herzberg, *Molecular Spectra and Molecular Structure: Infrared and Raman Spectra of Polyatomic Molecules* (Malabar, Florida, Krieger, 1991)
22. J.H. Seinfeld, S.N. Pandis, *Atmospheric Chemistry and Physics* (Wiley, New York, 1998)
23. J.H. Seinfeld, S.N. Pandis, *Atmospheric Chemistry and Physics* (Wiley, Hoboken, 2006)
24. M.L. Salby, *Physics of the Atmosphere and Climate* (Cambridge University Press, Cambridge, 2012)
25. https://en.wikipedia.org/wiki/Ozone-layer
26. https://en.wikipedia.org/wiki/Ozone

27. D.A. Zhilyaev, B.M. Smirnov, JETP **133** (2021)
28. B.M. Smirnov, J. Phys. D. Appl. Phys. **51**, 214004(2018) §6
29. G.R. North, K.-Y. Kim, *Energy Balance Climate Models* (Weinheim, Wiley, 2017)
30. B.M. Smirnov, Int. Rev. At. Mol. Phys. **10**, 39 (2019)
31. B.M. Smirnov, J. Atmos. Sci. Res. **2**, N4, 21 (2019)
32. D. Archer, R. Pierrehumbert (Ed.) *The Warming Papers* (Oxford, Wiley-Blackwill, 2011)
33. S. Arrhenius, Phil. Mag. **41**, 237 (1896)
34. G.S. Calendar, Weather **4**, 310 (1949)
35. G.N. Plass, Tellus **8**, 141 (1956)
36. G.N. Plass, D.I. Fivel, Quant. J. Roy. Met. Soc. **81**, 48 (1956)
37. W.M. Elsasser, Phys. Rev. **54**, 126 (1938)
38. Palaeosens Project Members, Nature **491**, 683 (2012)
39. L.B. Stap, P. Köhler, G. Lohmann, Earth Syst. Dynam. **10**, 333 (2019)
40. Intergovernmental panel on climate change. Nature **501**, 297;298 (2013). http://www.ipcc.ch/pdf/assessment?report/ar5/wg1/WGIAR5-SPM-brochure-en.pdf
41. J.T. Fasullo, K.E. Trenberth, Science **338**, 792 (2012)
42. N. Andronova, M.E. Schlesinger, J. Geophys. Res. **106**, D22605 (2001)
43. M.A. Snyder, J.L. Bell, L.C. Sloan, Geophys. Res. Lett. **29**, 014431 (2002)
44. J.D. Annan, J.C. Hargreaves, Geophys. Res. Lett. **33**, L06704 (2006)
45. A. Ganopolski, T. Schneider von Deimling, Geophys. Res. Lett. **35**, L23703 (2008)
46. M.E. Walter, Not. Amer. Mat. Soc. **57**, 1278 (2010)
47. A. Schmittner, N.M. Urban, J.D. Shakun et al, Science **334**, 1385 (2011)
48. B.M. Smirnov, EPL **114**, 24005 (2016)
49. B.M. Smirnov, *Microphysics of Atmospheric Phenomena* (Switzerland, Springer Atmospheric Series, 2017)
50. V.P. Krainov, B.M. Smirnov, *Atomic and Molecular Radiative Processes* (Switzerland, Springer Nature, 2019)
51. B.M. Smirnov, High Temp. **57**, 609(2019)+++++ §8

Chapter 8
Emission of Atmospheric Particles

Abstract The absorption cross section of infrared radiation by a water microdroplet is determined on the basis of the Mie model according to which a droplet is characterized by a sharp boundary. The mass of atmospheric condensed water in clouds in the form of water microdroplets follows from the analysis of passage of infrared radiation through the atmosphere, as well as from scattering of solar radiation by the atmosphere on the basis of the Fresnel theory. The total mass of water microdroplets per an atmospheric column is estimated as $\sim 10\,\mathrm{mg/cm^2}$ with the contribution of the same order of magnitude from cumulus clouds and rare clouds, though cumulus clouds cover a small part over the Earth's surface. Aerosols are particles of various chemical types of micron and submicron sizes which are suspended in atmospheric air. Aerosols affect an environment in chemical processes in such atmospheric phenomenon as a smog and also they are responsible partially for absorption and reflection of solar radiation. Aerosols of the specific mass of the order of $0.1\,\mathrm{mg/cm^2}$ are able to change remarkably the global temperature. If they are injected in the stratosphere as a result of volcano eruptions or nuclear explosions, they are capable to change the global temperature during several years.

8.1 Radiative Processes Involving Atmospheric Condensed Water

8.1.1 Emission of Cloud Microdroplets

Water microdroplets or microparticles which constitute the clouds are the third greenhouse component of the atmosphere, and below we consider radiative atmospheric processes involving condensed water. Since the temperature of atmospheric air is of the order of room one, emission in the infrared spectrum range subjects to the Wien law in the rough approximation. Indeed, the maximum of blackbody radiation at the temperature T according to the Wien law has the form [1]

$$\lambda T = 0.29\,\mathrm{cm} \cdot K$$

B. M. Smirnov, *Global Energetics of the Atmosphere*, Springer Atmospheric Sciences, https://doi.org/10.1007/978-3-030-90008-3_8

For room temperature this corresponds to a typical wavelength $\lambda \sim 10\,\mu$m. In the case of water molecules, an effective emission results from transitions between vibration–rotation and rotation states of molecules.

Because the individuality of molecules is conserved partially in the condensed water phase, the spectrum of emission of individual molecules may be conserved in the condensed water phase. We focus only on the liquid phase state of water because the basic part of condensed phase in atmosphere relates to liquid water microdroplets. Therefore, below we consider radiative transitions in the infrared spectrum range for liquid water microdroplets as the basic condensed component in the atmosphere.

We start from a free water molecule and present the infrared water spectrum of a free water molecule [2, 3]. The water molecule has the C_{2v}—symmetry. This means that the H_2O molecule consists of two segments OH with the length of 0.957 Å between atoms O and H, and the angle between two segments OH is equal 104.5^o for the ground molecule state. The structure of the H_2O molecule leads to three vibration types of the H_2O molecule. The first one, the symmetric vibration, results from motion of hydrogen atoms in identical directions with respect to the oxygen atom. The frequency of this vibration is $\nu_1 = 3657\,\mathrm{cm}^{-1}$ (the wavelength is $\lambda_1 = 2.734\,\mu$m).

In the antisymmetric vibration state hydrogen atoms move in opposite directions with respect to the oxygen atom, and its frequency is equal $\nu_2 = 3756\,\mathrm{cm}^{-1}$ (the wavelength is $\lambda_2 = 2.662\,\mu$m). The lengths of OH bonds do not vary in the third vibration with change of an angle between segments. The frequency of this torsion vibration state is equal $\nu_3 = 1595\,\mathrm{cm}^{-1}$ (the wavelength is $\lambda_3 = 6.269\,\mu$m). Emission created as a result of these vibration transitions is not of importance for thermal radiation of the atmosphere because these vibration transitions correspond to far wings of Earth's thermal radiation. These vibrations respect to an infrared spectrum range which is of importance for a wet gas at high temperatures.

The dipole moment of a water molecule is equal $1.85ea_o$ [4, 5], where e is the electron charge, a_o is the Bohr radius. This means that the electron charge is redistributed between hydrogen and oxygen atoms. Due to the electric dipole moment, radiative transitions as a result of the change of rotation states in water molecules are effective. Rotational constants of the H_2O molecule are $27.9\,\mathrm{cm}^{-1}$, $14.5\,\mathrm{cm}^{-1}$ and $9.3\,\mathrm{cm}^{-1}$, respectively.

If a weak signal penetrates from a vacuum into a uniform mater through a flat boundary, its intensity decreases with an increase of a distance z from the boundary. According to the Beer–Lambert law [6, 7] the intensity $J(z)$ of an electromagnet wave which propagates perpendicularly to the flat boundary at a distance z from it is given by

$$J_\omega(z) = J_o \exp(-k_\omega z) = J_o \exp(-z/\delta), \qquad (8.1.1)$$

where J_o is the intensity at the boundary, k_ω is the absorption coefficient, and δ is the penetration depth for an electromagnet wave.

Interaction of an electromagnet wave with a matter proceeds through displacement of electric dipoles. Liquid water is a weak electrolyte with ions H_3O^+ and OH^-. Therefore, along with a braked rotation of H_2O molecules inside liquid water,

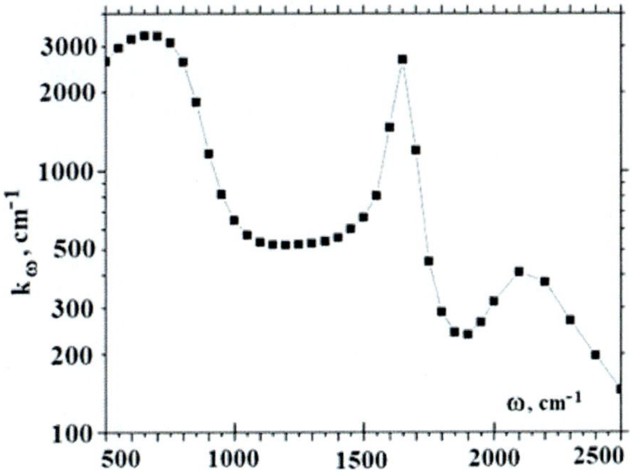

Fig. 8.1 Absorption coefficient for liquid water in the infrared spectrum range according to [12]

absorption of an infrared radiation may be resulted from displacement of the above ions inside liquid water. Hence, the atomic structure of water is of importance for interaction of an electromagnet wave with liquid water [8–11].

Figures 8.1 and 8.2 represent the dependence of the absorption coefficient k_ω for liquid water on the frequency or on the wavelength for different scales of wavelengths. One can see the resonance structure of the absorption coefficient. Moreover, the resonance at the photon frequency about $1600\,\mathrm{cm}^{-1}$ in Fig. 8.1 corresponds to the torsion vibration inside water molecules. Evidently, the next resonance of Fig. 8.1 corresponds to the torsion vibration for ions or radicals inside liquid water. We indicate in Fig. 8.2 the vibrations of free water molecules inside liquid water. In addition, the penetration depth δ of liquid water varies in the frequency range under consideration from $3\,\mu\mathrm{m}$ in the frequency range $(600 - 800)\,\mathrm{cm}^{-1}$ up $20\,\mu\mathrm{m}$ in the frequency range $(1000 - 1500)\,\mathrm{cm}^{-1}$.

Evidently, radiative transitions in liquid water for an infrared spectrum range below $1000\,\mathrm{cm}^{-1}$ are determined by braked rotations or electric interaction between neighboring molecules of liquid water. In this frequency range such an interaction is strong and leads to an effective absorption of an electromagnetic wave by liquid water. On contrary, there is the transparency window of liquid water in the visible range of frequencies, as it follows from Fig. 8.2. In the scale of this figure, the penetration depth in liquid water varies from of the order of 5 m for the visible spectrum range up to of the order of $10\,\mu\mathrm{m}$ for the infrared spectrum range. Transferring this to water droplets, one can obtain from this that microdroplets are opaque for infrared photons and are transparent for visible ones.

From this the Twomey effect [16, 17] follows according to which aerosols, as submicron particles in the atmosphere, are nuclei of condensation for formation and growth of water microdroplets in the atmosphere. Being distributed over all microdroplet, a bound water molecules provide a more strong scattering in the visible spectrum range compared with the case when they are free. In this manner, microdroplets provide a more strong scattering of visible light in the atmosphere. Because

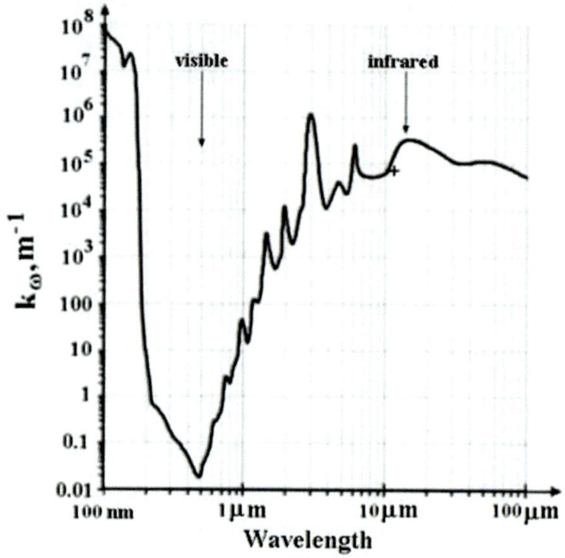

Fig. 8.2 Absorption coefficient for liquid water under normal conditions [13, 14]. Arrows indicate the visible and infrared spectrum ranges, and the cross refers to the measurement [15] for the cirrus-cumulus cloud

clouds consist of water microdroplets, they are transparent for visible light until they are poor, and are opaque for infrared radiation. But if clouds absorb admixture molecules or dust of a small concentration (their mass is less than $\sim 1\%$ of the water mass), clouds become visible. This physical fact explains an observed phenomenon, when clouds suddenly flare up among a clear sky. In reality, clouds consisting of water microdroplets arise below an observation time, but they consisted of pure water. Absorbing dust or other nontransparent particles, water microdroplets become visible.

8.1.2 *Absorption Cross Section of Water Microdroplets*

In analyzing the infrared atmospheric emission, we find above that radiation of clouds gives a remarkable contribution to the total radiative flux from the atmosphere. In determination of this contribution, we were based on the energetic balance of the Earth and atmosphere and used the model of a sharp cloud boundary. Within the framework of this model, the optical thickness of clouds is large, as well as its gradient. Because of the thermodynamic equilibrium between microdroplets of clouds, air molecules and the radiation field, emission of clouds is characterized by the temperature of the boundary cloud layer. In other words, a cloud layer depth which is responsible for cloud emission is small compared to an altitude range where the atmosphere temperature varies significantly. Under these conditions, clouds emit as a blackbody with the temperature of air where they are located. Other parameters of cloud water microdroplets are not important for the radiative fluxes created by the atmosphere in the infrared spectral range.

Considering scattering of electromagnetic waves on water droplets within the framework of the Mie theory [18] in the standard method [19–21], we assume the magnetic field of the wave to be relatively small because of a small water conductivity. Indeed, for the stationary electric field and the temperature of liquid water of 25 °C, its specific resistivity is equal $18 M\Omega \cdot \mathrm{cm}$ [22, 23] that corresponds to the conductivity $\Sigma = 5 \times 10^6\,\mathrm{s}^{-1}$. Since for the infrared spectrum range we deal with frequencies $\omega \sim 10^{14}\,\mathrm{s}^{-1}$, the criterion

$$\Sigma \ll \omega \tag{8.1.2}$$

holds true. This criterion allows one to neglect a magnetic field in the course of scattering of electromagnetic wave on a water microdroplet.

Let us apply the Mie theory for a water droplet of a small radius r

$$r \ll \lambda, \tag{8.1.3}$$

where λ is the wavelength of an electromagnet wave. Under the criterion (8.1.3), the absorption cross section σ_{abs} for a spherical drop and the scattering cross section σ_{sc} of an electromagnetic wave depend on the drop radius as [24]

$$\sigma_{abs} \sim r^3, \ \sigma_{sc} \sim r^6 \tag{8.1.4}$$

The Mie theory implies a sharp boundary for a water microdroplet located in atmospheric air. Then assuming the absence of interaction of an incident electromagnet wave with air and connecting the electric field strength inside the microdroplet with its value at the boundary, one can express the scattering cross section through electric parameter of liquid water as a microdroplet material at the wave frequency. As a result, one can obtain for the absorption cross section of a small particle through its dielectric constant $\varepsilon(\omega)$ at a given frequency ω as [24, 25]

$$\sigma_{abs} = \frac{12\pi\omega r^3}{c} \cdot \frac{\varepsilon''}{(\varepsilon' + 2)^2 + (\varepsilon'')^2}, \tag{8.1.5}$$

where the dielectric constant $\epsilon(\omega)$ at a given frequency has the complex form

$$\epsilon = \varepsilon'(\omega) + i\varepsilon''(\omega) \tag{8.1.6}$$

Then one can express the absorption cross section (8.1.5) through parameters of a plane electromagnet wave. One can use the dependence of an electric field strength for a monochromatic electromagnet wave on time t and coordinate $\mathbf{R}$ which has the form

$$\mathbf{E} = \mathbf{E}_o \exp(-i\omega t + i\mathbf{k}\mathbf{R}),$$

where $\mathbf{E}_o$ does not depend on time and coordinates. The wave vector k of this wave is given by

$$k = \sqrt{\epsilon} \cdot \frac{\omega}{c} = (n + i\kappa)\frac{\omega}{c} \quad (8.1.7)$$

This allows one to express the dielectric constant of liquid water in formula (8.1.5) through the refractive index n of bulk water. Accounting for the complex form of the refractive index $n + i\kappa$, where κ is the extinction coefficient. One can represent the connection between the refractive coefficient of a bulk matter with its dielectric constant as [24, 26]

$$n^2 = \frac{\sqrt{(\varepsilon')^2 + (\varepsilon")^2} + \epsilon'}{2}, \ \kappa^2 = \frac{\sqrt{(\varepsilon')^2 + (\varepsilon")^2} - \varepsilon'}{2}, \quad (8.1.8)$$

The inverse relation has the form

$$\varepsilon' = n^2 - \kappa^2, \ \varepsilon" = 2n\kappa \quad (8.1.9)$$

From this we have for the following connection between the absorption k_ω and attenuation κ_ω coefficients

$$k_\omega = \frac{4\pi\kappa}{\lambda}, \quad (8.1.10)$$

where λ is the wavelength. On the basis of the Beer–Lambert law (8.1.1) we have for the penetration depth

$$\delta = \frac{1}{k_\omega} = \frac{\lambda}{4\pi\kappa_\omega}, \quad (8.1.11)$$

The frequency dependencies for the refractive index n and attenuation coefficient κ_ω were measured in [12, 27–34]. These dependencies in the infrared frequency range are given in Fig. 8.3.

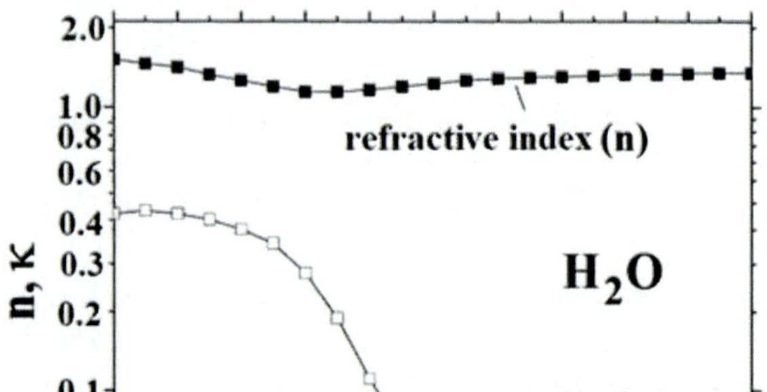

Fig. 8.3 Refractive index (n) and the extinction coefficient (κ) for liquid water in the infrared spectral range according to [12]

$$\sigma_{abs} = \frac{12\pi\omega r^3}{c} \cdot \frac{\varepsilon''}{(\varepsilon'+2)^2 + (\varepsilon'')^2}, \tag{8.1.12}$$

In considering another limiting case for the droplet size compared with (8.1.3), we assume a strong interaction of an electromagnetic wave with liquid water of a droplet. Then the absorption cross section for a large droplet is equal to its geometrical section

$$\sigma_{abs} = \pi r^2, \ r \gg \lambda \tag{8.1.13}$$

This corresponds to a blackbody model for an absorbing droplet. Combining formulas (8.1.5) and (8.1.13), one can represent the absorption cross section for all droplet sizes under the above assumptions in the following form

$$\sigma_{abs}(\omega) = \frac{\pi r^2}{1 + C(\omega)\frac{\lambda}{r}} \tag{8.1.14}$$

and the parameter $C(\omega)$ in formula (8.1.14) is given by

$$C(\omega) = \frac{(\epsilon'+2)^2 + (\epsilon'')^2}{24\pi\epsilon'} = \frac{(n^2+\kappa^2+2)^2 + 4n^2\kappa^2}{48\pi n\kappa} \tag{8.1.15}$$

Figure 8.4 contains the frequency dependence for parameters $C(\lambda)$ and $\lambda C(\lambda)$ in formula (8.1.14). As it follows from Fig. 8.4a, the transition from small droplet radii to large ones proceed in a wide range of radii from 3 μm to 16 μm depending on the frequency.

From this one can construct the absorption cross section (8.1.14) for a given droplet size as a frequency function. According to data of Fig. 8.5, this cross section depends strongly on a droplet radius. It is more convenient to operate with the specific absorption cross sections per unit mass. Indeed, the droplet mass m is equal

$$m = 4\pi r^3 \rho/3,$$

where $\rho = 1\,\mathrm{g/cm^3}$ is the mass density of liquid water. From this on the basis of formula (8.1.14) one can find the specific cross section σ_ω/m that is the ratio of the absorption cross section σ_ω to the droplet mass

$$\frac{\sigma_\omega}{m} = \frac{3}{4\rho[r + \lambda C(\omega)]} \tag{8.1.16}$$

Figure 8.6 gives the frequency dependence for the specific absorption cross section of an electromagnet wave by water microdroplets in the infrared frequency range.

Figure 8.6 contains also the result of the experiment [15] for stratocumulus clouds. In this experiment, the water content in the atmosphere ranges as $(0.02 - 0.3) g/m^3$ and the wavelength of radiation was $\lambda = (10 - 12)$ μm. According to the indicated measurements, the average specific cross section was 765 $\mathrm{m^2/g}$, but the range of

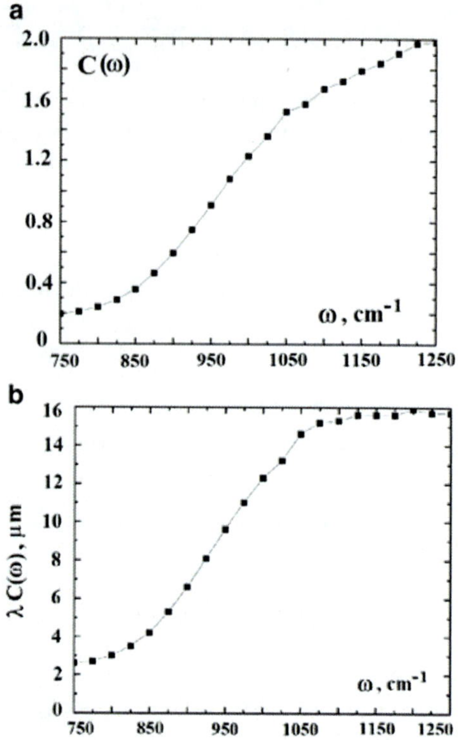

Fig. 8.4 Frequency dependence for the parameter $C(\omega)$ (**a**) according to formula (8.1.15) and such a dependence for the parameter $\lambda C(\omega)$ (**b**)

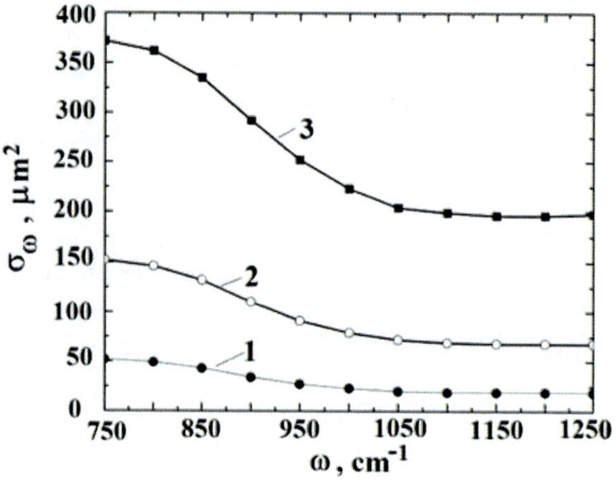

Fig. 8.5 Absorption cross section of a water microdroplet according to formula (8.1.14) for different radii r: 1- $r = 5\,\mu$m, 2- $r = 8\,\mu$m, 3- $r = 12\,\mu$m [35]

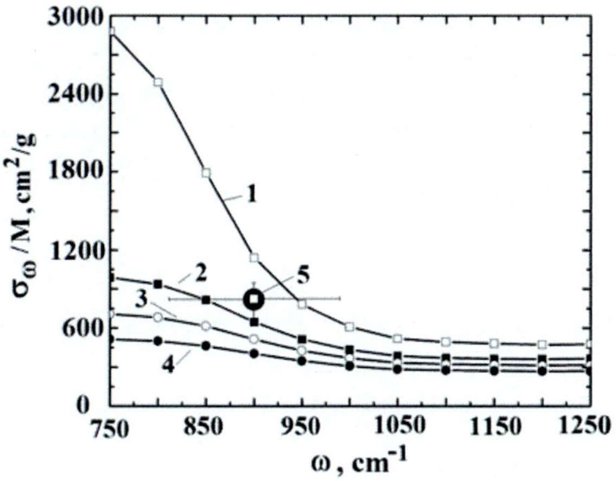

Fig. 8.6 Specific absorption cross section by a liquid water droplet according to formulas (8.1.14), (8.1.6), if a cloud consists of droplets of an identical indicated radius r: 1—small radius, 2—$r = 5\,\mu$m, 3—$r = 8\,\mu$m, 4—$r = 12\,\mu$m, 5—experiment [15] for stratocumulus clouds

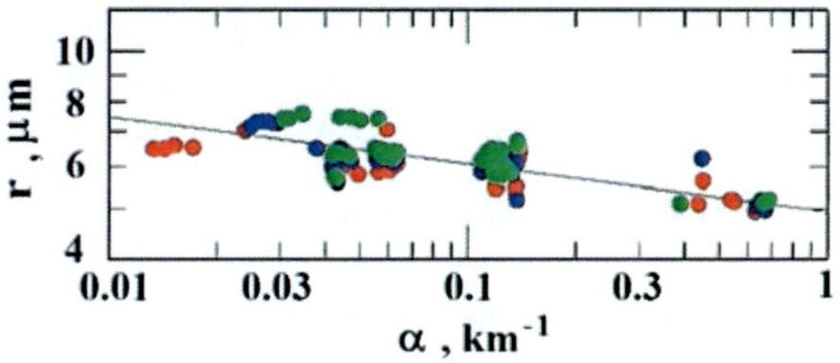

Fig. 8.7 Dependence of an effective radius r of water microdroplets of clouds on the aerosol extinction coefficient α in measurements on the basis of a Raman lidar and a microwave radiometer. In these measurements aerosols of nanometer sizes attach to microdroplets and determine the absorption in a visible spectral range where water is transparent [36]

location of this value was $(700 - 1000)\,\mathrm{m}^2/\mathrm{g}$. From comparison of the result of this experiment with formula (8.1.16), one can obtain for the radius of microdroplets which constitute a stratocumulus cloud in this experiment. From this one can obtain an average radius of water microdroplets in stratocumulus clouds

$$r = (4 \pm 1)\,\mu\mathrm{m} \tag{8.1.17}$$

As is seen, microdrops of stratocumulus clouds are less than those of cumulus clouds with the average radius $r = 8\,\mu$m according to formula (2.2.12).

In considering a size of cloud droplets, we note that this size depends on the density of atmospheric water where these droplets grow. Roughly speaking, the higher water density in the atmosphere, the larger is an average size of water microdroplets which form the clouds. In particular, we give in Fig. 8.7 dependence of the effective radius of microdroplets on the extinction coefficient of aerosols density of atmospheric water. In these measurements, ammonium sulfate aerosols are used. Being injected in the atmosphere, these aerosols are absorbed by water microdroplets, dissolve in them and are responsible for absorption of light by water microdroplets. This method allows one to determine the parameters of water microdroplets in the atmosphere on the basis of light measurements.

8.1.3 *Passing of Radiative Flux from Earth Through Clouds*

The energetic balance of the Earth given in Table 5.1 contains various information about atmospheric properties. We below use the radiative flux J_p passed through the atmosphere. Because absorption by optically active molecules may be evaluated on the basis of data of the HITRAN data bank, this allows us to determine absorption by clouds. In this manner one can determine the amount of water microdroplets over regions of the Earth's surface where radiation emitted by it goes outside the atmosphere. We assume the specific cloud mass per unit area of the Earth's surface to be identical for the globe and this part of the surface is the main part of the Earth's surface. According to data of Table 5.1, the average radiative flux which emits by the Earth's surface and goes outside the atmosphere is equal

$$J_p = (21 \pm 1)\,\mathrm{W/m^2} \tag{8.1.18}$$

We now determine the probability of passing through the atmosphere for thermal emission emitted by the Earth's surface. We assume clouds to be uniform over the surface and it includes the basic part of the Earth's surface.

Let us introduce the water mass ϱ per unit area of the atmospheric column that is contained in microdroplets. The optical thickness u_ω of the atmosphere with respect to water microdroplets is

$$u_\omega = \varrho \frac{\sigma_{abs}(\omega)}{nm_o}, \tag{8.1.19}$$

where ϱ is the mass density per an atmospheric column which is contained in water microdroplets of clouds and measured in mg/cm^2, σ_{abs} is the absorption cross section for one microdroplet, n is the number of water molecules per one microdroplet, $m_o = 3 \times 10^{-23}$ g is the mass of the water molecule. As it follows from Fig. 8.8, the main contribution to the outgoing radiative flux follows from the frequency range $(800 - 1400)\,\mathrm{cm^{-1}}$.

The radiative flux emitted by the standard atmosphere, that is a blackbody with the temperature $T_E = 288$ K, is equal according to the Stephane–Boltzmann law

$$J_E = \sigma T_E^4 = 390\,\mathrm{W/m^2}$$

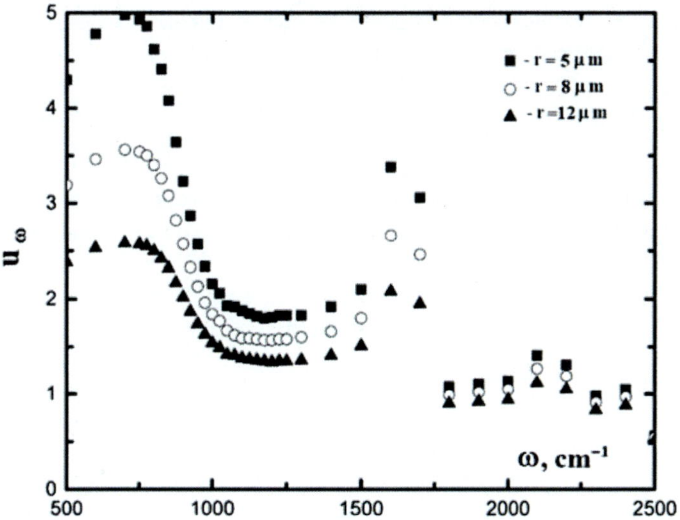

Fig. 8.8 Frequency dependence for the optical thickness in the infrared spectrum range for clouds consisting of water microdroplets of an indicated size if the total mass density of water microdroplets of an atmospheric column is 5 mg/cm^2 [35]

From this one can estimate the probability P for thermal radiation which passes through the atmosphere

$$P = \frac{J_p}{J_E} \approx 5\% \tag{8.1.20}$$

One can estimate also the probability of passing of radiative flux through clouds. Because the flux of $J_c \approx 120\,\mathrm{W/m^2}$ reaches the lower boundary of clouds, the part

$$p = \frac{J_p}{J_c} \approx 18\% \tag{8.1.21}$$

which attains clouds passes through them.

Since the optical thickness of clouds according to formula (8.1.19) is proportional to the density of atmospheric water, one can determine from this the average water density ϱ_o per unit area of the atmospheric column that is contained in microdroplets. In this analysis we take into account the distribution over the specific densities ϱ of clouds. For simplicity, we take the distribution function $f(\varrho)$ over specific densities ϱ in the form

$$f(\varrho) = \frac{1}{\varrho_o} \exp\left(-\frac{\varrho}{\varrho_o}\right), \tag{8.1.22}$$

where ϱ_o is the average specific density of condensed atmospheric water, and the distribution function satisfies to the normalization condition

$$\int f(\varrho)d\varrho = 1$$

Note that for large microdroplets $r \ll \lambda$, where the absorption cross section is independent of the frequency, the distribution function corresponds to that over the optical thicknesses.

In a general case, the outgoing radiative flux resulted from that emitted by the Earth's surface, in accordance with formula (8.1.19) is given by

$$J_p = \int_0^\infty I_E(\omega) g\left[u_\omega + \varrho\frac{\sigma_{abs}(\omega)}{m_o}\right]\frac{1}{\varrho_o}\left(-\frac{\varrho}{\varrho_o}\right)d\omega d\varrho$$

Here u_ω is the optical thickness of the atmospheric gap between the Earth's surface and clouds, and we take the total optical thickness of the atmosphere as the sum of those due to atmospheric molecules u_ω and due to clouds. If we use the approximation $g(u_\omega) = \exp(-1.6u_\omega)$, which takes place for $u_\omega \sim 1$ [35], one can obtain from this

$$J_p = \int_0^\infty \frac{I_E(\omega)\exp(-1.6u_\omega)d\omega}{1 + 1.6\sigma_\omega \varrho_o/m_o} \tag{8.1.23}$$

Figure 8.9 contains the dependence of the outgoing radiative flux on the specific mass density of condensed water which is located in the atmosphere in the form of

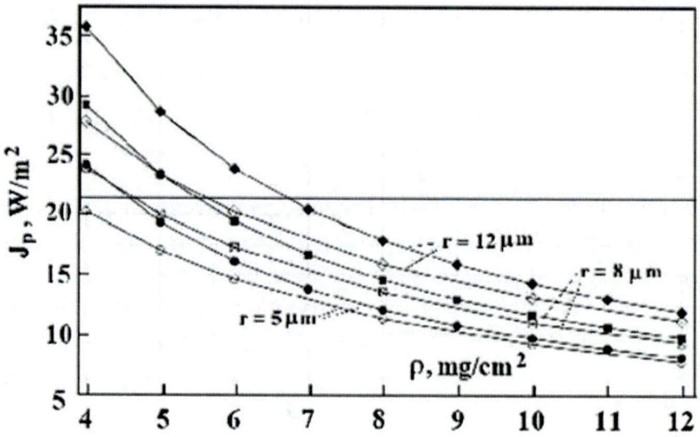

Fig. 8.9 Radiative flux which is formed at the Earth's surface and passes through the atmosphere as a function of the water mass in clouds per unit atmospheric column. Clouds consist of identical microdroplets of an indicated radius. Filled signs—calculations [35], open signs corresponds to results of [37]. Solid curve corresponds to Table 5.1 or formula (8.1.18)

microdroplets of the same radius. Sizes of water microdroplets taken in Fig. 8.9 are typical for clouds. Accepting the outgoing radiative flux in accordance with formula (8.1.18), one can find the average mass density of atmospheric condensed water in clouds

$$\varrho_o = (5 \pm 1)\,\mathrm{mg/cm^2} \tag{8.1.24}$$

According to this formula, the mass of condensed water in the atmosphere is approximately of 0.2% of the mass of atmospheric water in the form of free molecules.

It should be noted that in this consideration we are guided by globe regions with a small density of clouds. In reality, a larger density of water microdroplet corresponds to cumulus clouds which are responsible for atmospheric electricity. These clouds cover a small area over the Earth's surface and therefore give a small contribution to the passed radiative flux from the ground. But these clouds can give the contribution to the amount of condensed water in the atmosphere. For this reason, formula (8.1.24) may be considered as a lower limit for the amount of atmospheric condensed water.

8.1.4 Scattering of Visible Radiation on Water Droplets

The energetic balance of the Earth and atmosphere represented in Table 5.1 is connected with some processes in the atmosphere and its properties. As a result, this allowed us in the previous paragraph to determine the amount of atmospheric condensed water on the basis of the value of the flux of infrared radiation through the atmosphere. We now consider the connection between the amount of atmospheric condensed water and the flux of reflected solar radiation from the atmosphere. Because the basic part of atmospheric condensed water is located in the form of liquid microdroplets, this connection results from scattering of solar radiation on water microdroplets.

In considering this process for an individual droplet, we assume that a droplet is separated from air by a thin shell, where in accordance with the Mie theory the radiative transition involving a water microdroplet located in atmospheric air takes place in a narrow space region. In addition, a typical droplet size $\sim 10\,\mu\mathrm{m}$ exceeds significantly a wavelength of the electromagnetic wave that is $(0.4-0.7)\,\mu\mathrm{m}$ for visible radiation. This allows one to use the principles of optics [38, 39]. Then the trajectory of a beam of photons consists of straightforward lines for each matter, and the scattering character is determined by the Fresnel formulas [40, 41].

The geometry of light scattering on a spherical droplet is given in Fig. 8.10 in the case if the scattering intensity is small. The scattering takes place twice in the course of intersection of the droplet boundary by an electromagnet wave whose trajectory is described now by a straightforward line. Scattering at each intersection is characterized by the reflection coefficient R which is the ratio of the intensity of a scattered beam to an incident one. The reflection coefficient depends on the wave polarization [19, 24, 42]. Because Fresnel formulas are cumbersome, and

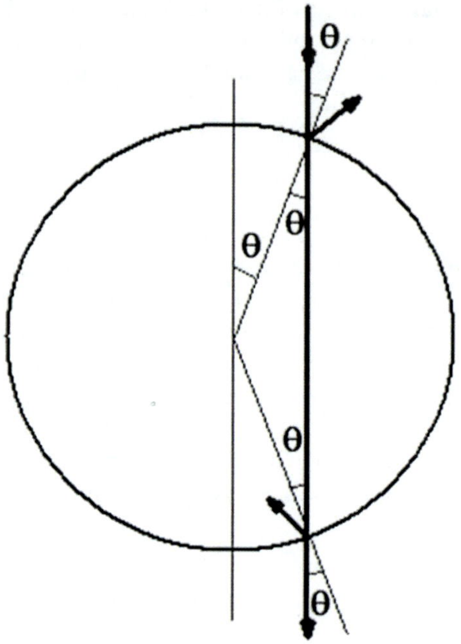

Fig. 8.10 Character of propagation of an electromagnet wave through a spherical droplet with a sharp boundary in a case of weak scattering. θ is the incident angle and the angle between a radius-vector and scattering wave

the final result is an estimation, below we restrict by the case where the electric field strength of an incident wave is perpendicular to air-water boundary. Then the reflection coefficients as the ratio of intensities of reflected and incident waves are given by [19, 24]

$$R_{aw}(\theta) = \left[\frac{\sqrt{\epsilon}\cos\theta - \sqrt{1-\epsilon\sin^2\theta}}{\sqrt{\epsilon}\cos\theta + \sqrt{1-\epsilon\sin^2\theta}}\right]^2, \quad R_{wa}(\theta) = \left[\frac{\cos\theta - \sqrt{\epsilon - \sin^2\theta}}{\cos\theta + \sqrt{\epsilon - \sin^2\theta}}\right]^2, \tag{8.1.25}$$

Here $R_{aw}(\theta)$ is the reflection coefficient for entrance of an incident electromagnet wave from air into liquid water from air, $R_{wa}(\theta)$ is the reflection coefficient for exit of a wave from water into air, and ϵ is the dielectric constant of water at an electromagnet wave frequency, the air dielectric constant is one, and θ is the angle between the direction of an incident wave and a normal to the droplet surface at a point of entering of the electromagnetic wave inside the droplet. We note that because we consider the reflection coefficient to be small, a direction of the electromagnet wave do not vary in the course of its propagation through the water droplet, i.e., an angle θ is identical for an entering and outgoing waves.

Being guided below by droplets consisting of liquid water, we represent the refractive index in the complex form $n + i\kappa$, and the connection between the refractive index and the dielectric constant is given by formulas (8.1.8), (8.1.9). Table 8.1 contains values of components of the refractive index in the visible spectrum range.

Table 8.1 Refractive index n, the extinction coefficient κ, and the absorption coefficient k_ω for liquid water under normal conditions as a function of the wavelength λ in the visible spectrum range [43]

λ, μm	n	κ, 10^{-8}	k_ω, 10^{-4} cm^{-1}
0.40	1.339	0.19	5.8
0.45	1.336	0.10	2.8
0.50	1.335	0.10	2.5
0.55	1.333	0.20	4.5
0.60	1.332	1.1	23
0.65	1.330	1.6	32
0.70	1.328	3.4	60

As it follows from data of Table 8.1, in the visible spectrum range one can neglect by absorption of an electromagnet wave by liquid water. In addition, the refractive index varies weakly in this spectrum range. Therefore, below we neglect by absorption and take $n = 1.33$. Correspondingly, according to formula (8.1.9) the dielectric constant of liquid water in the visible spectrum range is equal $\epsilon = 1.77$. In this case the reflection coefficient for the normal incidence ($\theta = 0$) formula (8.1.25) gives $R_1(\theta = 0) = R_2(\theta = 0) = 5.1 \times 10^{-3}$.

Let us introduce the scattering cross section σ_{sc} for a droplet of a radius r and the probability γ of radiation scattering, so that

$$\sigma_{sc} = \pi r^2 \gamma \tag{8.1.26}$$

This probability results from two events of intersection of the droplet boundary and therefore may be represented in the form

$$\gamma = \gamma_{aw} + \gamma_{wa}, ; \tag{8.1.27}$$

where γ_{aw} is the probability of beam scattering when it enters in water droplet, and γ_{wa} is the scattering probability when the beam leaves the droplet.

In accordance with formulas (8.1.25), the reflection probabilities are determined by the following expressions

$$\gamma_{aw} = \int_0^1 2x dx \left[\frac{x - \sqrt{\epsilon - 1 + x^2}}{x + \sqrt{\epsilon - 1 + x^2}}\right]^2, \quad \gamma_{wa} = \int_{x_o}^1 2x dx \left[\frac{\sqrt{\epsilon}x - \sqrt{1 - \epsilon + \epsilon x^2}}{\sqrt{\epsilon}x + \sqrt{1 - \epsilon + \epsilon x^2}}\right]^2,$$
$$x_o = \sqrt{1 - 1/\epsilon} \tag{8.1.28}$$

Here we use the variable $x = \cos\theta$ and the parameter x_o corresponds to an angle of entire reflection of the radiative flux. Taking $\epsilon = 1.77$ for water microdroplets, one can obtain from formulas (8.1.28)

Table 8.2 Average values of albedo for the atmosphere and Earth obtained on the basis of the energetic balance for the Earth and atmosphere taken from Table

Source	1	2	3	4	5	Average
Atmosphere	0.31	0.26	0.23	0.23	0.22	0.25 ± 0.04
Earth	0.08	0.09	0.12	0.12	0.13	0.11 ± 0.02

$$\gamma_{aw} = 0.10, \ \gamma_{wa} = 0.057, \ \gamma = 0.16 \tag{8.1.29}$$

Solar scattering by the Earth's surface or its atmosphere is characterized by albedo [44] α that is the part of reflected radiation. This value is zero for a blackbody and is one for a mirror. Values of albedo for certain surfaces are given in [45–48]. In particular, for a grass and forest in summer the albedo varies in the limits from 0.03 to 0.06, whereas for fresh snow it is equal to 0.9.

We below give in Table 8.2 the average values of albedo α for the Earth and its atmosphere. The basis of these values is the energetic balance of the Earth and atmosphere represented in Table 5.1. In this operation, we ignore the radiative flux reflected from the Earth and absorbed subsequently by the atmosphere. One can determine also the average value of albedo according to their definition with taking into account the data of Fig.5.4. The average radiative flux from the Sun entered in the Earth's atmosphere is $J_{sol} = 341\,\mathrm{W/m^2}$, and the reflected radiative flux from the Earth's atmosphere is equal $J_{sc} = 86\,\mathrm{W/m^2}$. This gives for the atmospheric albedo

$$\alpha = \frac{J_{sc}}{J_{sol}} = 0.25 \tag{8.1.30}$$

As is seen, the average value of albedo for the Earth's atmosphere according to formula (8.1.30) and Table 8.2 coincides.

One can connect the atmospheric albedo with its optical thickness u_ω in the case if a depletion of the radiative flux is determined by absorption, rather than by scattering. Indeed, according to the definition of the optical thickness, the probability P_ω to survive as a result of a pass through the layer

$$P_\omega = \exp(-u_\omega)$$

We assume the layer to be uniform in the horizontal direction, whereas the flux propagates in the vertical direction. One can generalize this expression to the case with an angle θ between the radiative flux and the layer surface, and then according to the definition for the albedo we have

$$\alpha = \langle \left(1 - \frac{\exp(-u_\omega)}{\cos\theta}\right) \rangle, \tag{8.1.31}$$

where triangle parentheses mean an average over frequencies. In the limit of a transparent layer this formula gives

$$\alpha = \frac{u}{\cos\theta}, \tag{8.1.32}$$

where $u = \langle u_\omega \rangle$ is the average optical thickness of the layer.

Defining the atmosphere albedo with respect to solar radiation according to formula (8.1.30), we have that the radiative flux at a given point of the Earth's surface located at the solar side of the planet is $J_{sol}\cos\theta$. Hence, the atmospheric albedo as the relative loss of the radiative flux from the Sun with respect to the incident flux is $J_{sol}u$ in the limit of a low optical thickness of the atmosphere coincides with its optical thickness on average

$$\alpha = u \tag{8.1.33}$$

Table 8.2 contains values of the average albedo for the Earth's atmosphere which follows from the energetic balance of the Earth given in Table 5.1. Evidently, formula (8.1.30) holds true in the case of reflection of an incident flux if the optical thickness of the atmosphere is defined with respect to beam depletion, rather than by formula (6.1.2) through the absorption coefficient. As is seen, the atmospheric albedo is not large and the reflection process may be analyzed on the basis of formula (8.1.33).

We now apply these results for determination of the average mass of clouds on the basis of scattering of solar radiation. One can connect the atmospheric albedo with the specific mass density of clouds consisting of water microdroplets. One can introduce the specific mass ϱ of atmospheric condensed water on the basis of formula (8.1.19). which is measured in g/cm^2. From its definition it follows for the number n_d of microdroplets per unit area of the Earth

$$n_d = \frac{\varrho}{m} = \frac{3\varrho}{4\pi r^3 \rho},$$

where m is the droplet mass, r is the radius of the microdroplet, $\rho = 1\,\text{g/cm}^3$ is the mass density of liquid water.

Under the assumption of the transparent atmosphere $u = n_d\sigma_{abs} \ll 1$, we have on the basis of formula (8.1.33)

$$\alpha = \sigma_{sc} n_d = \frac{3\gamma\varrho}{4r\rho} \tag{8.1.34}$$

where the scattering cross section is $\sigma_{sc} = \pi r^2 \gamma$ in accordance with formula (8.1.26). This allows one to express the specific mass ϱ of atmospheric condensed water through the albedo α of the atmosphere if it is determined by scattering of an electromagnet wave on microdroplets of clouds in accordance with formula (8.1.26)

$$\varrho = \frac{4\alpha r\rho}{3\gamma} \tag{8.1.35}$$

In the case, where radii of microdroplets are different, one can obtain the following expression instead of (8.1.35)

$$\varrho = \frac{m_o \alpha}{\pi \gamma r_W^3 \langle 1/r \rangle}, \tag{8.1.36}$$

where triangle parentheses means an average over droplet radii. Taking a typical droplet size $r = 8\,\mu\text{m}$ of clouds and the reflection probability $\gamma = 0.16$ for water droplets, one can obtain from formula (8.1.35) a typical specific density of condensed atmospheric water

$$\varrho = 17\,\text{mg/cm}^2 \tag{8.1.37}$$

in contrast to formula (8.1.24).

In spite of an estimation character of the above operations, the difference of ϱ values is not contradictive. Formula (8.1.24) relates to a clear atmosphere, so that infrared radiation can penetrate inside this atmosphere. This formula excludes regions with cumulus clouds which occupy a small part of sky over the Earth's surface. Cumulus clouds because of a large density of water microdroplets do not give a pass for infrared radiation. On contrary, formula (8.1.37) includes the basic part of water droplets located in the atmosphere, and cumulus droplets contain the most part of condensed water. Hence, comparing formulas (8.1.24) and (8.1.37), one can conclude that the significant part of condensed atmospheric water is concentrated in cumulus clouds.

One can add to this analysis that we assume that the atmospheric albedo is determined by condensed water as the main condensed component in the atmosphere. But the amount of atmospheric water droplets and particles is less than that of water molecules more than two orders of magnitude. Besides this, interaction of solar radiation with water microdroplets which is characterized by the parameter γ in formula (8.1.26) is weak. Hence, the assumption that water microdroplets dominate in scattering of solar radiation is questionable. Moreover, according to the total contribution of aerosols to the atmospheric albedo is approximately 0.1 [49]. From this it follows that the mass of atmospheric condensed water in the form of water microdroplets is

$$\varrho \sim 10\;\text{mg/cm}^2 \tag{8.1.38}$$

In addition, the mass of condensed water in cumulus clouds according to formula (8.1.24) is comparable with that in rare clouds which contain water microdroplets of a smaller size.

8.2 Small Particles in Atmospheric Processes

8.2.1 *Aerosols in the Earth Atmosphere*

Small particles formed in the atmosphere is named aerosols. Usually they do not include water and ice particles, i.e., aerosols are small particles other than water microdroplets which can fly in the atmosphere. Figure 8.11 represents various types of nanoparticles and microparticles located in atmospheric air together with a range of their sizes. This testifies both about a variety of small particles in the atmosphere, and various types aerosols act on different atmospheric properties. Aerosols are distributed in the atmosphere nonuniformly, as well as their occurring in the atmosphere has a random character in time. Information for aerosols and their influence on various atmospheric properties is contradictory, The goal of the analysis of aerosols now is to give a general representation about various aerosols as a physical object and to make estimations for parameters atmospheric processes under consideration with participation aerosols.

The first type of studied aerosols is Aitken particles [51–54] which were studied earlier than other aerosols. They are located at high altitudes, above clouds, and their basis are radicals of sulfur compounds which result from vaporization of meteorites at high altitudes and from processes near the Earth surface at low altitudes, including processes of ocean evaporation. The number density of Aitken particles at altitudes $10 - 20\,\mathrm{km}$ is $10^2 - 10^4\,\mathrm{cm}^{-3}$ [55]. Because of a small size (below $0.1\,\mu\mathrm{m}$), the Aitken particles are located in the atmosphere during a long time, and a low concentration prevents their from condensation. A small size of Aitken particles is also the reason of a blue sky color because shortwave photons scatter on these particles effectively.

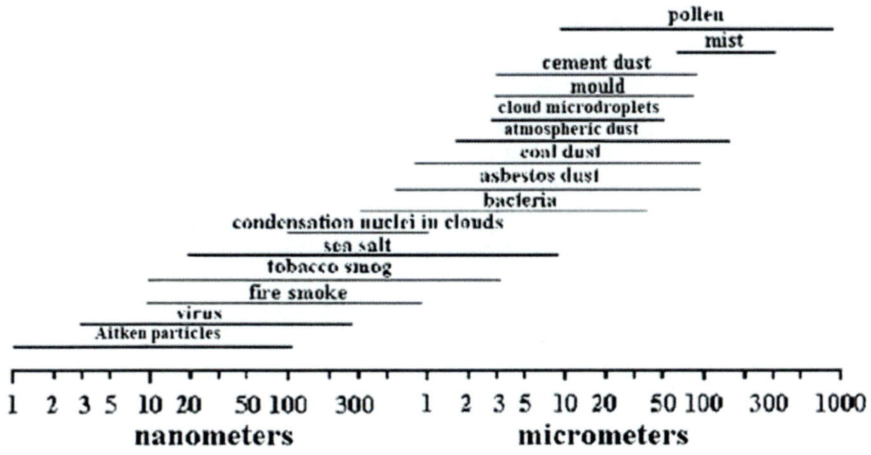

Fig. 8.11 Typical size of aerosols and microparticles located in the Earth's atmosphere [50]

Because of a chemical activity of aerosols, they may be nuclei of condensation. Indeed, under a high concentration water molecules attach to aerosols, and hence under appropriate conditions, aerosols promote formation and growth of water droplets. As optically active particles, aerosols influence optical properties of the atmosphere which may be realized as in a direct manner with absorption of solar radiation and in the indirect one, where aerosols are nuclei of condensation. The latter is named as the Twomey effect [16, 17] which enforces interaction with solar radiation. Subsequent dissolution of aerosols in water microdroplets causes absorption of light by water microdroplets, whereas this absorption is absent for pure water microdroplets.

In considering aerosols which partake in absorption of solar radiation, we divide them in several groups, as this is usually accepted. Sulfate aerosols consist of compounds of the sulfur acid. Their salts contain sodium, potassium, ammonium and other compounds as anions. Another type of aerosols, black carbon, is a soot formed as a result of the pyrolysis, where chemical bonds are formed between carbon atoms during combustion of fossil fuels. Organic aerosols are formed from organic material mostly of the biological origin, which are located in the atmosphere. Dust particles rise in the atmosphere by winds and consist mostly from SiO_2. Sea salts are formed in the atmosphere from evaporation of open water, mostly, from the surface of oceans. These types of aerosols are formed or injected in the atmosphere. They give the contribution to absorption of solar radiation by the Earth's atmosphere.

Usually the contribution of each channel to the Earth's heat balance is characterized by the so-called radiative forcing [56], which is the energy flux in the atmosphere due to this channel of the Earth's energy balance. The radiative flux from the Sun penetrated in the Earth's atmosphere is equal 1365 W/m^2. Averaging this flux over the globe surface, one can obtain the value 341 W/m^2 which is given in Table 5.1. The total radiation effect due to aerosols compared to the pre-industrial epoch, calculated on the basis of modern programs of the energy balance of the atmosphere, is $-0.5 \pm 0.4\,\mathrm{W/m^2}$ [57].

The radiation effect due to aerosols of a certain type averaged over season and geographical coordinates is $(-0.4 \pm 0.2)\,\mathrm{W/m^2}$ for sulfates, $(-0.05 \pm 0.05)\,\mathrm{W/m^2}$ for fossil organic compounds, $(0.2 \pm 0.15)\,\mathrm{W/m^2}$ for black carbon (coal and soot), $(0.03 \pm 0.12)\,\mathrm{W/m^2}$ for burned biomass, $(-0.1 \pm 0.1)\,\mathrm{W/m^2}$ for nitrates and $(-0.5 \pm 0.4)\,\mathrm{W/m^2}$ for mineral dust [58]. Leaving aside the question of the reliability of the computer programs used, we note that these results are published after the discussion at the International Panel on Climate Change (IPCC) in its proceedings [57]. In terms of the processes considered here, it is essential that, for the one hand, the uncertainty of the results is relatively large and does not allow us to reliably determine the sign of action of each component. For the other hand, the quantities themselves are small. We remark that that a temperature change of $0.1K$ corresponds to the radiation effect of $0.2\,\mathrm{W/m^2}$ according to formula (5.3.5). In addition, the temperature change of 0.1 K is the minimum temperature change that can be recorded with the existing fluctuations in the parameters of the atmosphere. Therefore, the contribution of radiative processes involving aerosols to the energy balance of the Earth and its atmosphere can be ignored.

Table 8.3 Rounded global parameters of aerosols in the Earth's atmosphere

Parameter	Sulfates	Soot	Organic	Dust	Sea salt
Injection mass, Tg/year	200	10	100	1700	4000
Atmospheric mass, Tg	2	0.2	2	20	6
Residence time, days	3	1	4	9	6
Absorption cross section at 550 nm, m^2/g	10	9	6	1	3
Optical thickness at 550 nm	0.034	0.004	0.018	0.032	0.032

Table 8.3 contains parameters of these aerosols which are associated with their injection and residence in the Earth's atmosphere. These data are taken from the USA Program of Climate Change Science of 2009 [49] and International Panel on Climate Change [57]. Values of parameters of Table 8.3 are rough, and therefore, we make rounded them. Hence, Table 8.2 gives the scales of global parameters of atmospheric aerosols. We express the total mass of aerosols in Tg (1 Tg=10^{12} g = 10^6 tons), the average cross section of absorption and scattering of light is taken at the wavelength of 550 nm, as well as the average optical thickness of the atmosphere with respect to aerosols. According to data of Table 8.3, the total optical thickness of the atmosphere with respect to aerosols is 0.13. Comparing it with the atmospheric albedo which is equal $\alpha = 0.25$ according to Table 8.2, one can conclude that aerosols give a remarkable contribution to scattering and absorption of solar radiation by the Earth's atmosphere.

Along with aerosol properties associated with absorption or scattering of solar radiation in the course of its entering in the atmosphere, their presence in the atmosphere causes a chemical action on an environment, influencing on its chemical properties. According to the character of their formation, radicals and nonsaturated chemical bonds are typical for aerosols. Indeed, chemical processes in the atmosphere proceed with participation of sulfur oxides SO_x, nitrogen NO_x oxides and atmospheric ions which resulted from electric processes in the atmosphere at the Earth surface. Nitrogen oxides and soot particles are formed in combustion processes at the Earth surface. In addition, underground processes and volcanoes are the source of some radicals which are injected in the atmosphere. Therefore, the processes involving aerosols in atmospheric air a branch of atmospheric physics and chemistry [59–63]. Note that these processes have a nonregular character, and hence the distribution of aerosols in a space and time in strongly nonuniform.

Dividing aerosols in several groups, one can keep in mind that each group includes many chemical compounds. In particular, soot which results from combustion of organic materials in the pyrolysis channel [64] that is a chemical process at high temperature involving organic material in absence or lack of oxygen. This process leads to bonding of carbon atoms and to formation various organic molecules and radicals. In particular, polycyclic aromatic hydrocarbons (PAH) [65] formed in this process are a toxic material in the atmosphere and are the most danger compound for inhabitants of the Earth. In the atmosphere, soot and other chemically active molecules or radicals are captured subsequently by water aerosols during rain.

Atmospheric water microdroplets with attached to them toxic aerosols are of a danger for the human health if they are formed in absence of winds. Penetrating in lungs together with inhaled air, these compounds can cause pneumonia, bronchitis, tuberculosis and heart failure. A risk to cancer catch follows from a long location in impure air. The most strong result of such an action of atmospheric pollution took place in a London smog in December 1952 [66]. An inverse temperature distribution was observed in that time, i.e., the temperature at the Earth's surface was below that at higher altitudes. Under such conditions, motion of air was promoted, and a mist in air with smoke from chimney was collected near the Earth's surface.

As a result, pollution from chimney was accumulated in that time in London during 5 days until the inverse temperature distribution was maintained. In that time the visibility does not exceed 500m during 5 days and 50 m during 2 days, and the amount of sulfur oxides SO_x in the atmosphere exceeded the standard value ($700 ppb$) in 7 times. Next, the mass density of particulate matter for particles of diameter below 10 μm PM_{10} was in the limits from 3000 up to 14000 $\mu g/cm^3$, whereas their content in a quiet atmosphere was approximately 30 $\mu g/cm^3$. A strong air pollution during a London smog led to tragic consequences. Namely, 4000 persons were dead during the smog and 8000 persons died after the smog. Specific conclusions were accepted in 1956. Then methane became the fuel together with electricity and smokeless coal in London. This allowed one to escape such consequences subsequently.

8.2.2 Volcanoes as a Source of Atmospheric Aerosols

Though volcano eruptions [67–69] has a local character, they are the power source of aerosols, and therefore, we consider the volcano problem separately. Therefore, they give a certain contribution to formation of atmospheric aerosols which are distributed over all atmosphere. We below consider volcano eruption as a strong natural method of generation of atmospheric aerosols. In turn, these aerosols influence the atmosphere state and also the action on people through a smoke which is formed as a result of volcano eruptions. Some examples of volcano eruptions are presented in Figs. 8.12 and 8.13. Along with an ash and smoke, volcanoes are a source of blocks and massive particles which are located in the atmosphere for a small time. Small particles of the low atmosphere are washing out by rains, while small particles located above clouds can be found there during several months [70]. In accordance with the character of volcano processes, a volcano ash contains a large amount of toxic substances, and the basic part of them are sulfur-containing substances. Therefore, a volcano ash has a negative influence on the environment nature and people.

Volcano eruptions lead to atmosphere pollution and polluted air may influence on the health and live of people. In particular, volcano eruptions proceeded in Iceland in 1783 led to the death of 20% habitants of Iceland. A toxic cloud with sulfur-containing compounds moved to Europe and led to the death of 23 thousands of persons in the

Fig. 8.12 Eruption of Grimsvotn volcano in Iceland in May 2011 [71, 72]. The eruption flow reaches up to 11 km and along with gases and small particles contains stones and fragments of rocks

Great Britain [70]. The remembering eruption of the Krakatau volcano in 1883 [73, 74] caused 36 thousands of victims. This eruption is equivalent to explosion of 200-megaton bomb that exceeds the action of the Hirocima atomic bomb in 1945 in 13,000 times. A volcano ash acts not only on persons, but also on the equipment, and can put out of operation the aircraft engine. For this reason 400 flights were canceled in May 2010 during volcano eruption in Iceland, when a volcano ash was spread over the Europe North [75].

One can compare a smoke of volcano eruption with that formed in combustion process. Combustion of forests and peatbogs creates a smoke which propagates by analogy with volcano, but the power of this process is low, and this smoke does not reach the stratosphere. Hence, this smoke is washed out by rain. In addition, along with carbon dioxide, the product of combustion of forests and peatbogs is soot formed as a result of pyrolysis, while sulfur-containing compounds or silicon compounds are the main components of volcano eruptions. A combustion smoke is characterized by narrowed sizes of formed particles, and during their residence in the atmosphere a smoke propagates on distances up to 100 km. A smoke resulted from combustion of forests and peatbogs creates a haze or mist in a daytime with a specific smell.

Apparently, the most powerful eruption of the last millennium occurred on April 10–11, 1815 from the Tambora volcano (Sumbawa Island, Zond Archipelago,

Fig. 8.13 Eruption of St-Elena volcano (north-west of USA) in 1980 [76]

Indonesia). As a result of the explosion, approximately 150 km^3 of rocks was emitted in the atmosphere, so that the surrounding areas were covered with a thick layer of ash. As a result of this eruption, the house of the governor of this colony, located 111 km from the volcano, was destroyed under the weight of ash covered the roof. The eruption of this volcano was accompanied by earthquakes, tidal waves, hurricanes and led to the death of about 80 thousand people.

However, the most memorable was the eruption of the Krakatau volcano in 1883 in the Zond strait between the islands of Java and Sumatra in Indonesia. This eruption occurred on August 27, 1883 [73, 74]. In its effect, it was equivalent to the explosion of a 200-megaton bomb, which is 13 thousand times greater than the effect of the atomic bomb dropped on Hiroshima in 1945. This eruption led to the death of 36 thousand people. Approximately 18 km^3 of rock was lifted into the air, which is eight times less than in the eruption of the Tambora volcano. But an emitted material was highly dispersed and the ash layer spread over a long distance up to 1000 km. Bright sunrises and sunsets were observed after the eruption of the Krakatoa volcano due to the ash which reached the stratosphere and was resided there.

The change in the optical properties of the atmosphere after the eruption of the Krakatau volcano led to some cooling of the planet during several months, because a part of solar radiation was reflected by emitted particles. In the course of falling of the emitted dust to the ground, this effect disappeared. Another feature of the Krakatoa eruption was the exceptionally high force of the explosion, which was heard in Australia and on the island of Rodriguez located in the Indian Ocean at a distance of almost 5000 km from the volcano. This explosion caused a huge tidal wave with a height of up to 40 m, which hit the shores of islands of Indonesia. The tidal wave from the explosion as a result of eruption of the Krakatoa volcano was recorded even in the strait English Channel.

The Krakatau eruption lead to the death of 36,417 people, where about 90% were killed by the tsunami. The remaining 10% were killed as a result of a falling volcanic debris and under the action of fast-moving masses of hot volcanic ash and gas. A huge ash cloud covered Indonesia, the Sun was not visible for three days, and the ash cloud spread to 360 km. The ash rose to such a height that had led to the complete darkness in Java and Sumatra. There was so much ash that in Nicaragua, located at the other side of the Pacific Ocean, the Sun turned blue. After the eruption, the thickness of the volcanic foam floating in the ocean reached 3 m. It filled the seaports and counteracted to movement of ships. Before the explosion, the island of Krakatoa had a height of about 800 m, while after the explosion, almost the entire island went under water. In December 1927, fishermen discovered that a new volcano had grown out of the Krakatoa caldera, which remains active up to this time.

Products of the eruption in the form of hot gases with dust are a great danger to humans. In this regard, the large number of victims was resulted from the volcano eruption, which occurred on May 8, 1902 in Mont Pele on the island of Martinique (Lesser Antilles Islands in the Caribbean Sea). This catastrophe led to destroy of the Saint-Pierre city with killing of 30 thousand of its inhabitants. In this case the top of the mountain cracked with a terrible crash, and a huge scorching cloud burst out a wall of fire, rushing down the slope with incredible speed. In a few seconds it reached the Saint-Pierre city, and all living creatures perished in its flames.

A volcanic eruption can be accompanied by the formation of toxic particles in the atmosphere, so that volcanic dust has a negative impact on the environment and humans. For example, the volcano eruption in Iceland in 1783 led to the death of 20% of the island's population. Toxic cloud with sulfur-containing compounds was transported to Europe and led to the death of 23 thousand people in the Great Britain. Volcanic ash acts not only on people, but also on equipment, and can disable the engine of the aircraft. Just for this reason more than 4000 flights were cancelled in Europe during the volcanic eruption in Iceland in 2010, when volcanic ash spread across northern Europe [75].

It should be noted that our ground consists of the material which was the result of volcano eruptions. Indeed, if the material of the Tambora eruption in 1816 is distributed over the Earth's surface uniformly, it forms a layer of the thickness of 0.3 mm which covers the Earth's surface. One can represent what amount of the volcano material participates during the Earth's existence. One can see that in the course of volcano eruption its material is taken from the Earth's interior and is transported to the Earth's surface.

Volcano eruptions accompany by the weather change during a long time. In past, it was remarked repeatedly that some volcano eruptions were accompanied by a cold summer in subsequent years. We consider almost detective history when a building in London was discovered as burial place where of the order of 50 thousands of people were found [77]. The carbon analysis shown that this burial relates to the second half of 13th century. In that time approximately 150 thousand of habitants resided in London. The subsequent study of the Greenland ice and the Antarctice one allowed one to establish that the Samalas volcano (Lombok Island in Indonesia) is responsible for this eruption which occurred in 1257 [77].

Due to injection of nanoparticles in the stratosphere, the eruption of this volcano resulted in a decrease of the Earth's temperature because an additional part of solar radiation was reflected by these particles. The analysis of the Greenland ice shows that this eruption led to injection in the stratosphere of $(158 \pm 12) \times 10^6$ tonnes of sulfur dioxide, $(227 \pm 18) \times 10^6$ tonnes of chlorine, and $(1.3 \pm 0.3) \times *10^6$ tonnes of bromine [78]. Residing in the stratosphere, nanoparticles of volcanic origin spread throughout the hemisphere and, possibly, penetrate into the other hemisphere, i.e., the decrease in temperature applies to the entire surface of the Earth. In particular, in the case of the eruption of the Salamas volcano, the decrease in global temperature is estimated from 0.6 to 5.6 °C. This lasted for 4–5 years. As a result of this cooling in London, snow fell in the summer, and this state was conserved during subsequent several years. This led to a crop failure and caused death of about 15 thousand residents of London out of 50 thousand of this city at that time [77].

As can be seen, volcanoes are one of the ways of influencing the atmosphere, leading to the observed change in its state [68, 78]. Sulfates, as one of the products of volcano eruption, form the nanometer and submicron-sized aerosols which reduce the flux of solar radiation toward the Earth's surface. As a result, the temperature of the Earth decreases. This effect is similar to nuclear winter [79, 80], which occurs during a nuclear explosion, when nanometer-sized particles are injected into the stratosphere and absorb or reflect solar radiation. If they were to fall into the troposphere, they would be washed away by rain water within a few days. When these particles penetrate the stratosphere, they can remain there for several months and years. This effect is typical for eruption of other volcanoes. In particular, release of sulfates during the eruption of the Peaktu volcano in 946 is estimated as 90×10^6 tonnes of sulfur dioxide, and the eruption of Mount Tambora in 1815 was accompanied by injection of $(73 - 91) \times 10^6$ tonnes of sulfur dioxide. The effect of these emissions is the absorption or reflection of solar radiation by these particles.

It should be noted that the volcano effect in absorption or reflection of solar radiation by aerosols resulted from volcano eruption has a temporary character. Figure 8.14 contains the time dependence for the optical thickness of the atmosphere due to volcano eruption as a result of different treatments of observations obtained on the basis [81–83] As is seen, some time after the corresponding volcano eruption the change in the atmosphere albedo becomes remarkable during some time.

A strong season change of the weather under the action of volcano eruptions is observed several times in the millennium. In addition to the above case with the Samalas volcano, we give one more example of the volcano eruption in Peru in 1600 [84]. A large amount of snow falls was observed next winter in Europe, and snow in Russia occurred even in the summer. This change caused crop failure and the famines lasted from 1601 to 1603. Such a situation led to political consequences that resulted in "time of troubles" from 1605 to 1613 as a hard period in the Russian history.

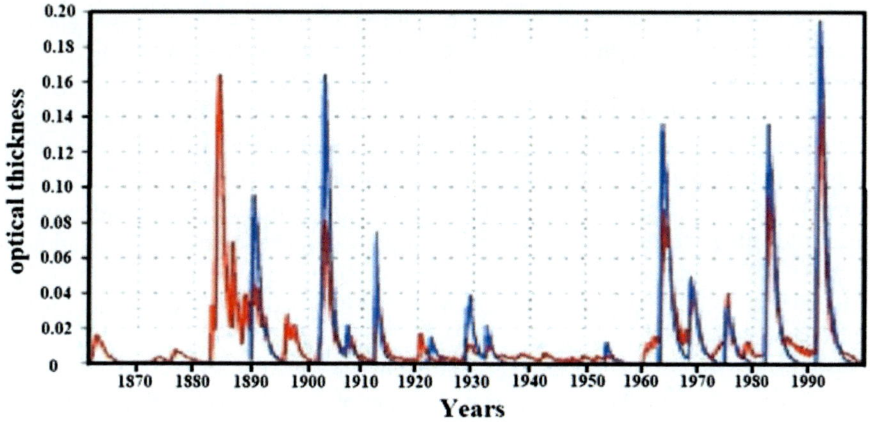

Fig. 8.14 Character of the change of the atmospheric optical thickness for the wavelength of 0.55 μm due to volcano eruptions [57] with using of different scheme of treatment

8.2.3 Aerosols in Atmospheric Phenomena

In considering radiative processes of the atmosphere due to solid and liquid particles, we divide these particles in water microparticles and aerosols. In any case, the specific mass of water particles in the atmosphere exceeds significantly that of aerosols, and hence in the case where water particles may be responsible for a certain property or process, the presence of aerosols in the atmosphere may be ignored. This takes place for emission or absorption of the atmosphere in the infrared spectrum range, so that participation of aerosols in these processes may be neglected. On contrary, since absorption by water in the visible spectrum range is weak (see Fig.8.2), the role of aerosols may be significant for radiative processes in the visible spectrum part.

Aerosols contribute to reflection of solar radiation from the atmosphere. This is enforced by the fact that approximately 10% of atmospheric aerosol is of an anthropogenic origin [85]. The optical thickness of the atmosphere for the solar radiation flux resulting from the presence of aerosols in the atmosphere was measured by [86] during the period from 2003 to 2012. The average contribution of aerosols to the optical thickness of the atmosphere is 0.126, and the average value for land is 0.131, and for the oceans it is 0.124. For the Northern and Southern Hemispheres, these values differ significantly, amounting to 0.151 for the Northern Hemisphere and 0.101 for the Southern Hemisphere. Comparing these values with data of Table 8.1 for the atmospheric albedo, one can conclude that scattering of solar radiation on aerosols gives a remarkable contribution to the albedo of the atmosphere.

Contemporary study of aerosol processes and their contribution of each aerosol component to interaction of solar radiation with the atmosphere includes satellite measurements and complex computer models. In spite of a uncertainty of results and some contradictions from various studies, they give the understanding what is the role of each component in indicated processes. We leave aside these studies because the goal of this book relates to global processes, we leave aside detailed studies of aerosols. In addition, in the real atmosphere processes involving aerosols are nonregular and nonuniform, that complicates their studies.

Aerosols act on an environment twofold. In the first place, aerosols as chemically active particles partake in atmospheric chemical processes. In this respect, it is important participation of aerosols in chemical processes which lead to formation some danger compounds for the human health. In particular, aerosols of organic origin together with ultraviolet radiation and ozone are the basis in formation of a smog. As the industry develops, a number of cities with often smogs increases. Unfortunately, the progress in study of smogs is weak, and we do not consider below this phenomenon.

Another aspect of aerosol influence on an environment associates with their interaction with solar radiation. In general, because of a strong interaction of infrared radiation with water microdroplets, condensed water of the atmosphere dominates in processes involving infrared radiation. Therefore, one can ignore aerosol processes in this interaction. On contrary, for solar radiation the aerosol processes are important because of a weakness of interaction of solar radiation with water microparticles. One can extract two types of processes in propagation of solar radiation through the atmosphere, namely, reflection and absorption of solar radiation. Roughly, the basic contribution to reflection of solar radiation from the atmosphere gives clouds in the form of water microdroplets or microparticles. But, the contribution of aerosols in this process is also remarkable, so that up to one half of the reflected radiative flux from the Sun is created by aerosols. Absorption of solar radiation by the atmosphere proceeds mostly due to aerosols.

Because of the importance of atmospheric processes involving aerosols in the energetic balance of the Earth, we now estimate the amount of the aerosol mass in the atmosphere that is able to influence on this balance. In this analysis, we assume a strong interaction of solar radiation with an aerosol and obtain a minimal aerosol mass which can influence the energetic balance of the Earth. In this case, if the wavelength of radiation λ is small compared to its radius r, the cross section σ_r of absorption and reflection of an electromagnet wave by the aerosol is πr^2. Being guided by solar radiation, we take in this estimation the aerosols radius $r = 1\,\mu\mathrm{m}$, that corresponds to the mass $m \sim 10^{-11} g$ of an individual aerosol particle at the mass density of the aerosol material $\rho \sim 2\,\mathrm{g/cm^3}$, and to the cross section $\sigma_r = 3 \times 10^{-8}\,\mathrm{cm^2}$.

For definiteness, we assume a significant influence of this process on the energetic balance of the Earth if a change of the global temperature under this process exceeds $5\,^\circ C$. Taking the climate sensitivity for solar radiation to be identical to that for infrared radiation which is equal $S = 0.5\,\mathrm{K} \times \mathrm{m^2/W}$ in accordance with formula (5.3.5), one can obtain that this corresponds to a change in the radiative flux of $\Delta J = 10\,\mathrm{W/m^2}$. Namely such change of the solar radiative flux we take below as the value which characterizes a significant change due to its interaction with aerosols. The change in the solar radiative flux ΔJ is estimated as

$$\Delta J \sim J_s n_{aer} \sigma_r,$$

where $J_j = 341\ \mathrm{w/m^2}$ is the average solar radiative flux, n_{aer} is the number density of aerosols per unit area of the atmospheric column. From this one can obtain for the specific number density of aerosols

$$n_{aer} \sim 1\times!0^6\,\mathrm{cm}^{-2} \tag{8.2.1}$$

Correspondingly, this gives for the mass density of aerosols per unit of the atmospheric column

$$\varrho = mn_{aer} \sim 10^{-5}\,\mathrm{g/cm^2} \tag{8.2.2}$$

Comparing this formula with (8.1.38), one can see that the minimal mass density of aerosols which causes a significant increase in the global temperature is three orders of magnitude smaller than that of condensed water in the atmosphere. This shows that a not large amount of a material which injected in the atmosphere in the form of aerosols is able to influence on the thermal state of the Earth. If this material is distributed uniformly over the total Earth's surface, this corresponds to the total mass of aerosols M located in the Earth's atmosphere

$$M = \varrho S \sim 10^{13}\,\mathrm{g} \tag{8.2.3}$$

One can compare this with the total carbon mass in fossil fuels extracted from the Earth's interior annually, that is equal 10^{16} g (10 billion tons).

In order to understand the scale of the above values, we consider the following situation. Let us assume that all the world coal which is extracted from the Earth's interior is transformed in soot and is injected in the atmosphere. Subsequently this soot is washed out by rain, and it establishes the equilibrium for atmospheric soot. Let us assume that all extracted coal is transformed in soot injected in the atmosphere. Let us determine the mass of soot in the atmosphere under these conditions. We take the annual coal production to be 3 billion tonnes ($3 \times 10^{16}\,\mathrm{g/yr}$) and a residence time of soot in the atmosphere to be equal to that of water molecules that is 10 days according to formula (2.1.10). This gives for the total soot mass $M_s \sim 10^{15}$ g that is two orders of magnitude above the aerosol mass of formula (8.2.3). This shows that the minimal mass (8.2.3) of aerosols in the atmosphere is available.

Let us consider the above situation from the standpoint of the nuclear winter if aerosols injected in the stratosphere can reside there during a long time and reduces the solar radiative flux. Because aerosols cannot be washed out the stratosphere, they remain there during a long time and leave the stratosphere as a result of falling under the action of the gravitation field of the Earth. According to formula (2.2.9) the falling velocity of aerosols of a size $r = 1\,\mu\mathrm{m}$ is equal 0.01 cm/s. From this it follows that the residence time of aerosols in the stratosphere is several years. Above it was indicated that the amount of sulfur dioxide injected in stratosphere as a result of eruption of the volcano Samalas in 1259 and volcano Tambora in 1815 was of the order of 100 millionen tonnes (10^{14} g). If molecules of sulfur dioxide are transformed partially into particles of a submicron size, the above situation may be realized during, at least, several years.

From this analysis one can conclude that injection of the available amount of a chemically active material into the stratosphere which form there aerosols of micron and submicron sizes can cause a remarkable decrease of the Earth's temperature because of absorption of solar radiation by formed aerosols. This explains a decrease of the global temperature after strong eruption of volcanoes in the course of several years if an injected material forms aerosols in the stratosphere of micron and submicron sizes. Along with this, one can influence on a local temperature in this method, on the basis of injection of a suitable amount of aerosols in the atmosphere.

One can estimate the change of the solar radiative flux as a result of scattering by strongly reflected particles if their amount would be corresponded to that after eruption of the Salamas volcano. If particles of sulfur dioxide are distributed uniformly over the Earth's surface, their mass per unit air column will be $3 \times 10^{-5}\,\mathrm{g/cm^2}$. If these particles are pressed in a compact material, its value exceeds $0.1\,\mu\mathrm{m}$. Hence, in the course of a strong interaction with solar radiation, these particles may scatter a remarkable part of solar radiation which wavelength ranges approximately from $0.4\,\mu\mathrm{m}$ up to $0.8\,\mu\mathrm{m}$. Hence, eruption of this material in the form of micron-sized particles in the stratosphere can lead to a remarkable change of solar radiation which is directed to the Earth.

References

1. W. Wien, Wied. Ann. Phys. Chem. **58**, 662 (1896)
2. G. Herzberg, *Molecular Spectra and Molecular Structure* (Van Nostrand Reinhold, Princeton, 1945)
3. http://www1.lsbu.ac.uk/water/water-vibrational-spectrum
4. A.A. Radzig, B.M. Smirnov, *Reference Data on Atoms, Molecules, and Ions* (Springer, Berlin, 1985)
5. S.V. Khristenko, A.I. Maslov, V.P. Shevelko, *Molecules and Their Spectroscopic Properties* (Springer, Berlin, 1998)
6. A. Beer, Annalen der Physik und. Chemie **86**, 78 (1852)
7. J.H. Lambert, *Photometry, or, on the measure and gradations of light, colors, and shade.* (Eberhardt Klett, Augsburg, 1760)
8. O. Boucher, *Atmospheric Aerosols. Properties and Climate Impacts.* (Springer, Dordrecht, 2015)
9. D. Eisenberg, W. Kauzmann, *The Structure and Properties of Water* (Oxford University Press, New York, 1969)
10. J.B. Hasted, *Aqueous Dielectrics* (Chapman and Hall, London, 1973)
11. M.N. Afsar, J.B. Hasted, Infrared Phys. **18**, 835 (1978)
12. H.D. Downing, D.W. Williams, J. Geoph. Res. **80**, 1656 (1975)
13. https://en.wikipedia.org/wiki/Electromagnetic-absorption-by-water
14. http://www1.lsbu.ac.uk/water/water-vibrational-spectrum.html
15. C.M.R. Platt, Quart. J. Roy. Meteorolog. Soc. **102**, 553 (1976)
16. S. Twomey, Geofis. Pure Appl. **43**, 227 (1959)
17. S. Twomey, J. Atmos. Sci. **34**, 1149 (1977)
18. G. Mie, Annalen der Physik **330**, 377 (1908)

19. J.A. Stratton, *Electromagnetic Theory* (McGraw-Hill, New York, 1941)
20. H.C. van de Hulst, *Light Scattering by Small Particles*. (Wiley, New York, 1957)
21. C.F. Bohren, D.R. Huffmann, *Absorption and Scattering of Light by Small Particles* (Wiley, New York, 2010)
22. https://en.wikipedia.org/wiki/Properties-of-water
23. T.S. Light, et al., Electrochem. Solid State Lett. **8**, E16 (2005)
24. L.D. Landau, E.M. Lifshitz, *Electrodynamics of Continuous Media* (Pergamon Press, Oxford, 1984)
25. B.M. Smirnov, *Clusters and Small Particles in Gases and Plasmas* (Springer, New York, 1999)
26. https://en.wikipedia.org/wiki/Refractive-index
27. D.A. Draegert, N.W.B. Stone, B. Curnutte, D. Williams, J. Opt. Soc. Am er. **56**, 64 (1966)
28. W.M. Irvine, J.B. Pollack, Icarus **8**, 324 (1968)
29. M.R. Querry, B. Curnutte, D. Williams, J. Opt. Soc. Am er. **59**, 1299 (1969)
30. V.M. Zolatarev, B.A. Mikhailov, L.I. Aperovich, S.I. Popov, Opt. Spectrosc. **27**, 430 (1969)
31. C.W. Robertson, D. Williams, J. Opt. Soc. Am. **61**, 1316 (1971)
32. A.N. Rusk, D. Williams, M.R. Querry, J. Opt. Soc. Am. **61**, 895 (1971)
33. P.S. Ray, Appl. Opt. **11**, 836 (1972)
34. C.W. Robertson, B. Curnutte, D. Williams, Mol. Phys. **26**, 183 (1973)
35. B.M. Smirnov, *Transport of Infrared Atmospheric Radiation* (de Gruyter, Berlin, 2020)
36. G. Feingold, W.L. Eberhard, D.E. Veron, M. Previdi, Geophys. Res. Lett. **30**, 1287 (2003)
37. D.A. Zhilyaev, B.M. Smirnov, JETP. **133** (2021)
38. M. Born, E. Wolf, *Principles of Optics* (Pergamon Press, Oxford, 1970)
39. F.A. Jenkins, H.E. White, *Fundamentals of Optics* (McGraw-Hill, New York, 1976)
40. A. Fresnel, Annales de Chimie et de Physique **17**, 102 (1821)
41. A. Fresnel, Annales de Chimie et de Physique **17**, 167 (1821)
42. https://en.wikipedia.org/wiki/Fresnel-equations
43. https://en.wikipedia.org/wiki/Optical-properties-of-water-and-ice
44. https://en.wikipedia.org/wiki/Albedo
45. B. Briegleb, V. Ramanathan, J. Appl. Meteor. **21**, 1160 (1982)
46. R.E. Dickenson, Adv. Geophys. **25**, 305 (1983)
47. B.P. Briegleb, et al., J. Climate Appl. Meteor. **25**, 214 (1986)
48. G.E. Thomas, K. Stamnes, *Radiative Transfer in the Atmosphere and Oceans* (University Press, New York, Cambridge, 1999)
49. W.J. Brennan, P.A. Schulz, et al., *Atmospheric Aerosol, Properties and Climate Impacts*. (US Climate Change Science Program, New York, 2009)
50. B.M. Smirnov, *Nanoclusters and Microparticles in Gases and Vapors* (De Gruyter, Berlin, 2012)
51. J. Aitken, Nature **23**(583), 195–197 (1880)
52. J. Aitken, Nature **23**(588), 311–312 (1881)
53. J. Aitken, Nature **23**(591), 384–385 (1881)
54. J. Aitken, Trans. Roy. Soc. Edinburgh **35**(1), 1–19 (1888)
55. S.H. Harris, in *Encyclopedia of Physics*. ed. by R.G. Lerner, G.L. Trigg. (VCH Publ, New York, 1990, p.30; Weinheim, Wiley, 2005, p. 61)
56. V. Ramaswamy, A.W. Collins, B.J. Haywood, et al., *American Meteorological Society. Monograph*, chapter **14** (2019). https://doi.org/10.1175/AMSMONOGRAPHS-D-19-0001.1
57. P. Forster, V. Ramaswamy, P. Artaxo et al., *Changes in Atmospheric Constituents and in Radiative Forcing* (Cambridge University Press, Cambridge, 2007)
58. I. Lagzi, et al., *Atmospheric Chemistry*. (Institute of Geography and Earth Science, Budapest, 2013)
59. R.G. Fleagle, J.A. Businger, *Introduction to Atmospheric Physics* (Academic Press, San Diego, 1980)
60. M.L. Salby, *Fundamentals of Atmospheric Physics* (Academic Press, San Diego, 1996)
61. J.H. Seinfeld, S.N. Pandis, *Atmospheric Chemistry and Physics* (Wiley, New York, 1998)

62. L.S. Ivlev, *Chemical Composition and Structure of Atmosphere Aerosols*. (Izd. LGU, Leningrad, 1982; in Russian)
63. I.V. Petryanov-Sokolov, A.G. Sutugin, *Aerosols*. (Nauka, Moscow, 1989; in Russian)
64. http://en.wikipedia.org/wiki/pyrolysis
65. http://en.wikipedia.org/wiki/soot
66. http://www.eoearth.org/article/London-smog
67. H. Rast, *Vulkane und Vulkanismus* (Teubner, Leipzig, 1980)
68. R. Decker, B. Decker, *Volcanoes* (Freeman, New York, 1989)
69. T. Simkin, L. Siebert, *Volcanoes of the world* (Geoscience Press, Arizona, Tucson, 1994)
70. http://en.wikipedia.org/wiki/volcanic-ash
71. http://www.liv.ac.uk/science-eng-images/earth/research/VolcanicAsh.jpg
72. http://www.liv.ac.uk/info/research/microstruc-litosphere
73. R.D.M. Verbeck, Nature **30**, 10 (1884)
74. S. Self, M.R. Rampino, Nature **294**, 699 (1981)
75. http://news.bbc.co.uk/2/hi/8621407.stm
76. http://static.howstuffworks.com/gif/volcanic-ash-2
77. https://www.theguardian.com/uk/2012/aug/05/medieval-volcano-disaster-london
78. C.M. Vidal, N.M. Trich, J.Ch. Komorowski, et al., Sci. Rep. **6**, 34868 (2016)
79. P.J. Crutzen, J.W. Birks, Ambio **11**, 114 (1982)
80. R.P. Turco, O.B. Toon, T.P. Ackerman, J.B. Pollack, C. Sagan, Science **222**, 1283 (1983)
81. M. Sato, J.E. Hansen, M.P. McCormick, J.B. Pollack, J. Geophys. Res. **98**, 987 (1993)
82. C. Ammann, G. Meehl, W. Washington, C. Zender, Geophys. Res. Lett. **30**, 1657 (2003)
83. G. Schmidt, et al., Geosci. Model Dev. **4**, 33 (2011)
84. https://www.sciencenewsforstudents.org/article/explainer-volcano-basics
85. C. Textor, et al., Atmos. Chem. Phys. **6**, 1777 (2006)
86. K.B. Mao, et al., Atmospheric Environ. **94**, 680 (2014)

Chapter 9
Equilibrium Radiation of Earth and Venus

Abstract In the first approximation, the powerful radiative atmospheric processes assume to be stationary that means the equilibrium state of the atmosphere stationarity. In addition, in the first approximation, atmospheric regions which are responsible for atmospheric emission to the Earth's surface and outside are separated. The power of solar radiation absorbed at each point of the globe is characterized by day and season oscillations. The equilibrium between solar radiation penetrated in the atmosphere and outgoing infrared radiation is analyzed for the Earth and Venus. Emission of the Venusian atmosphere to its surface is formed close to its surface. The difference between them is determined by transparency windows in the spectrum of CO_2 molecules and is approximately 10%, as it follows from data of the HITRAN bank. The outgoing radiative flux created in the Earth's and Venusian atmospheres in the infrared spectrum range compensates the solar radiative flux penetrated in the atmosphere and is formed to a greater extent by clouds.

9.1 Energetic Equilibrium in Standard Atmosphere

9.1.1 Character of Heat Equilibrium in Standard Atmosphere

In analyzing the energetic balance of the atmosphere, we are based on the model of standard atmosphere which assumes it to be a stationary system. This means that the atmosphere is found in the equilibrium with both the Earth's surface and an environment. In addition, an interaction between the near-surface part of the atmosphere and its outer part is relatively weak, so that energetic processes for these atmospheric parts must be analyzed independently. These facts simplify the analysis of atmosphere energetics and allow one to construct a simple physical picture of atmospheric processes. This analysis assumes the stationarity of the atmosphere and separation of its near surface and outer parts. In addition, the equilibrium takes place between the Earth and lower atmosphere, as well as between the outer part of the troposphere and an environmental space. These assumptions hold true in the first approximation. We below analyze the degree of violation of these assumptions.

B. M. Smirnov, *Global Energetics of the Atmosphere*, Springer Atmospheric Sciences, https://doi.org/10.1007/978-3-030-90008-3_9

Table 9.1 Average energy fluxes involving the Earth and atmosphere as a whole according to different sources

Channel, process	1	2	3	4	5	Average
S.R. absorbed by A *)	57	68	78	75	79	72 ± 9
I.R. outgoing emission by A	200	215	217	220	218	215 ± 8
Outgoing IR from E	20	22	22	20	21	21 ± 1
Total outgoing IR	220	237	239	240	239	235 ± 8
Transport of heat from E to A	102	106	98	104	98	102 ± 4
Emitted by E and passed through A	20	22	22	20	21	21 ± 1

*) S.R. is solar radiation in the visible spectrum range, I.R. is thermal infrared radiation, A is the atmosphere, and E is the Earth. The energy fluxes for an indicated process are expressed in W/m^2. Sources of data are the same as in Table 5.1

Let us consider the equilibrium at the top of the troposphere which results from equality between the flux of solar radiation penetrated into the Earth's atmosphere and the total flux of outgoing infrared radiation in accordance with Table 5.3. It should be noted a small violation of this equilibrium that has the form of an imbalance in fluxes of radiation at the top of the troposphere. Values of this imbalance for the average radiative fluxes expressed in W/m^2 are 0.9 ± 0.5 according to [1], 0.64 ± 0.11 according to [2], 0.8 ± 0.2 according to [3], 0.5 ± 0.4 according to [4] and 0.6 ± 0.4 according to [5].

As is seen, this imbalance is below 1 W/m^2, and we above ignore this effect. In addition, there is the season dependence for energy fluxes under consideration. Because of a larger error for energy fluxes, we also neglect the season changes of energy fluxes. One can note also that the average energy flux emitted by the Earth $J_E = 392\,\mathrm{W/m^2}$ exceeds slightly that of a blackbody at the temperature of standard atmosphere $T_E = 288K$ that leads to the radiative flux of $J_{bl} = 390\,\mathrm{W/m^2}$ for a blackbody. A smallness of the imbalance for the atmosphere energetics, as well as other small violations of the equilibrium for the atmosphere, allows one to ignore it in construction of the physical picture of powerful atmospheric processes.

We below consider the energy equilibrium for the troposphere top. From this equilibrium, it follows the energy equilibrium for the upper part of the troposphere. Basing on data of Table 5.1, as well as on formula (5.1.14), we give in Table 9.1 total energy fluxes involving the atmosphere. References for sources of these data are represented in Table 5.1.

Note that dividing the atmosphere into the upper and lower parts of as an independent object requires the separate equilibrium for the upper part of the troposphere with an environment and the lower part of the troposphere. This is represented in Table 9.1 where the energy loss of the upper part of the troposphere results from its emission, whereas it obtains an energy through thermal conductivity of the atmosphere, as well in the form of infrared radiation from lower layers of the atmosphere. The thermal conductivity is connected with convective transport of air in the atmosphere and according to formula (5.1.14). From this, one can find the radiative flux which penetrates in the upper part of the atmosphere and is presented in Table 9.1. This table contains for comparison the average radiative flux which is emitted from the Earth's surface and leaves the atmosphere. In accordance with the scheme under consideration, this radiative flux is small compared to that transported from other parts of the atmosphere and is absorbed by upper atmosphere layers.

9.1.2 Temporary Changes of Atmospheric Fluxes

In considering emission of the atmosphere, we assume a stationary atmosphere state, so that the altitude distribution of the temperature establishes fast. Since the Earth is moving in a space around the Sun and rotates, the average temperature of the Earth's, that is the global temperature, varies according to the harmonic law

$$C\frac{\partial T}{\partial t} = J_o(1 + \cos\omega t) - j(T) \tag{9.1.1}$$

Here T is the global temperature, i.e., the average temperature of the Earth's surface, C is the specific heat capacity of the atmosphere which is the heat capacity per unit area of the atmospheric column and is measured in $\mathrm{J}/(\mathrm{m}^2 \cdot \mathrm{K})$, J_o is the average solar radiative flux, $j(T)$ is the energy flux which includes other channels of the Earth's energetic balance in accordance with data of Table 5.1, t is time, and ω is the frequency of the Earth's rotation, so that $2\pi/\omega = 24$ h. We also denote by J_o the radiative flux which is absorbed by the Earth or its atmosphere. In accordance with data of Table 5.1, the outgoing radiative flux of the atmosphere is equal $J_o = (236 \pm 9)\,\mathrm{W/m^2}$.

Let us solve equation (9.1.1) on the basis of the perturbation theory on the basis of the assumption that the average Earth's temperature T_E is the zero-th approximation. This approximation gives

$$J_o = j(T_E) \tag{9.1.2}$$

Taking into account the first approximation and consider as a perturbation the terms which depend on time, we obtain for the Earth's temperature

$$T = T_E + \frac{\Delta T}{2}\sin\omega t, \;\; \Delta T = \frac{2J_o}{C\omega} \tag{9.1.3}$$

The criterion of validity of the used perturbation theory has the form

$$T_E \gg \frac{J_o}{C\omega} \tag{9.1.4}$$

According to the version under consideration, in the course of evolution of the Earth's atmosphere system, the temperature of its various layers of the atmosphere follows for the solar flux. We use also the assumption that absorption of solar radiation

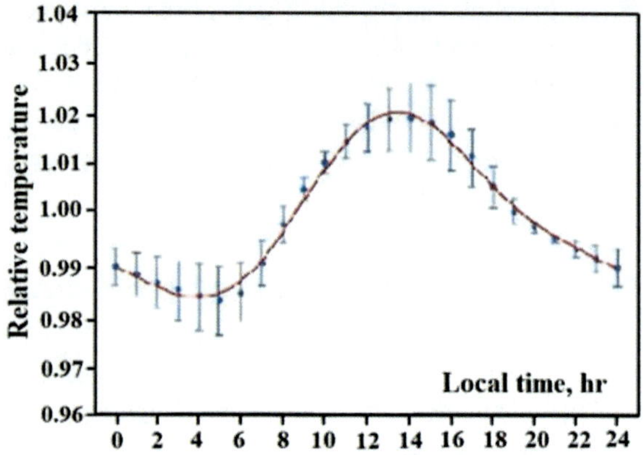

Fig. 9.1 Daily variations of the relative global temperature according to the NASA Global Reference Atmosphere Model [7]

is independent of its intensity and other factors. As a result, the global temperature varies harmonically with the amplitude ΔT. Formula (9.1.3) gives also on the basis of parameters of standard atmosphere [6]

$$C\Delta T = 3.0 \times 10^6\,\text{J/m}^2 \tag{9.1.5}$$

Let us assume that the atmosphere partake in heat processes as a whole, i.e., it is heated or is cooled as a whole under the action of energetic processes represented in Table 5.1. In this case, the parameter C in formula (9.1.1) is the heat capacity of air located in atmospheric column. Taking the air temperatures to be of the order of room one, we use the parameters of standard atmosphere [6], so that the heat capacity equals $1\,\text{J/(g}\cdot\text{K)}$, and the density of atmospheric air per unit column area is $1\,\text{kg/cm}^2$. From this, one can obtain $1 \times 10^7\,\text{J/m}^2$ for the specific capacity of atmospheric (mostly, tropospheric) air. Correspondingly, formula (9.1.5) gives for the amplitude of daily oscillations of the average Earth's temperature

$$\Delta T = 0.3\,\text{K} \tag{9.1.6}$$

It is clear that the value (9.1.6) is small compared with observed daily temperature change or daily variations of the Earth's temperature which according to data of Fig. 9.1) is equal $\Delta T \approx 10$ K.

Figure 9.1 contains the time dependence for the relative global temperature. As is seen, the amplitude of daily oscillations of the global temperature exceeds 10 K that is large compared to the result of formula (9.1.6). From this, it follows that only a part of atmospheric air partakes in the heat process. Let us determine the effective thickness of the air layer L that partakes in the heat process as

$$L = \frac{C}{c_p} = 50\,\text{m}, \tag{9.1.7}$$

where $c_p = 1.0\,\mathrm{J}/(\mathrm{g}\cdot\mathrm{K})$ is the specific heat capacity.

One can determine the effective altitude H of heat propagation from another standpoint on which the heat propagates for one half of a day $t = 12\,\mathrm{h} \approx 5\times10^4\,\mathrm{s}$. We have

$$H = 2\sqrt{D_L t} \approx 700\,m \tag{9.1.8}$$

Taking the diffusion coefficient of air for its displacement at large distances due to its convection motion as $D_L = 5\times10^4\,\mathrm{cm}^2/s$ according to formula (3.2.38). From comparison on the values (9.1.7) and (9.1.8), one can conclude that the main contribution to the heat capacity of the Earth's surface and near-surface air follows from the ground.

9.2 Emission of Venus

9.2.1 Properties of the Venus Atmosphere

We above consider the Earth's atmosphere as a thick layer of atmospheric air in which molecules are found in thermodynamic equilibrium with a radiation field. Infrared emission of the Earth's atmosphere and its absorption of solar radiation and infrared radiation from the Earth's surface, as well as heat transfer from the Earth to the atmosphere due to convective transport of atmospheric air and condensation of water molecules evaporated from the Earth's surface constitute the energy balance of the Earth and its atmosphere. Infrared or thermal emission of greenhouse components of the atmosphere plays an important role in the energy balance of the Earth and its atmosphere. Evidently, similar situation takes place in atmospheres of other planets, and we below consider from this standpoint the energetic balance of the Venus.

First of all, we analyze emission of the Venusian atmosphere, and for this, it is necessary to represent its parameters. The Venusian atmosphere consists of carbon dioxide (96.5%) and nitrogen (3.5%) [8]. The atmospheric pressure at the Venus surface is 92 atm, and the temperature of its surface is 737 K [8, 9] that corresponds to the number density of molecules of carbon dioxide near the Venus surface $N = 9.2\times10^{20}\,\mathrm{cm}^{-3}$. The temperature gradient is about -8 K/km for the altitude range h from 0 to 60 km [9]. On the basis of these parameters, we construct the energetic balance of the Venus similar to that of the Earth in which parameters are given in Table 5.1.

It is convenient also to introduce the scale parameter Λ for the number density of carbon dioxide molecules N of the Venusian atmosphere similar to that (2.1.1) for atmospheric air

$$\Lambda = \frac{d\ln N}{d(1/h)},$$

where h is a distance from the Venusian surface. We have near the Venusian surface $\Lambda = 19\,\mathrm{km}$.

In constructing the energetic balance of the Venus and its atmosphere, we model the Venus surface as a blackbody for infrared radiation. We have that an absolutely blackbody with a temperature of 737 K creates an infrared radiative flux $J_o = 16.7\,\mathrm{kW/m^2}$. On the other hand, the average flux of solar radiation which penetrates in the Venus atmosphere is $2.6\,\mathrm{kW/m^2}$ [10]. The Venus albedo is 0.80 ± 0.02 according to [11, 12] and 0.76 ± 0.01 in accordance with [10]; we take this value to be 0.78. It follows that the average flux of solar radiation absorbed by the atmosphere and surface of the Venus is equal to $0.14\,\mathrm{kW/m^2}$ per unit surface area of the Venus. As is seen, the radiative flux emitted by the Venus surface exceeds by two orders of magnitude the solar radiative flux penetrated and absorbed by the surface and atmosphere of the Venus. This means that the total flux of Venus atmospheric radiation is relatively large. One can expect that the main contribution to the radiative atmospheric flux gives CO_2 molecules. Therefore, below we evaluate the radiative flux emitted by CO_2 molecules toward the Earth and outside the Venus. Though according to [13] only a small part of the solar radiation of $17\,\mathrm{W/m^2}$ is absorbed by the Venusian surface, we check the reliability of this value by the subsequent analysis.

Note that we use old information about the Venusian atmosphere in this analysis of its emission. Now this information is improved (e.g., [14–17]), so that we have more detailed data for parameters of the Venusian atmosphere. Nevertheless, an old description of parameters of the Venusian atmosphere is enough for this analysis.

9.2.2 *Infrared Radiation of Venus and Its Atmosphere*

In the analysis of the energetic balance of the Venus, one can restrict ourselves by solar radiation J_{sol} which attains the Venusian surface and is absorbed by it. We also account for emission of the Venusian surface J_{surf} and emission of the Venusian atmosphere J_a to its surface with absorption by it. Then the energy balance of the Venus has the form

$$J_{\mathrm{sol}} + J_a = J_{\mathrm{surf}}, \qquad (9.2.1)$$

In the first approximation, modeling the Venusian surface by a blackbody, we have for the radiative energy flux at a given frequency according to the Planck formula [18, 19]

$$I_\omega(T) = \frac{\hbar\omega^3}{4\pi^2 c^2 \left[\exp\left(\frac{\hbar\omega}{T}\right) - 1\right]} \qquad (9.2.2)$$

Figure 9.2 presents the frequency dependence of the radiative flux of a blackbody at the temperature of 737 K which is the temperature of the Venus surface.

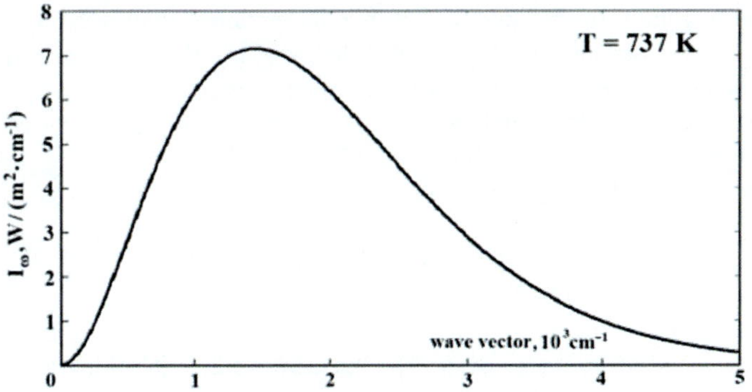

Fig. 9.2 Radiative flux of the blackbody with the temperature $T = 737$ K of the Venus surface as a function of the wave number $k = 1/\lambda$, where λ is the wavelength

From this, the radiative flux $J_b(T)$ of a blackbody of a temperature T according to the the Stephan–Boltzmann law [18, 19]

$$J_b(T) = \int I_\omega(T)d\omega = \sigma T^4, \qquad (9.2.3)$$

where $\sigma = 5.67 \times 10^{-8}\mathrm{W}/(\mathrm{m}^2 \cdot \mathrm{K}^4)$ is the Stephan–Boltzmann constant. In particular, at the temperature of the Venusian surface, the radiative flux is equal to $J_b = 17\,\mathrm{kW/m}^2$. Note that the temperature T is expressed in energetic units, and the connection between these units is $1\,\mathrm{K} = 1.381 \times 10^{-23}\,\mathrm{J}$. In the same manner, the spectroscopic unt cm^{-1} is used for the photon energy $\hbar\omega$ and frequency ω. The connection between the energy units is $1\,\mathrm{cm}^{-1} = 1.986 \times 10^{-23}\,\mathrm{J}$, and $1\,\mathrm{K} = 0.695\,\mathrm{cm}^{-1}$.

Introducing the gray coefficient γ for surface emission, one can represent the total radiative flux $J_{\rm sur}$ from the Venusian surface in the form

$$J_{\rm surf} = \gamma\sigma T^4 \qquad (9.2.4)$$

According to definition, $\gamma < 1$ as the gray coefficient of the surface characterizes the difference of a radiating surface from a blackbody. Note that by modeling the Venusian surface by a blackbody in evaluation of its radiative flux, we use the experience of emission of various materials in the infrared spectrum range [20]. As it follows from Fig.5.3, the gray coefficient for some object located at the Earth's surface is close to one in the infrared spectrum range.

The basic approach to the analysis of the energetic balance of the Venus which is determined by emission processes is similar to that for the Earth. We below evaluate the difference of the radiative fluxes from the Venusian atmosphere J_a and its surface $J_{\rm surf}$ modeling radiation of the surface by a blackbody. Emission of the atmosphere at a given frequency is determined by the radiative temperature T_ω which is close to the surface temperature. In this analysis, we use the parameters of radiative transitions of CO_2 molecules taken from HITRAN database [21–23] with some additions [24, 25]. The HITRAN database uses some parameters which are convenient to describe the radiative transitions in molecules. Namely the cross section σ_ω of molecule absorption due to a given vibrational transition and different rotation numbers j at a certain frequency ω is given by formula [26]

$$\sigma_\omega = \sum_j S_j a(\omega - \omega_j), \tag{9.2.5}$$

where S_j is the intensity for $j-$th spectral line and for this vibration transition which values are contained in the HITRAN database, ω_j is the frequency of the line center for this transition, and $a(\omega - \omega_j)$ is the frequency distribution function which is normalized by the condition.

$$\int a(\omega - \omega_j)d\omega = 1 \tag{9.2.6}$$

In particular, in the case of the impact mechanism of broadening of spectral lines, the frequency distribution function is determined by formula

$$a(\omega - \omega_j) = \frac{\nu}{2\pi[(\omega - \omega_j)^2 + (\nu/2)^2]}, \tag{9.2.7}$$

where ν is the spectral line width.

Let us assume the rotation number j to be a continuous variable and introduce the transition intensity $S(\omega) \equiv S(\omega_j)$ to be a continuous function of the frequency ω. This holds true for a small intensity variation as a result of transfer to a neighboring rotation number that is governed by the criterion

$$\Delta\omega \cdot \frac{d \ln S(\omega)}{d\omega} \ll S, \tag{9.2.8}$$

where $\Delta\omega$ is the frequency difference for neighboring spectral lines. For carbon dioxide molecules consisting of isotopes ^{12}C and ^{16}O, which population exceeds 98%, we have $\Delta\omega = 4B = 1.56\,\mathrm{cm}^{-1}$ for not strongly excited vibrational states. Here $B = 0.39\,\mathrm{cm}^{-1}$ is the rotational constant for the ground vibrational state.

We now take into account a high gas pressure of the Venusian atmosphere that leads to overlapping of neighboring spectral lines. Hence, in contrast to the spectroscopy of the Earth's atmosphere, where the absorption coefficient at the spectral line center exceeds by the order of magnitude that at the middle between two neighboring spectral lines, these values are close for the Venusian atmosphere. Thus, the criterion

$$\nu \gg \Delta\omega \tag{9.2.9}$$

holds true for the Venusian atmosphere near its surface. On the basis of this and the criterion (9.2.8), an average absorption cross section σ_ω for a given vibration transition which is averaged over rotation states j has the form

$$\overline{\sigma_\omega} = \frac{S(\omega)}{\Delta\omega} \tag{9.2.10}$$

Let us take the dependence on the altitude h over the Venusian surface for the number density of molecules $N(h)$ in the Venusian atmosphere as

$$N(h) = N_o \exp\left(-\frac{h}{\Lambda}\right), \tag{9.2.11}$$

and the parameters of this formula are $N_o = 9.2 \times 10^{20}\,\mathrm{cm}^{-3}$, $\Lambda = 19\,\mathrm{km}$ at altitudes near its surface. On the basis of these parameters, one can obtain the following expression for the optical thickness u_ω of the atmospheric column

$$u_\omega = \overline{\sigma_\omega} \int N(h)dh = \frac{S(\omega)N_o\Lambda}{\Delta\omega} \tag{9.2.12}$$

Taking the above values of these parameters, one can obtain

$$u_\omega = \frac{S(\omega)}{S_o}, \quad S_o = \frac{\Delta\omega}{N_o\Lambda} \tag{9.2.13}$$

This gives the following criterion of a large optical thickness of the Venusian atmosphere

$$S(\omega) \gg S_o, \quad S_o = 9 \times 10^{-28}\,\mathrm{cm} \tag{9.2.14}$$

9.2.3 *Transparency Windows of Venusian Atmosphere*

We now evaluate the radiative flux from the Venusian atmosphere to its surface within the framework of the "line-by-line" method [27, 28] that means calculation of radiative fluxes at each frequency and subsequent summation of partial fluxes. In this case, we use the intensities of radiative transitions taken from the HITRAN data bank. Some values of the transition intensities due to CO_2 molecules located near the Venusian surface are given in Figs. 9.2, 9.3, 9.4 and 9.5. According to these data and formulas (9.2.13), (9.2.14), the Venusian atmosphere is optically thick for almost the entire spectrum range which is responsible for surface emission. Hence, the total radiative flux from the Venusian atmosphere toward its surface is close to that according to formula (9.2.4), i.e., the Venusian atmosphere emits almost as a blackbody in which temperature is equal to that of the Venusian surface.

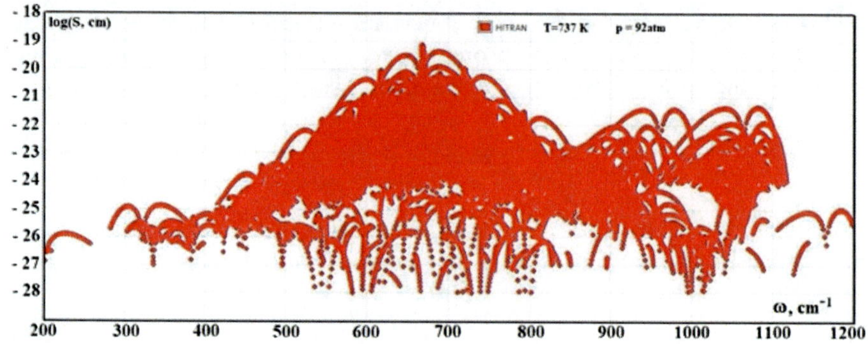

Fig. 9.3 HITRAN bank data [23] for the intensity of emission of CO_2 molecules of the Venusian atmosphere near its surface, where the temperature is $T = 737\,\mathrm{K}$ and the pressure is $p = 92\,\mathrm{atm}$. Frequencies of emitted photons are ranged from $200\,\mathrm{cm}^{-1}$ to $1200\,\mathrm{cm}^{-1}$

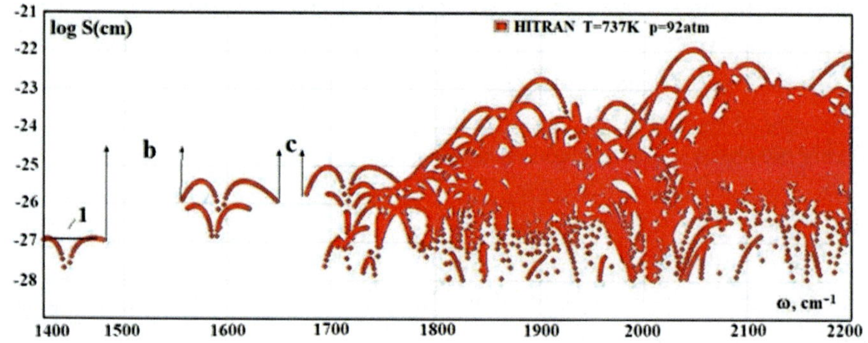

Fig. 9.4 HITRAN bank data [23] for the intensity of emission due to CO_2 molecules of the Venusian atmosphere near its surface as a function of photon frequency. Arrows indicate boundaries of transparency windows, and solid lines correspond to a rough approximation of intensities where they are small. Frequencies under consideration are ranged from $1400\,\mathrm{cm}^{-1}$ to $2200\,\mathrm{cm}^{-1}$

One can represent the total radiative flux from the Venusian atmosphere to its surface in the form

$$J_a(T) = \alpha(T)\sigma T^4, \tag{9.2.15}$$

where $\alpha(T) \approx 1$. Our goal is to determine the difference $1 - \alpha(t)$ that is small.

The main contribution to the difference $1 - \alpha(T)$ follows from the frequency ranges where the optical thickness u_ω of the Venusian atmosphere is not large, i.e., for transparency windows. The optical thickness u_ω of the Venusian atmosphere is given by formulas (9.2.13), (9.2.14). In the most part of frequencies, the optical thickness is large, and regions of the atmosphere which are responsible for atmospheric emission at a given frequency are located near the Venusian surface. Therefore, the radiative temperature for these frequencies coincides with that of the Venusian surface. Hence,

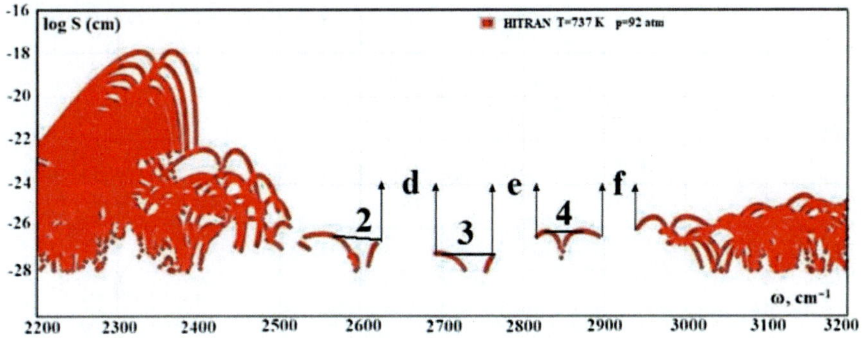

Fig. 9.5 HITRAN bank data [23] for the intensity of emission due to CO_2 molecules of the Venusian atmosphere near its surface. Arrows indicate boundaries of transparency windows, and solid lines correspond to a rough approximation of intensities where they are small. Frequencies under consideration are ranged from 2200 cm^{-1} to 3200 cm^{-1}

our task is to extract frequency ranges with not large values of the optical thickness. If the total optical thickness at such frequencies is large, due to local thermodynamic equilibrium, an effective altitude h_ω which is responsible for emission to the surface at a given frequency follows from the relation (6.1.10)

$$u_\omega \cdot \frac{h_\omega}{\Lambda} = \frac{2}{3} \tag{9.2.16}$$

Correspondingly, the radiative temperature T_ω at a given frequency ω is given by

$$T_\omega = T_V - \frac{2\Lambda S_o}{3S_\omega} \cdot \frac{\mathrm{d}T}{\mathrm{d}h}, \tag{9.2.17}$$

where $\mathrm{d}T/\mathrm{d}h = 8$ K/km is the temperature gradient for the Venusian atmosphere in the vicinity of its surface.

We assume the radiative temperature T_ω to be close to the temperature of the Venusian surface T_V, i.e.,

$$T_V - T_\omega \ll T_V$$

and represent the radiative flux in the form

$$J_\omega = I_\omega(T_V)\left[1 - \frac{\hbar\omega/T_V}{1-\exp\left(-\hbar\omega/T_V\right)} \cdot \frac{2S_o\Lambda}{3S_\omega T_V} \cdot \frac{\mathrm{d}T}{\mathrm{d}h}\right]$$

by expansion over a small parameter. Let us use formula (9.2.2) for the equilibrium radiative flux and the expression for the Stephan–Boltzmann constant

$$\sigma = \frac{\pi^2}{60c^2\hbar^3}$$

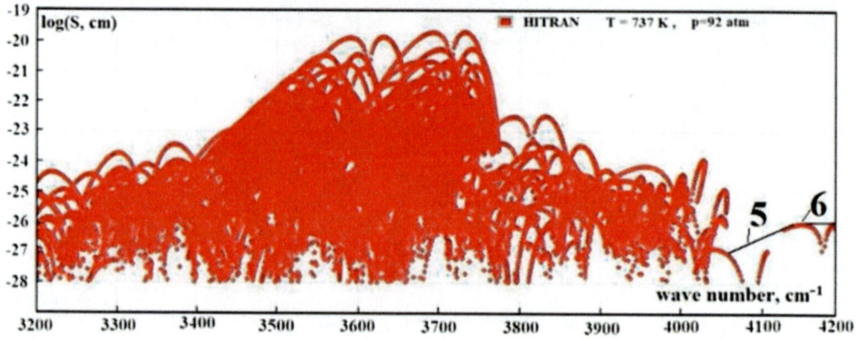

Fig. 9.6 HITRAN bank data [23] for the intensity of emission due to CO_2 molecules of the Venusian atmosphere near its surface. Solid lines relate to transparency windows, and a rough approximation is used in ranges 5 and 6 for intensities where they are small. Frequencies under consideration are ranged from 3200 cm^{-1} to 4200 cm^{-1}

This leads to the following expression for the correction with respect to the blackbody Venusian atmosphere

$$1-\alpha(T_V)=\frac{15}{\pi^4}\int_0^\infty\frac{x^4e^xdx}{(e^x-1)^2}\cdot\frac{s_o}{S_\omega},\ x=\frac{\hbar\omega}{T_V},\ s_o=\frac{2\Lambda S_o}{3T_V}\cdot\frac{dT}{dh} \tag{9.2.18}$$

For parameters of the Venusian atmosphere near its surface, we have $s_o=1.2\times 10^{-28}$ cm (Fig. 9.6).

If we take for simplicity $S_\omega=const$, this formula gives

$$1-\alpha(T_V)=\frac{4s_o}{S_\omega} \tag{9.2.19}$$

Formula (9.2.19) may be used for estimations. Let us assume that frequencies with the ratio s_o/S_ω below 1% do not give the contribution to the correction $1-\alpha(T_V)$. Hence, we ignore in the integral (6.1.38) frequencies for which $S_\omega>5\times10^{-26}$ cm.

First on the basis of parameters of the HITRAN data bank, we extract frequency ranges which are transparency windows for emission of CO_2 molecules located in the Venusian atmosphere. Boundaries of these ranges are marked in Figs. 9.2, 9.3, 9.4 and 9.5 by arrows, and also these data are collected in Table 9.2. The radiative flux ΔJ is lost in transparency windows with boundary frequencies ω_1 and ω_2. Hence, the change of the radiative flux due to a given window is determined by the expression

$$\Delta J=\int_{\omega_1}^{\omega_2}I(\omega)d\omega\approx I(\overline{\omega})\Delta\omega, \tag{9.2.20}$$

Table 9.2 Spectral ranges of transparency windows of the Venusian atmosphere due to CO_2 molecules and values ΔJ of the radiative flux which passes through this window

Range	Frequency range, cm^{-1}	ΔJ, W/m^2
a	0–200	44
b	1480–1556	540
c	1648–1677	202
d	2620–2690	277
e	2760–2820	211
f	2890–2940	157
Sum	–	1433

where $\overline{\omega} = (\omega_1 + \omega_2)/2$ is an average frequency of the transparency window, $I(\omega)$ is the partial radiative flux at the frequency ω which is determined by formula (9.2.2), and $\Delta\omega$ is the width of the transparency window. Values of the radiative flux ΔJ for each transparency window are presented in Table 9.2. As is seen, the total decrease of the radiative flux of the Venusian atmosphere due to CO_2 molecules decreases by almost 9%.

Another contribution to this decrease of the radiative flux from the Venusian atmosphere toward its surface due to atmospheric CO_2 molecules follows from frequency ranges where the intensity S_ω of spectral lines is not large. The decrease of the radiative flux for i-th frequency range of a small intensity with boundary frequencies ω_1 and ω_2 is given by

$$\Delta J_i = \int\limits_{\omega_1}^{\omega_2} I_\omega(T_V)d\omega \cdot \frac{x}{1-\exp(-x)} \cdot \frac{s_o}{S_\omega}, \quad x = \frac{\hbar\omega}{T_V} \tag{9.2.21}$$

Here we assume $S_\omega \gg s_o$, i.e., this correction is small compared to the blackbody one at the Venusian surface temperature T_V. We include in Figs. 9.3, 9.4 and 9.5 rough approximations for ranges of small intensities of spectral lines. Following to these approximations, one can obtain corrections to the blackbody radiative fluxes.

One can see that ranges of low intensities are narrow, i.e., $\omega_2 - \omega_1 \ll \omega_i$, where $\omega_i = (\omega_1 + \omega_2)/2$ is the average frequency for i−th range. Then formula (9.2.21) takes the form

$$\Delta J_i = I_\omega(T_V) \cdot \frac{x_i}{1-\exp(-x_i)} \cdot \frac{s_o(S_{max} - S_{min})(\omega_2 - \omega_1)}{S_{min}S_{max}\ln(S_{max}/S_{min})}, \quad x_i = \frac{\hbar\omega_i}{T_V} \tag{9.2.22}$$

In the case where intensities of spectral lines for a used approximation are nearby, i.e., $S_\omega = const$ in a given range of a small intensity of spectral lines, the radiative flux due to this frequency range is determined by the expression

Table 9.3 Spectral ranges of small intensities in the Venusian atmosphere due to CO_2 molecules and values ΔJ of the radiative flux which passes through this window

Number	Frequency range, cm^{-1}	S_ω, 10^{-28} cm	ΔJ, W/m^2
1	1430–1460	10	8
2	2530–2620	19	124
3	2690–2730	7–12	260
4	2830–2870	70	13
5	4056–4140	10–90	2
6	4140–4190	90	7
Sum	–	–	414

$$\Delta J_i = I_\omega(T_V) \cdot \frac{x_i}{1 - \exp(-x_i)} \cdot \frac{s_o(\omega_2 - \omega_1)}{S_\omega} \tag{9.2.23}$$

In the other limiting case $S_{max} \gg S_{min}$, it follows from formula (9.2.22)

$$\Delta J_i = I_\omega(T_V) \cdot \frac{x_i}{1 - \exp(-x_i)} \cdot \frac{s_o(\omega_2 - \omega_1)}{S_{min} \ln(S_{max}/S_{min})} \tag{9.2.24}$$

We give in Table 9.3 values of the radiative flux decrease compared to the blackbody one due to ranges of a small intensity of spectral lines for indicated approximations. As it follows from data of Table 9.3, the range of a low intensity of spectral lines gives a small contribution to the decrease of the blackbody radiative flux compared to that due to transparency windows.

In summarizing these evaluations for the radiative flux in the infrared spectrum range from the Venusian atmosphere, one can note the important role of the HITRAN data bank for parameters of radiative transitions due to CO_2 molecules. Next, the decrease in the atmosphere radiative flux compared to that of a blackbody is determined by spectrum ranges where the intensity of spectral lines is below 10^{-27} cm, whereas the maximum intensity of spectral lines due to CO_2 molecules is of the order of 10^{-18} cm. In other words, the results are determined by weak overtones. The accuracy of radiative parameters for such transitions is less than that for basic radiative transitions. As a result of this analysis, one can conclude from data of Tables 9.2 and 9.3 that the parameter α of formula (6.1.14) for the Venusian atmosphere is equal

$$\alpha = 0.9 \tag{9.2.25}$$

with an accuracy of several percent.

Let us analyze on the basis of formula (9.2.25) the energy balance for the Venus and its atmosphere. According to [13], the solar radiative flux which is absorbed by the Venusian surface is 17 W/m^2 that is approximately 1% of the radiative flux from the Venusian surface. Accounting for the accuracy of the above analysis, one can represent the energetic balance of the Venus on the basis of formulas (9.2.4) and (9.2.15) as

$$\alpha(T_V) = \gamma(T_V) \qquad (9.2.26)$$

Other channels of the energy balance are weak compared to these ones. Nevertheless, they may give the contribution to the energetic balance of the Venus. Along with solar radiation absorbed by the Venusian surface, these channels include convective transport in the Venusian atmosphere and emission of a hypothetical dust located in the Venusian atmosphere. Let us analyze the emission of dust particles under optimal conditions if its size r is of the order of a typical wavelength $\lambda \sim 5\mu m$ of infrared radiation for the Venus, that is $\lambda \sim r$. Then the absorption cross section of photons by dust particles is $\sim \pi r^2$ [29] in the case of their effective interaction with the radiation field.

Let us assume that dust particles of the Venusian atmosphere absorb infrared radiation effectively, and the optical thickness of an atmospheric dust is of the order of one. In this case if we collect all this dust on the surface, its thickness is of the order of λ. Being guided by sand as a dust material in which mass density is $2g/cm^3$, one can obtain the total dust mass per unit area of the surface as $\rho \sim 10^{-3}$ g/cm^2 for this optimal case. Note that the mass per unit area of the Venusian surface due to carbon dioxide and nitrogen molecules is $\sim 10^7$ g/cm^2. This value is ten orders of magnitude larger than the minimal mass density of an atmospheric dust which is able to influence the energy balance of the Venus. Note that this dust is located close to the Venusian surface.

We note that the convective transport of the Venusian atmospheric gas proceeds in the Venusian atmosphere by analogy with that in the Earth's atmosphere. The heat flux due to convective transport in the Venusian atmosphere is less than that in the Earth's atmosphere because of a higher gas pressure there and is comparable with the temperature gradients. Therefore, convection gives a small contribution to the Venusian energy balance. But dust particles may be arisen from the Venusian surface due to convection. The dust presence in the atmosphere can increase the radiative flux to the surface in transparency windows and spectrum ranges of a small intensity of spectral lines. This effect varies weakly the radiative flux from the atmosphere toward the Venusian surface because according to formula (6.1.27) it acts in a small part of the emission spectrum. But this effect may be of importance on the stage of establishment of the Venusian temperature.

Thus, the experience in the analysis of the greenhouse effect in the Earth's atmosphere may be used for other planets. In the case of the Venus under consideration, the energetic balance of the Venusian surface is determined mostly by emission of atmospheric CO_2 molecules. In this analysis, data from the HITRAN bank play the key role because this bank contains radiative parameters of CO_2 molecules in the spectral range of transparency windows.

9.3 Equilibrium Between Atmosphere and Environment

9.3.1 *Outgoing Atmospheric Radiation from Earth*

In considering emission of atmospheres of the Earth and Venus, we take into account that the lower part of these atmospheres and their upper part which are responsible for infrared emission in the infrared spectrum range are separated. Above we consider emission of the Earth and Venus toward the Earth which is determined mostly by regions near the planet surface. Our task now to determine the outgoing radiative flux which is created by upper layers of planets. Using the experience in the analysis of the greenhouse effect in the Earth's atmosphere, we are based on the energetic balance of the Earth and environment given in Table 5.1. From the fact of separation of regions for emission creation, it follows that the optical thickness of clouds, i.e., the atmosphere region between indicated ones, is large.

In considering thermal outgoing emission of the atmosphere, we use the same method of evaluation of radiative fluxes as well as that for emission toward the Earth. Namely formula (6.1.12) is taken for the outgoing radiative flux with accounting for the contribution to this flux that follows from emission of H_2O and CO_2 molecules, as well as from water microdroplets and microparticles. Analyzing the energetic balance of the Earth and atmosphere represented in Table 5.1 and Fig.5.4, one can conclude a large optical thickness of the atmosphere u that allows one to separate radiative fluxes toward the Earth and outside. It should be noted that the optical thickness of the atmosphere is not so large when one can use the value $1/u$ as a small parameter in this analysis. In particular, the model of identical absorption coefficient over frequencies gives $u = 2.4$ according to formula (6.2.8). Nevertheless, it is possible to separate the atmosphere in independent regions for this optical thickness.

As before, in this analysis we use as a basis the model of standard atmosphere and at first will be guided by a simple model with averaging the absorption coefficient over frequencies. Then on the basis of formulas (6.2.4) and (6.2.5), we have for the effective altitude of outgoing radiation $h_\uparrow = 6.1$ km and the average radiative temperature for outgoing radiation $T_\uparrow = 248$ K. We have at this altitude $N(H_2O) = 2 \cdot 10^{16} cm^{-3}$ and $N(CO_2) = 2 \times 10^{16}\,\text{cm}^{-3}$. Next, from the model of standard atmosphere [6], the atmospheric temperature decreases with an increasing altitude, and the temperature gradient is equal to $\mathrm{d}T/\mathrm{d}h = -6.5$ K/km up to altitude $h = 11$ km, while the atmospheric temperature is constant $T = 217$ K between altitudes of 11 and 22 km

Similar to atmospheric emission toward the Earth, outgoing radiation of the atmosphere is determined by H_2O and CO_2 molecules, as well as by water droplets or water particles. But a lower number density of molecules and their contribution to the outgoing radiative flux are less than that for the radiative flux to the Earth's surface. Next, the overlapping of spectral lines is not so important as for radiation to the Earth's surface. As before, emission of CO_2 molecules proceeds mostly inside the absorption band between 580 cm^{-1} and 760 cm^{-1}, while water molecules create radiation at lower frequencies. Figure 9.7 [30] contains the optical thickness of the atmosphere in the range of the absorption band of CO_2 molecules for the model of

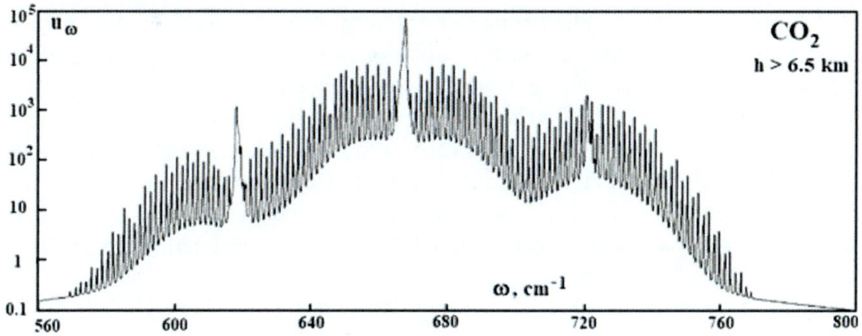

Fig. 9.7 Optical thickness of the atmosphere for outgoing radiation due to radiation of CO_2 molecules above the clouds located at the altitude $h = 6.1$ km

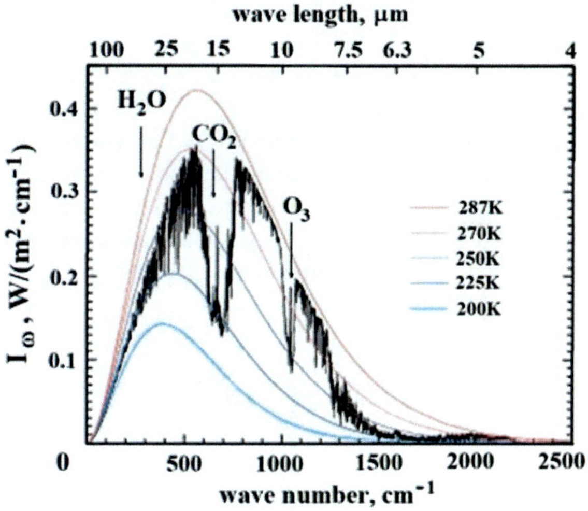

Fig. 9.8 Outgoing atmospheric radiative flux for the case of the clear atmosphere [31]

standard atmosphere. One can see that it is similar to the absorption coefficient near the Earth's surface inside the absorption band for CO_2 molecules according to its form.

The problem of emission of outgoing radiation from the atmosphere for a clear sky was considered in [32–34] in the same method as for that directed to the Earth's surface. Figure 30 represents the results of some satellite measurements for a clear sky. In this case, the radiative flux for the transparency window at frequencies between 800 cm^{-1} and 1200 cm^{-1} corresponds to the temperature of the Earth's surface. Absorption by ozone molecules in the frequency range of the transparency window proceeds in the stratosphere and has no relation to tropospheric emission. It is clear that the contribution of molecules to the outgoing radiative flux of the atmosphere is less and from clouds is larger than that for the radiative flux toward the Earth.

We now evaluate the radiative flux of outgoing radiation in accordance with the analysis [35]. In considering outgoing emission created by CO_2 molecules, we are restricted by the absorption band of these molecules in the spectrum range from $\omega_1 = 560\,\mathrm{cm}^{-1}$ up to $\omega_2 = 760\,\mathrm{cm}^{-1}$. Taking the concentration of CO_2 molecules in atmospheric air $c = 4 \times 10^{-4}$ with respect to air molecules, one can obtain that practically for all frequencies of the absorption band the effective altitude of radiation ranges between $h = 11$ km and $h = 22$ km with temperature $T_c = 217$ K. Taking the radiative temperature in this frequency range $T_\omega = T_c$, one can determine the radiative flux due to these frequencies

$$\int_{\omega_1}^{\omega_2} I_\omega(T_\omega) g(\omega) d\omega = 24\,\mathrm{W/m}^2, \quad \int_{\omega_1}^{\omega_2} I_\omega(T_\omega) d\omega = 27\,\mathrm{W/m}^2, \tag{9.3.1}$$

where $I_\omega(T)$ is the equilibrium radiative flux at frequency ω from a blackbody radiator of a temperature T which is given by the Planck formula (6.1.3). As it follows from formulas (9.3.1), inside the absorption band for CO_2 molecules the average opaque factor $g(\omega) \approx 0.9$.

In order to correct formula (9.3.1) for the outgoing radiative flux created by CO_2 molecules, we take into account that radiation created by CO_2 molecules screens that created by clouds, and because emission of clouds is not absorbed by CO_2 molecules, it partially goes outside. Then the total radiative flux created at frequency inside the absorption band of CO_2 molecules is given by

$$J_\uparrow(CO_2) = \int_{\omega_1}^{\omega_2} I_\omega(T_\omega) g(\omega) d\omega + \int_{\omega_1}^{\omega_2} I_\omega(T_{cl})[1 - g(\omega)] d\omega = 30\,\mathrm{W/m}^2, \tag{9.3.2}$$

and we take here $T_{cl} = 244$ K. One can see that this result is close to that according to formula (9.3.1).

We give in Fig. 9.8 the frequency dependence for the outgoing radiative flux of the atmosphere according to satellite measurements. This dependence consists of the Planck curve due to radiation of clouds and downfalls which correspond to emission of molecules. Emission of clouds outside is characterized by the temperature near 250 K that is the atmosphere temperature at altitudes above 6 km. Each downfall describes radiation of indicated molecules for which the effective altitude of emission is higher and the radiative temperature is lower than that for clouds. Note that we study emission of the troposphere assuming that the radiative flux is fixed at the tropopause, whereas measurements of a type of Fig. 9.8) were fulfilled at higher altitudes above the stratosphere. Therefore, being guided by emission of the troposphere, it is necessary to exclude from the consideration the downfall in the absorption band of ozone molecules which are concentrated in the stratosphere mostly.

In this consideration, we assume atmospheric emission outside at frequencies below $\omega_1 = 580\,\mathrm{cm}^{-1}$ to be created by water molecules, whereas at frequencies

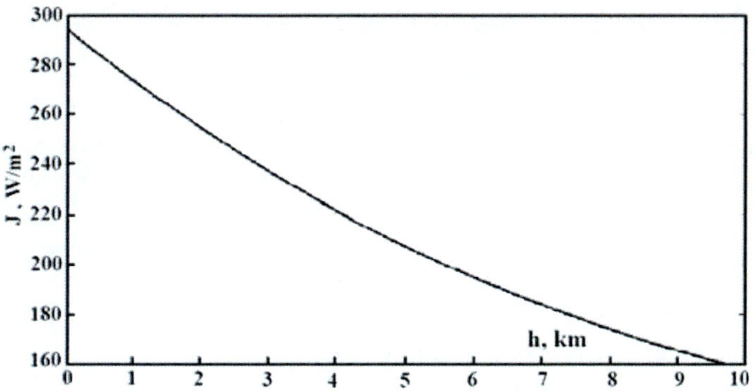

Fig. 9.9 Outgoing radiative flux of the troposphere created by clouds depending at the altitude h where cloud emission is formed

above ω_2, it is determined by cloud radiation. In evaluation the radiative flux due to water molecules, we are based on Fig. 9.9 assuming that the radiative flux varies linearly with frequencies and is independent of the cloud temperature. We then obtain for the radiative flux from this frequency range

$$J_(H_2O) = \int_0^{\omega_1} J_\omega d\omega = 98\,\text{W/m}^2, \tag{9.3.3}$$

As a result, one can obtain for the outgoing radiative flux at the tropopause $J_\uparrow$ for the model of standard atmosphere [6] with accounting for this radiation to be formed in the troposphere

$$J_\uparrow = J_\uparrow(H_2O) + J_\uparrow(CO_2) + \int_{\omega_2}^{\infty} I_\omega(T_{cl})d\omega \tag{9.3.4}$$

As is seen, the total outgoing radiative flux depends on the cloud temperature, i.e., on an altitude from which clouds emit. We give in Fig. 9.9 the dependence of the effective altitude of cloud radiation for the outgoing radiative flux of standard atmosphere, where the connection between the layer temperature T and its altitude h has the form

$$T = T_E - \frac{dT}{dh}h, \; T_E = 288\,\text{K}, \; \frac{dT}{dh} = 6.5\,\text{K/km} \tag{9.3.5}$$

In particular, on the basis of the energetic balance of the Earth given in Table 5.1, the average outgoing radiative flux from the atmosphere is equal $J_\uparrow = 200\,\text{W/m}^2$. This radiative flux and Fig. 9.9 lead to the altitude $h = 6\,\text{km}$ of cloud emission, and the cloud temperature is equal $T_{cl} = 249\,\text{K}$ at this altitude.

9.3.2 Outgoing Radiation of Venus

Our aim is to determine peculiarities of the outgoing radiative flux from the Venus. By analogy with the Earth case, it results to a greater extent from emission of aerosols consisting of sulfur acid mostly. This radiative flux determines the energy balance involving the Venusian atmosphere and environment that differs in principle from the energy balance between the Venusian atmosphere and its surface. Indicated equilibria are determined by different atmosphere regions. Therefore, they are not connected, and we consider below the energy balance between the Venusian atmosphere and environment independently of that for the Venusian surface.

As in the case of the Earth, solar radiation is the basis of the energetic balance of the Venus. The Venusian albedo is 0.80 ± 0.02 according to [11, 12] and 0.76 ± 0.01 according to [10], so that we use the average value 0.78. This gives for the solar radiative flux $J_\uparrow = 140\,\mathrm{W/m^2}$ absorbed by the Venusian atmosphere, and the same radiative flux is emitted by it in the infrared spectrum range. If infrared radiation of the Venusian atmosphere forms by one source of this emission, its temperature is $T_\uparrow = 223\,\mathrm{K}$. Because of thermodynamic equilibrium between the atmospheric molecular gas and radiation, the emitting atmosphere layer is located at the altitude of $h_\uparrow = 72\,\mathrm{km}$ above the Venusian surface. The atmospheric pressure at this altitude is $p = 0.5\,\mathrm{atm}$, and the number density of CO_2 molecules is equal there $N(CO_2) = 1.2 \times 10^{18}\,\mathrm{cm^{-3}}$.

Evidently, basic radiators of the Venusian atmosphere are aerosols and CO_2 molecules. These aerosols form clouds at altitudes of 60–70 km [36–38]. Clouds cover the Venus by a thick layer and create outgoing infrared radiation of the Venus. Taking into account that aerosols give the main contribution to the outgoing radiative flux of the Venus, one can consider two limiting cases for their emission. In the first case, we ignore the contribution of emission of CO_2 molecules to the outgoing radiative flux. In this case, the effective atmospheric layer, where aerosols create outgoing emission, is located at the altitude $h = 72\,\mathrm{km}$ above the Venusian surface and the radiative temperature, which is equal to the temperature of this layer, is $T_\uparrow = 223\,\mathrm{K}$.

In other limiting case, we assume that emission of CO_2 molecules dominates inside the absorption band of these molecules which ranges from $\omega_1 = 580\,\mathrm{cm^{-1}}$ up to $\omega_2 = 760\,\mathrm{cm^{-1}}$. Because effective altitude for emission of CO_2 molecules lies higher than that for aerosols, the radiative flux inside the absorption band of CO_2 molecules is lower than that in the absence of CO_2 molecules. Let us take it to be zero. Then the outgoing radiative flux is given by

$$J_\uparrow = \int_0^{\omega_1} I_\omega(T_\uparrow)d\omega + \int_{\omega_2}^{\infty} I_\omega(T_\uparrow)d\omega, \tag{9.3.6}$$

where $I_\omega(T)$ is the equilibrium radiative flux per unit frequency at a given frequency and temperature which is determined by formula (6.1.5). The solution of this equation for $J_\uparrow = 140\,\mathrm{W/m^2}$ leads to $T_\uparrow = 235\,\mathrm{K}$.

Basing on these results, we are guided by altitudes $h = (70 - 90)$ km, where the altitude dependence of the atmospheric temperature $T(h)$ may be approximated as

$$T(h) = T_1 + T_2 \exp\left(-\frac{H}{\lambda}\right) \qquad (9.3.7)$$

Here we use a new altitude variable $H = h - 70$ km, and $T_1 = 150$ K, $T_2 = 80$ K, $\lambda = 15$ km. Hence, the temperature $T = 235$ K corresponds to the altitude $h = 69$ km, and the atmospheric temperature $T = 223$ K relates to a layer at the altitude $h = 72$ km. Thus, the lower and upper limits for an effective altitude of emission of the Venusian atmosphere outside are nearby.

From this, one can estimate also the upper limit for emission of CO_2 molecules. Indeed, from formula (9.3.6) we have that the radiative flux from the Venusian atmosphere due to aerosols is $J_\uparrow = 113$ W/m^2 in the limit where the radiative temperature for CO_2 molecules coincides with that for aerosols, i.e., if the effective radiative temperature is $T_\uparrow = 223$ K. This flux is 0.81 part of the outgoing radiative flux $J_\uparrow = 140$ W/m^2 of the Venusian atmosphere. From this, it follows that the contribution from CO_2 molecules to the total radiative flux of outgoing radiation does not exceed 19%. Thus, simple models allow us to describe the character of outside emission of the Venus enough precisely.

Nevertheless, using an experience in outgoing emission for the Earth's atmosphere, we below analyze emission of the Venusian atmosphere outside more accurately by taking the approximation for the number density $N(CO_2)$ of CO_2 molecules at altitudes near $h = 70$ km in the form

$$N(CO_2) = N_o \exp\left(-\frac{H}{\Lambda}\right), \qquad (9.3.8)$$

$H = h - 70$ km, $N_o = 1.2 \times 10^{18}$ cm^{-3} and $\Lambda = 4.7$ km. If we assume the upper part of the Venusian atmosphere to be optically thick with respect to emission of CO_2 molecules, the atmospheric radiative temperature follows from the relation [39, 40]

$$u_\omega(h_\omega) = \kappa_\omega \lambda = 2/3 \qquad (9.3.9)$$

Here u_ω is the optical thickness of the atmospheric layer which is located between an altitude h_ω and infinity, κ_ω is the absorption coefficient at an altitude h_ω, and the radiating temperature T_ω for the outgoing flux is the atmospheric temperature at this altitude

$$T_\omega = T(h_\omega) \qquad (9.3.10)$$

We here use the "line-by-line" model [27, 28] where parameters of emission are evaluated for each frequency independently.

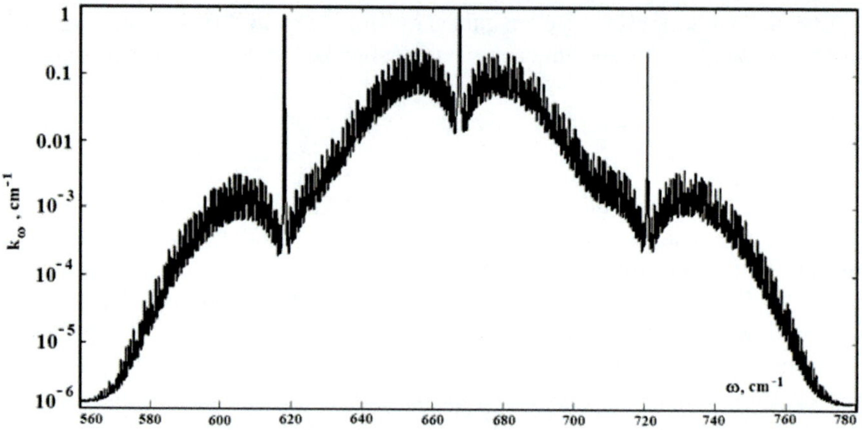

Fig. 9.10 Absorption coefficient k_ω of the Venus atmosphere at the temperature is $T = 273$ K and pressure $p = 0.5$ atm. Radiative parameters from the HITRAN data bank [23] are used

Taking the radiative temperature in accordance with the absorbed flux of solar radiation $T_\uparrow = 223$ K which refers to the altitude $h_\uparrow = 72$ km, one can obtain from formula (9.3.9) the average absorption coefficient

$$\kappa_o = 1.2 \times 10^{-6}\,\text{cm}^{-1} \tag{9.3.11}$$

One can determine the absorption coefficient of the Venusian atmosphere due to CO_2 molecules on the basis of radiative parameters which are taken from the HITRAN data bank [23]. In this case, the criterion (9.2.9) is violated, and the emission spectrum of CO_2 molecules consists of separate broaden spectral lines. Using a general method to evaluate the absorption coefficient of a gas [30], we give in Fig. 9.10 an example of the absorption coefficient due to CO_2 molecules for atmospheric parameters at the temperature $T = 273$ K. Though this temperature differs from the radiative one, it allows us to determine the effective temperature at each frequency inside the absorption band of CO_2 molecules. From this, we have that approximately 10% of the outgoing radiative flux from the Venusian atmosphere is created by CO_2 molecules.

References

1. K.E. Trenberth, J.T. Fasullo, J.T. Kiehl, Bull. Am. Meteorol. Soc. **90**, 311 (2009)
2. J.M. Lyman, S.A. Good, VV. Gouretski, et al., Nature. **465**, 334 (2010)
3. J. Hansen, M. Sato, P. Kharecha, K. von Schuckmann, Atmos. Chem. Phys. **11**, 13421 (2011)
4. N.G. Loeb, J.M. Lyman, G.C. Johnson, et al., Nature Geosci. **5**, 110 (2012)
5. G.L. Stephens, J. Li, M. Wild, et al., Nature Geosci. **5**, 691 (2012)
6. U.S. Standard Atmosphere. (Washington, U.S. Government Printing Office, 1976)

7. F.W. Leslie, C.G. Justus. https://ntrs.nasa.gov/search.jsp
8. https://en.wikipedia.org/wiki/Venus
9. https://en.wikipedia.org/wiki/Atmosphere-of-Venus
10. V.I. Moroz, et al., Adv. Space Res. **5**, 197 (1985)
11. M.G. Tomasko, et al., J. Geophys. Res. **85**, 8187 (1980)
12. D.V. Titov, et al., https://asp.colorado.edu-espoclass/ASTR-5835-2015...Notes/Titov
13. D. Crisp, D.V. Titov. In: *Venus-II*, ed. by S.W. Bougher, D.M. Hunten, R.J. Phillips. (University Arizona Press, Tucson, Arizona, 1997; p 35396384)
14. R. Haus, et al., Icarus **272**, 178 (2016)
15. P.L. Read, et al., Quart. J. Roy. Meteorol. Soc. **142**, 703 (2016)
16. S.S. Limaye, D. Grassi, A. Mahieux, et al., Space Sci. Rev. **214**, 102 (2018)
17. N.M. Johnson, R.R. de Oliveira, Earth Space Sci. **6**, 1299 (2019)
18. F. Reif, *Statistical and Thermal Physics* (McGrow Hill, Boston, 1965)
19. L.D. Landau, E.M. Lifshitz, *Statistical Physics*, vol. 1 (Pergamon Press, Oxford, 1980)
20. M. Wendisch, P. Yang, *Theory of Atmospheric Radiative Transfer* (Wiley, Singapore, 2012)
21. https://www.cfa.harvard.edu/
22. http://www.hitran.iao.ru/home
23. http://www.hitran.org/links/docs/definitions-and-units/
24. L.S. Rothman, I.E. Gordon, Y. Babikov, et al., JQSRT. **130**, 4 (2013)
25. I.E. Gordon, L.S. Rothman, C. Hill, et al., JQSRT. **203**, 3 (2017)
26. M. Simeckova, D. Jacquemart, L.S. Rothman, et al., JQSRT. **98**, 130 (2006)
27. R.M. Goody, *Atmospheric Radiation: Theoretical Basis* (Oxford University Press, London, 1964)
28. R.M. Goody, Y.L. Yung, *Principles of Atmospheric Physics and Chemistry* (Oxford University Press, New York, 1995)
29. L.D. Landau, E.M. Lifshitz, *Electrodynamics of Continuous Media* (Pergamon Press, Oxford, 1984)
30. B.M. Smirnov, *Transport of Infrared Atmospheric Radiation* (de Gruyter, Berlin, 2020)
31. https://en.wikipedia.org/wiki/Outgoing-longwave-radiation
32. R.T. Pierrehumbert, *Principles of Planetary Climate* (Cambridge University Press, Cambridge, 2010)
33. R.T. Pierrehumbert, Phys. Today **64**, 33 (2011)
34. W. Zhong, J.D. Haigh, Weather **68**, 100 (2013)
35. B.M. Smirnov, *Global Atmospheric Phenomena Involving Water* (Springer Atmospheric Series, Switzerland, 2020)
36. L.V. Zasova, V.I. Moroz, L.V. Esposito, C.Y. Na, Icarus **105**, 92 (1993)
37. L.V. Zasova, N.I. Ignatiev, I.V. Khatuntsev, V.M. Linkin, Planet Space Sci. **55**, 1712 (2007)
38. Y.J. Lee, D.V. Titov, S. Tellmann, et al., Icarus. **217**, 599 (2012)
39. B.M. Smirnov, *Physics of Ionized Gases* (Wiley, New York, 2001)
40. B.M. Smirnov, *Microphysics of Atmospheric Phenomena* (Springer Atmospheric Series, Switzerland, 2017)

Chapter 10
Physical Aspects of Climate Change

Abstract The analysis of power atmospheric processes depending on its parameters and external conditions allows one to describe the thermal state of the Earth. For simplicity, one parameter is used usually for this goal—the global temperature that is the temperature of the Earth's surface averaged over the globe and time. An important factor of a change of the global temperature in time is the intensity of solar radiation that penetrates in the Earth's atmosphere. Its variation due to a change of the Earth's position with respect to the Sun causes the periodicity in the global temperature with the period of the order of hundred thousand years. Two regimes exist for evolution of the global temperature, with and without glaciers, because of different values of the Earth's albedo for these regimes. The local temperature in past resulted from the isotope analysis of sediments allows one to restore evolution of the Earth's temperature. In particular, the difference between the maximal and minimal temperatures in the course of Earth's evolution is almost 20 °C that by one order of magnitude exceeds the corresponding value that is the basis of the Paris agreement 2015 on climate. The influence of the human activity is stronger for a megapolis with a heightened density of population. It is shown that artificial cooling of the megapolis temperature may be resulted from injection in the atmosphere the water microdroplets with a dissolved absorbed additivity. This technology is analogous to aerial firefighting. In addition, it is shown that the power of hydroelectric plants and wind turbine is compared with that of greenhouse radiation due to water evaporated from the ground under the action of these systems. As it follows from our experience, the start of the Earth's warming begins from eighties of twentieth century, as well as the anthropogenic change of the climate.

10.1 Natural Factors in Earth's Thermal State

10.1.1 Solar Radiation in Earth's Climate

Solar radiation penetrated in the Earth's atmosphere is the driving factor which determines its energetics. Indeed, the power of absorbed solar radiation causes a chain of power processes which proceed on the Earth's surface and in its atmosphere.

B. M. Smirnov, *Global Energetics of the Atmosphere*, Springer Atmospheric Sciences, https://doi.org/10.1007/978-3-030-90008-3_10

But the Sun is not a permanent source of energy, as well as the position of the Earth with respect to the Sun. Therefore, the solar energy absorbed by the Earth and its atmosphere varies in time. Such variations lead to the change of the Earth's global temperature as the basic parameter which characterizes the Earth's thermal state. We below consider the character of real variation of these parameters in the course of evolution of the Sun and the Earth.

We first consider the influence of solar radiation on the Earth's thermal state during contemporary time which includes the last several centuries. Short-term oscillations of the intensity of solar radiation are characterized by a period of 11 years. Direct satellite measurements show that the change in the solar flux in the course of the previous two cycles does not exceed 0.1%, which corresponds to the variations in the flux itself of about 0.3 W/m^2 [1, 2] that according to formula (5.3.5) gives the value of 0.15 K for variation of the global temperature during one solar period. This corresponds to the estimation [3] of (0.18 ± 0.08) K [3] for the difference of global temperatures at the maximal and minimal solar irradiance of the same solar period.

However, the observed change in the global temperature for recent decades is larger by several times than that due to variations in the solar irradiance during one period. In addition, the observed change of the global temperature is not oscillatory, but monotonous in time. One can emphasize that for last centuries, starting from the Maunder period, the solar irradiance for current cycle minima increases approximately by 0.5 W/m^2, i.e., by 0.04% of its value [4]. Therefore, short-term changes in the solar activity do not explain the observed warming of the planet. The restoration of the solar irradiance from 7000 BC for 7500 years [1] gives its variations from 1358 up to 1370 W/m^2. This corresponds to variations of the global temperatures up to 1.5 K. From this, one can conclude that though variations in the solar irradiance do not determine observed changes in the global temperature, solar processes are of importance in establishment of the global temperature.

One can understand that variation of solar irradiance is not the reason of an observing variation of the global temperature. According to Fig. 5.1, the solar radia-

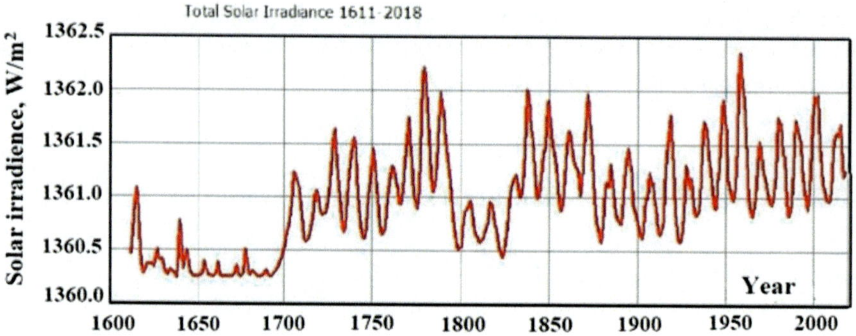

Fig. 10.1 Evolution of the radiative flux generated by the Sun and entering the Earth's atmosphere over the past centuries [5, 6]

tive flux penetrated in the Earth's atmosphere correlates with a number of sunspots. Figure 10.1 represents the character of evolution of the solar irradiance over the four last centuries. This curve is restored on the basis of the number of sunspots at the corresponding time. As is seen, values of the solar irradiance during last four centuries are concentrated in the range of 2 W/m^2 that corresponds to deviations of 0.15%. Being guided by the climate sensitivity of $S = 0.5\ \mathrm{m}^2 \cdot \mathrm{K/W}$ according to formula (5.3.5), one can find the temperature change $\Delta \approx 1$ K in this time range. This value is comparable with an observed one in contemporary time.

Since solar radiation is the basis of powerful processes in the atmosphere and the thermal state of the Earth, change of the intensity of solar radiation is responsible for the long-term changes in the thermal state of the Earth. When the Earth moves around the Sun along its orbit, the intensity of solar radiation to the Earth's surface varies that leads to a variation in the global temperature. Such changes of the global temperature have an oscillate character, as it follows from Fig. 5.11.

Let us analyze these oscillations in more detail in the framework of the Milankovich theory [7, 8]. Indeed, the elliptical orbit of the Earth itself, along which it moves around the Sun, is stable, so that it is characterized by small shifts during one period. Small perturbations lead to a small change of the trajectory. Three types of oscillations can occur for the elliptical orbit along which the Earth moves. Namely, these oscillations include the oscillation of the plane of the trajectory, of the angle between the Earth's axis and the plane of its orbit and of the eccentricity of the orbit, which characterizes the minimum and maximum distance between the Sun and the Earth as they move in their orbits. The analysis of long-term variations of the Earth's trajectory around the Sun allows us to study climate changes over a long period of time.

The character of change in radiative fluxes to the Earth's surface in the course of oscillations of the Earth's orbit is shown in Fig. 10.2. Note that the elliptical orbit of

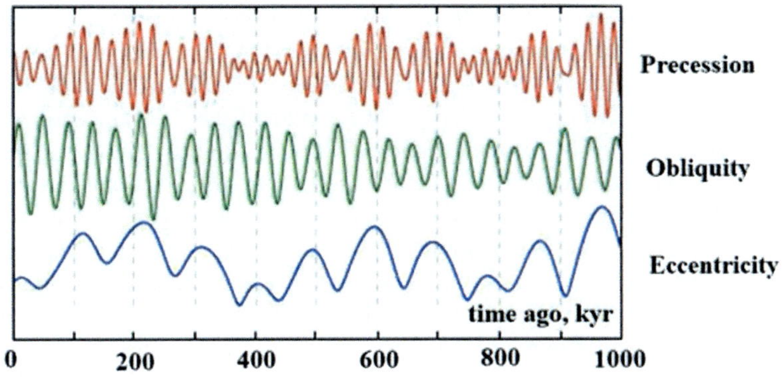

Fig. 10.2 Evolution of the solar radiative flux in the visible region of the spectrum as a result of oscillations of the Earth's orbit with respect to the Sun. The time dependence for the solar radiative flux entered the Earth's atmosphere is presented for each of the types of oscillations of the Earth's orbit relative to the Sun in the framework of the Milankovich model [9]

the Earth is close to a circular one, and from the point of view of the Earth's climate change, the most important are the oscillations of the seasonal average distance from the Earth to the Sun. The oscillation period for this parameter is approximately one hundred thousand years, which more or less corresponds to the period of recurrence of glacial stages. The most rapid type of oscillations due to changes in the Earth's orbit corresponds to a change in the angle of inclination of the Earth's axis to the plane of its orbit, which varies from 22.1° to 24.5°, and the period of such oscillations is 41 thousand years. Currently, the angle of inclination of the Earth's axis to the plane of the orbit is 23.44°. It is essential that in the process of oscillation of the Earth's orbit, the trajectory of the Earth's motion can be repeated only in the absence of other interactions, except for the interaction between the Earth and the Sun. The inclusion of other interactions (in particular, the interaction of the Earth with other planets of the solar system) violates the deterministic nature of the Earth's movement, and a typical time of transformation of this movement into chaotic is the longer, the weaker this interaction is.

The characteristic of interaction of solar radiation with the Earth as a planet is the albedo (see §8.1.4) that is a part of the radiative flux reflected from the Earth's surface. The albedo varies from zero for a blackbody to one for a high-quality mirror. Along with the solar irradiance, thermal state of the Earth's surface depends on its albedo. As a demonstration of influence of the character of interaction between solar radiation and the Earth's surface, we consider the case when the surface is covered by ice. If ice is melted, another regime of thermal equilibrium is realized. In this example, we are guided by Greenland which was free from the ice in eleventh century when vikings occupied it. Let us take for definiteness that the albedo is $\alpha = 0.6$, when it is covered by a glacier and is equal to $\alpha = 0.3$ if it is free from ice. The energy balance equation has the form

$$J_{\downarrow} = (1 - \alpha)\sigma T_s^4, \tag{10.1.1}$$

where $J_{\downarrow}$ is the total energy flux absorbed by the surface, and T_s is its temperature. Here, we assume that a region under consideration has a large size, so that the horizontal transfer is not essential, and the total energy flux $J_{\downarrow}$ includes both radiative and heat transport of energy. We have two regimes of the thermal regime for the surface thermal state, if the surface is covered by ice or by grass. As is seen, there are two thermal regimes of the surface, if it is covered by ice or grass. Transition between these regimes proceeds with a delay, so that transitions in the course of heating and cooling proceed at different temperatures. In reality, this situation is realized also for lakes. In this case, the lake surface is covered by ice in winter during the heating process and is covered by grass autumn. Thus, a different character of evolution is realized at the same ground temperature. These examples show that the thermal balance of the Earth's surface depends not only on the solar irradiancy, but also on the character of interaction between incident solar radiation and the Earth's surface.

It should be noted the role of the Earth's atmosphere in interaction of solar radiation with the Earth. Let us remove the atmosphere in ours mind. On the basis of definition for the albedo of the Earth's surface α, we have

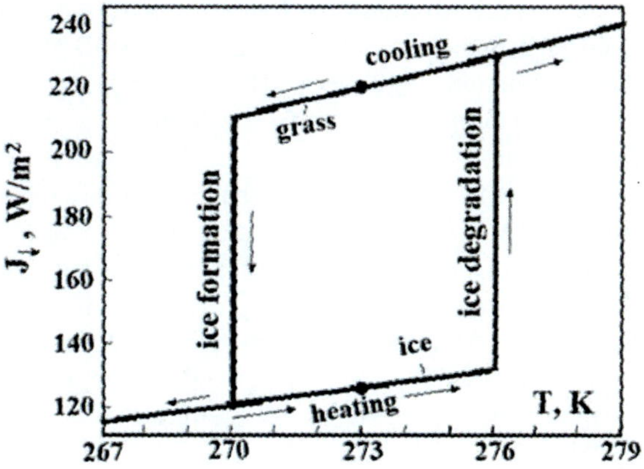

Fig. 10.3 Character of evolution of the thermal state for the Earth's surface in the course of an increase (heating) of the heat flux $J_\downarrow$ to the Earth's surface for its certain region and a flux decrease (cooling), as it is shown by arrows. T is the surface temperature averaged over a short-term variations

$$J = J_r(1 - \alpha)$$

Here, J is the flux absorbed ty the Earth's surface, and J_r is the incident solar radiative flux. Assuming radiation of the Earth's surface to be identical to that of a blackbody in the infrared spectrum range, one can obtain in accordance with Fig.5.4 the flux J_E from the Earth's surface according to the Stephan–Boltzmann law

$$J_E = \sigma T_E^4$$

From this, one can obtain the connection between the flux of incident solar radiation J_r and the surface temperature T_E of the Earth

$$J_r(1 - \alpha) = \sigma T_E^4 \qquad (10.1.2)$$

Taking the incident solar radiative flux $J_r = 341$ W/m^2 according to data of Table 5.1 and the albedo is $\alpha = 0.11$ according to Table 8.2 data, one can obtain for the Earth's temperature $T_E = 270$ K, whereas within the framework of the model of standard atmosphere, it is equal $T_E = 288$ K. As is seen, the processes with participation of the atmosphere lead to an increase of the global temperature T_E.

10.1.2 Instability of Global Temperature

In the analysis of the energetic balance for the Earth (Chap. 5), we assume a certain chain of powerful processes which establish the contemporary energetic balance of the Earth and atmosphere. Variation of values of parameters in this balance leads to a shift of the energy equilibrium of the Earth and changes the global temperature, if we discuss within the framework of the model of standard atmosphere [10]. In particular, an observed warming of the Earth now assumes to be a result of the change of the atmospheric composition that leads to enforcing of the greenhouse effect. It is evident that the energetic equilibrium of the Earth exists at a certain limit. If this limit will be reached or the system under consideration approaches to this limit, two scenarios are possible.

In the first case of an energetic balance, this limit is a critical point, and approaching to this point is caused by an increase of factors which increase the global temperature. Evidently, in the case of the Earth's energetic balance, another regime of the Earth's thermal balance is similar to the process of a thermal explosion in chemical physics. In this case, the temperature dependence of the rate of heat release is sharper than that for heat remove [11–13]. If the system reaches the critical temperature, a new energetic equilibrium is realized with a higher temperature, and this transition is irreversible.

We demonstrate the thermal explosion process on the example of a burning process for a fuel particle located in atmospheric air. In this case, the thermal equilibrium of the particle results from heat extraction as a result of the chemical reaction of particle oxidation, and the particle transfers an energy to surrounding air through the air thermal conductivity. As a result of these processes, the particle temperature T_o differs from the temperature T of surrounding air far from it. Let us take the rate $k(T)$ of the heat release in the form of the Arrenius dependence [14]

$$k \sim \exp\left(-\frac{E_a}{T_o}\right)$$

Then, the heat balance equation has the form of the Zeldovich equation [15–17]

$$T_o - T = A\exp\left(-\frac{E_a}{T_o}\right) \tag{10.1.3}$$

The left-hand side describes a heat loss through the thermal conductivity. If the difference of the particle and air temperatures is small, this term is proportional to the temperature difference. The right-hand side of this equation includes heat extraction as a result of combustion which is characterized by the activation energy E_a, and the factor A of Eq. (10.1.3) includes numerical factors of rates of these processes which are independent of the temperature. The nature of the thermal explosion that the rate of thermal conductivity depends weakly for the particle temperature, whereas the rate of combustion has the exponential dependence. Therefore, the thermal equilibrium is violated starting from a certain power of the combustion process.

Figure 10.4 contains the graphic representation of the Zeldovich equation. Curves 1 and 2 of Fig. 10.4 correspond to the left-hand and right-hand sides of the Zeldovich Eq. (10.1.3) at a stable solution of this equation. Then, the point of intersection of these curves gives the particle temperature which differs from the air temperature corresponded to the origin. If the air temperature increases, the curve 1 displaces right, and the instability of this regime starts if a new curve position $1'$. This means

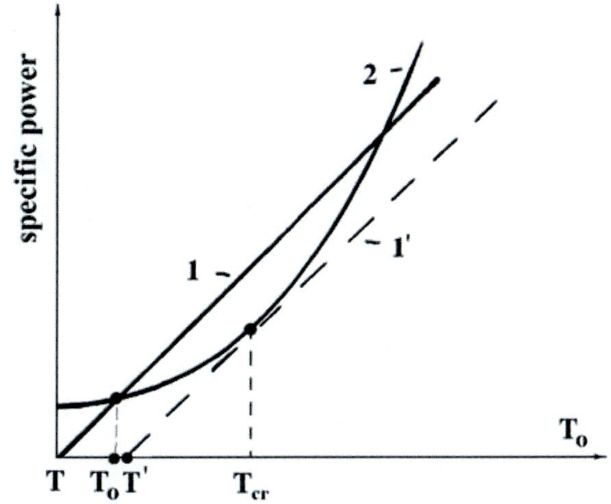

Fig. 10.4 Dependence on the particle temperature T far from the particle for the left-hand side and right-hand side of the Zeldovich Eq. (10.1.3) (solid line). The solid line is this dependence for the critical point [18]

that at larger shifts the thermal loss due to air thermal conductivity cannot provide the heat balance. This instability, that is the thermal explosion, starts if these two curves touch each other.

The equation of this thermal instability is determined by the following equation which results from the Zeldovich Eq. (10.1.3)

$$T_{cr} - T' = A \exp\left(-\frac{E_a}{T_{cr}}\right), \ \frac{T_{cr}^2}{E_a} = A \exp\left(-\frac{E_a}{T_{cr}}\right), \tag{10.1.4}$$

and because of a large value of the activation energy $E_a \gg T_{cr}$, one can obtain from Eq. (10.1.4)

$$T_{cr} - T' \ll T_{cr} \tag{10.1.5}$$

From this, it follows that the temperature difference between the particle temperature and that of surrounding air is small. This means that the thermal instability proceeds at a small difference between the temperatures of the particle and surrounding air. As a result of this instability, the particle temperature increases strongly, until a new character of the energetic balance of the particle is established. In particular, in the case of combustion of a fuel particle, heat removal in a new regime of the thermal particle evolution in the course of combustion results in radiation of the particle.

The critical character of thermal processes means that the global temperature may be found up to a certain limit. The above mechanism, applied to the Earth, means the possibility of two thermal regimes of its evolution. Along with the above explosive regime of evolution, another approach to the maximum temperature for the system is possible as a result of continuous development of this system. In this case, an approach to the maximum temperature has the character of saturation. We give in Fig. 10.5 possible regimes of evolution the system depending on the temperature.

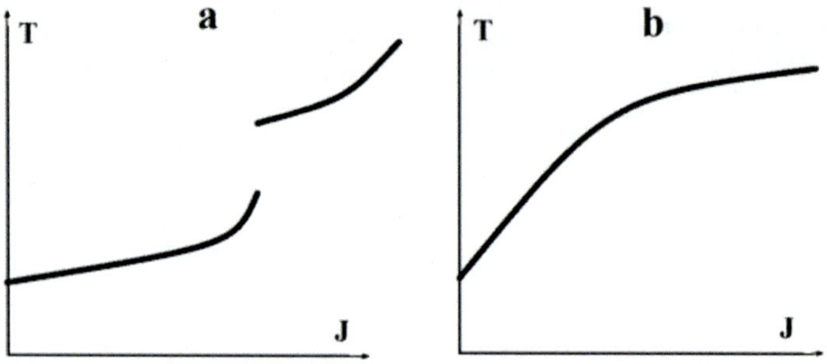

Fig. 10.5 Two regimes of evolution of the temperature depending on character of the process: **a** -explosion process, **b** -process of saturation

One can consider from another standpoint occurrence of the explosive character for evolution of the thermal state of an object as a result of a continuous change of its temperature from another standpoint. Namely, an extracted heat cannot be removal as a result of a given mechanism of heat absorption. Then, an extracted heat remains inside the object that causes an explosive character of an increase of the temperature. This character takes place in the case of the greenhouse instability (see §6.2.4). Under conditions of the standard atmosphere [10], this instability starts if the global temperature increases by approximately 10 °C. This instability finishes when a water vapor becomes saturated. The increase of the air moisture under conditions of the standard atmosphere leads to the Earth's heating by 2 °C.

10.1.3 Earth's Climate in Past

We now consider the Earth's energetic balance from this standpoint. The analysis of the energetic regime of the Earth under contemporary conditions is represented in Chap. 5. We characterize the thermal state of the Earth by the global temperature. But determination of the global temperature is possible only last 150 years, when meteorological stations arised over all the globe and the possibility occured to have simultaneously the local temperature at various points of the globe. In considering a longer period of time, we are based on data related to certain areas of the Earth. Nevertheless, a large number of such examples allows one to understand the character of evolution of the Earth's temperature, as well as limits in which it can be found.

Let us consider first methods of determination of the temperature in past. Some information about the temperature of the Earth follows from the solar activity expressed in terms of the number of sunspots. There is information on the number of sunspots dating back to the seventh century, and it is believed that the solar activity correlates with the total solar radiation flux. In this regard, the coldest time refers to the period 1645–1715, the so-called Maunder minimum, when sunspots were completely absent. More significant are the written records of the Earth's temperature and the climate of certain areas of the Earth in recent millennia, although they are often fragmentary. These data indicate a warming in northern Europe in the 11th–14th centuries, when grapes ripened in England, and the Vikings took over Greenland. After that, there was a period of cold weather, when the River Thames in London froze every year, and the Vikings left Greenland.

A universal and reliable method for determining the climate in past uses the isotope analysis of sediments. In this case, the temperature of the planet at a moment when the layer in question was deposited follows from the ratio of the concentrations of stable isotopes ^{18}O and ^{16}O, as well as from the ratio of the concentrations of stable isotopes ^{13}C and ^{12}C (e.g., [19–22]). The time of formation of this object can be determined both by the depth of the layer under study and on the basis of the so-called geochronological method (e.g., [23, 24]), which allowed us to look deep into the past. First of all, this method applied to the analysis of the radioactive isotope of carbon ^{14}C, for which half-decay time is 5730 years. In this method [25–27], the assumption is used that cosmic rays act on a sample, and this action ceases when this sample is transformed into a sediment. In this case, formation of the radioactive isotope proceeds according to the scheme

$$n +^{14} N \to p +^{14} C \tag{10.1.6}$$

This process has became an important tool of archeology and has played a significant role in creation and development of geochronology.

The radioactive carbon formed in the process (10.1.6) then passes into carbon dioxide in the atmosphere and, as a result of photosynthesis, becomes part of the plant, and the residence time of a carbon dioxide molecule in the atmosphere is approximately 4 years [28]. When the plant dies, this carbon goes underground and does not participate in atmospheric nuclear processes. Since this carbon isotope constantly decays, its concentration can give the residence time of the radioactive isotope ^{14}C in the sample. This method is widely used in archeology to determine the lifetime of objects, but it is also convenient for finding the time of formation of the corresponding layers in the pit, if this time is of the order of thousands of years. For other times scales, other radioactive isotopes are used, just as other stable isotopes can be used to determine the sample temperature. In particular, Fig. 5.19 represents the temperature change in the past for almost a million years based on the ratio of the deuteron and proton concentrations in the sample.

The use of radioactive methods for study of radioactive isotopes of heavy elements in a sample allows us to look into the much more long past and to estimate a long time range where the global temperature varies. In this case, the sample temperature at moment, when its state becomes stable, follows from the analysis of the concentrations of stable isotopes. The most wide method for this purpose is the analysis of the ratio of the amount of stable oxygen isotopes ^{18}O and ^{16}O, the occurrence of which in the Earth's crust is 0.2% and 99.76%, respectively. In addition, the concentration of stable carbon isotopes ^{13}C and ^{12}C in the Earth's crust equals 1.10% and 98.90%, correspondingly. Next, in studying the compositions of air bubbles in glaciers and clays, the temperature of the object at time of its occuring inside the

Earth is based on the ratio of the concentration of the stable isotopes of argon ^{38}Ar and ^{40}Ar, the occurrence of which in nature is 0.06% and 99.60%. Sometimes for this purpose, the ratio of the concentration of stable isotopes of hydrogen—deuteron ^{2}H and protons ^{1}H is used with the occurrence in nature of 0.015% and 98.985%, respectively. Isotopes of various elements allow one to study different time ranges.

This set of tools provides information about the temperature at the time of formation of objects located in the pits. This method also is a basis of the geochronology as part of the paleontology (e.g., [19, 29–33]). The combination of various radioactive isotopes allows one to fulfill this analysis in a wide range of times on the basis of methods of geochronology (e.g., [23, 24, 34, 35]. This combination gives the possibility to determine the evolution of the Earth's temperature for various epochs, as it is given in Fig. 10.6.

Along with the temperature, these methods give the evolution of the ice covering of the Earth. Formation of ice at the Earth's surface changes the Earth's albedo and in this manner acts on the energy balance of the Earth. As an example, Fig. 10.7 shows the changes in the total volume of glaciers in past, as follows from the analysis of soil deposits [36]. This figure shows that ice ages follow with a periodicity of about one hundred thousand years, and the Earth is currently in an interglacial relatively warm period. At the same time, during warm periods, most of the ice melts and then accumulates again.

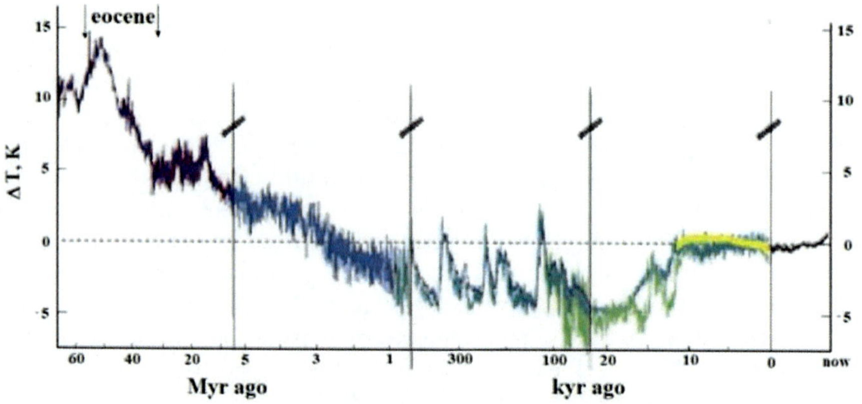

Fig. 10.6 Natural evolution of the Earth's temperature from Eocene up to end of the BC epoch

Fig. 10.7 Change in the total mass of glaciers on the Earth in past [36]

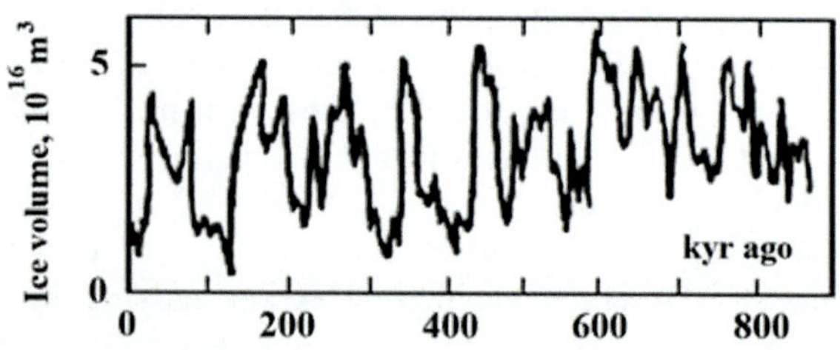

Let us estimate the power that is spent or released as a result of ice melting or its formation during the transition from cooling to warming or vice versa. As follows from Fig. 10.7, the characteristic change in the volume of ice over a single period is on the order of 10^{16}m^3 of ice or of the order of 10^{16} tons, and a typical timescale over which this occurs is on the order of ten thousand years. From the ratio of these values, one can obtain an estimate for a typical power of the corresponding heat processes as $\sim 10^{13}$ W. As is seen, this value is four orders of magnitude less than the power obtained by the Earth from the Sun, and it is compared with the power of the human activity.

If we look back at long times, one can see that the warmest period on the planet was the Eocene, which refers to the interval between 56 and 34 million years ago [38] when the Earth's temperature exceeded the current value of the global temperature of the planet by about 10 °C. Note that it would take approximately 500 years to achieve it, starting from the current temperature and based on the current rate of warming. The temperature of the planet during the Eocene period mainly follows from the analysis of the ratio of the concentrations of stable carbon isotopes ^{13}C and ^{12}C in fossil leaves [39–41]. Figure 10.8 presents the evolution of the Earth's temperature during the Eocene period. During the Eocene period, the temperature near the north pole at the beginning of this period was (23–24) °C [42, 43]. Further, during this period, the temperature near the south pole increased by (3–5) °C [44], i.e., the warming was global in nature. This warming influences the environment, and, in particular, mammoths appeared on Earth widely [37].

The peak in Fig. 10.8 is of interest, where a sharp increase in temperature was observed in the initial Eocene period over ten to twenty thousand years [37, 45, 46], i.e., over a relatively short time interval compared to geological scales. The simplest explanation of this fact is related to the increase in the greenhouse effect as a result of the injection of a large amount of greenhouse gases into the atmosphere [42]. However, this explanation contradicts the current understanding of this effect, since the time spent by greenhouse gases in the atmosphere does not exceed several years. It is more likely that the sharp increase in the temperature of the planet during this

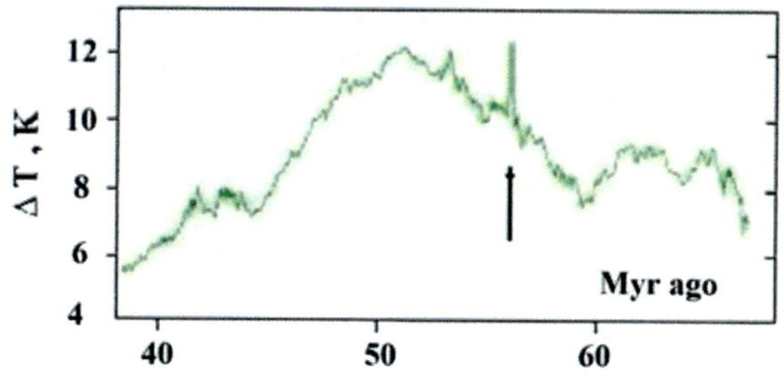

Fig. 10.8 Evolution of the difference of the Earth's temperature near the North Pole and contemporary global temperature during the Eocene period [37]. The arrow indicates the time when the temperature of the Earth's surface increased dramatically and, probably, results from the greenhouse instability (§6.2.4)

time is due to the greenhouse instability (see §6.2.4), to thermal processes inside the Earth, including the volcanic activity, or due to changes in the radiation flux from the Sun. Note that the basic method to study the temperature evolution during the Eocene period is based on the analysis of plants. Note that the study of the Earth's climate in past provides a rich experience for the analysis of the climate in the future millennia a result of natural processes [47–50].

Isotope methods allow us to study not only the evolution of the planet's temperature, but also the phenomena that lead to a sharp change in the state of the planet. One of the most important events of this kind is the fall to the Earth's surface of an astronomical body, which occurred on the Yucatan peninsula in Mexico [51, 52]. The size of this body was 10–15 km, and its fall led to the release of energy that exceeds the energy of the atomic bombs in Hiroshima and Nagasaki by billion times [51, 53]. This event occurred approximately 66 million years ago, and this time is restored to an accuracy of ten thousand years [51, 54]. As a result of this event, there was a strong release of dust into the troposphere and stratosphere, so that the Earth's surface temperature fell by about 7 °C, since almost 50% of solar radiation did not reach the Earth's surface due to this dust [55]. This event led to significant changes on our planet, in particular, to the disappearance of large dinosaurs, while flying dinosaurs were transformed into birds. This caused also the development of mammoths and a number of mammals on the planet, filling the vacant niches [56–58].

As it follows from Fig. 10.3, the presence of ice on the Earth's surface is of importance for evolution of global temperature. Evidently, the periodicity of this process is governed by the Earth's position with respect to the Sun, and at least during last epochs time may be divided in so called ice periods each of them lasts during of the order of hundred thousands years. Then, one can extract in each ice period a time of cooling, where glaciers grow, and warming, where glaciers melt. There are more than 132,000 glaciers on the Earth's surface now, and they cover an area of 740,000 km^2 [59]. At this warming period, the surface area covered by glaciers is reduced by half [60] in about 50 years. This characterizes a degree of warming now.

10.1.4 Action of Aerosols and Condensation Processes on Climate

We above analyze methods of the analysis of the climate change in past and give some results for the time dependence of the global temperature many years ago. Our goal now is to analyze the reasons of these variations and mechanisms which govern by these variations now. One can expect that the Sun causes the global temperature variation through the change of the Earth's position with respect the Sun which variations a change the intensity of irradiance. This effect takes place, but it is realized at larger timescales. Therefore one can accept that for a timescale under consideration variations in solar radiation do not lead to change of the climate.

In analyzing the weather, one can consider cyclones and anticyclones, atmospheric vertical and horizontal winds, condensation in the atmosphere and precipitation of water, ice and snow as the basic meteorological processes. Being guided by a long-term period, which exceeds, at least, 10 years, we average over these processes. Restricting by a time range of (10–10,000) years, one can ignore below the climate change due to solar processes with larger times. Therefore, we below consider only the climate change due to the greenhouse effect and to the presence of particles in the atmosphere.

The residence time for greenhouse molecules is small in this timescale. Indeed, the tropospheric residence time for water molecules is approximately 9 days, and the residence time of carbon dioxide molecules in the atmosphere is approximately 4 years. After this time, these molecules transfer on the Earth's surface. Hence, the global temperature changes due to the greenhouse effect in the total system, the atmosphere–Earth's surface. As for atmospheric droplets and particles, their residence time is restricted in the lower troposphere which is responsible for infrared radiation to the Earth. Therefore, they partake in the greenhouse effect only in the case of the stable change of their amount in the troposphere. But if they occur in the stratosphere, they can reflect solar radiation, and in this manner, they decreases the global temperature. We above analyze the action of volcanoes which throw out a dust in the atmosphere. Now, we analyze other examples of this type.

Along with injection of small particles in the atmosphere as a result of eruption of volcanoes, other ways are possible for occuring of small particles in the atmosphere. According to evaluations [61], a dust which is injected from the Halley comet with a period 72 ± 5 yr is able to cool the Earth's surface by 0.08 °C on average. Cooling of the Earth's surface results also from the cosmic dust of interplanetary and interstellar origin [62]. Meteorites form a dust during their pass through the atmosphere [63]. In all these cases, dust particles act not only as reflectors of solar radiation, but they are also nuclei of condensation. In addition, charged particles resulted from the pass of meteors it through the atmosphere change the atmospheric resistance up to several percent. This may influence the condensation processes in the atmosphere.

Being guided by eruption of volcanoes with injection of aerosols into the stratosphere, where they are not washed by rain, we estimate the action of aerosols on the Earth's energetic balance under favorable conditions where these aerosols reflect solar radiation in the strongest manner. Let us take a typical size of these aerosols to be $r = 1\mu\text{m}$ that weakly exceed the wavelength of solar radiation and the cross section of reflection of solar radiation to be

$$\sigma = \pi r^2 \sim 3 \times 10^{-8}\ \text{cm}^2$$

Taking the density of an aerosol material as $\rho = 2\text{g/cm}^3$, we have that its mass is

$$m = \frac{4\pi r^3 \rho}{3} \sim 1 \times 10^{-11}\text{g}$$

Next, the optical thickness of the atmosphere with respect to aerosols is

$$u = n\sigma,$$

where n is the number density of aerosols per unit area of the atmospheric column. For definiteness, we give estimations for the case if a decrease in the global temperature $\Delta T = 2\,^{\circ}\mathrm{C}$, or according to formula (5.3.5), the reflected flux of solar radiation is 4 W/m^2, i.e., approximately 1% of the incident radiative flux from the Sun which according to Table 5.1 is equal 340 W/m^2. This gives that the optical thickness of the atmosphere with respect to reflection of solar radiation by aerosols is $u \sim 0.01$. For aerosols under consideration, this leads to the following number density of aerosol per unit area of the atmospheric column

$$n \sim 5 \times 10^5 \mathrm{cm}^{-2}$$

Since the area of the Earth's surface is equal $S = 5.1 \times 10^{18} \mathrm{cm}^2$, the total mass of aerosols M in this case is equal

$$M = nmS \sim 3 \times 10^{13} \mathrm{g},$$

i.e., this mass is $M \sim 30$ million tons. In these estimations, we assume aerosols to be distributed uniformly over the Earth's surface. In addition, the velocity v of aerosol falling under these conditions is $v \sim 0.01$ cm/s. Hence, the residence time of these aerosols in the stratosphere is several years.

Cosmic rays are the source of ionization in the atmosphere, and formed charged particles are nuclei of condensation. According to investigations [2, 64, 65], an increase of the intensity of cosmic rays, as well an increase of their energy, leads to growth of the rate for formation of condensation nuclei. From this, it follows that variations in the intensity of cosmic rays which ionize atmospheric air, influences on the global temperature. The main part of cosmic rays are protons and the Earth's magnetic field does not allow them to penetrate in the Earth's atmosphere if the proton energy is below a certain value that is names the cutoff rigidity. The contemporary value of the cutoff rigidity near the Earth's equator is about 14 GeV. But this value may be varied in time that can be the reason of the climate change because of the change of the rate of cloud formation [2]. During last 3.4 million years, the Earth's magnetic pole changed its polarity nine times [2] that is the evidence of the fickleness of the cutoff rigidity. Therefore, this property of the Earth's magnetic field may be a reason of the climate change [2].

On the way toward the Earth, cosmic rays interact with the solar wind which intensity depends on the solar activity. Because of the periodicity of the solar activity, this periodicity may be observed in the rate of condensation in the atmosphere, and then, the cloudiness must contain a periodic component [66–68]. Figure 10.9 exhibits this connection for 22 cycle of the solar activity from 1981 to 1992. But during the subsequent 23 cycle of the solar activity cycle, this correlation is absent according to Fig. 10.9. One can explain this fact that during the 22 solar cycle cosmic

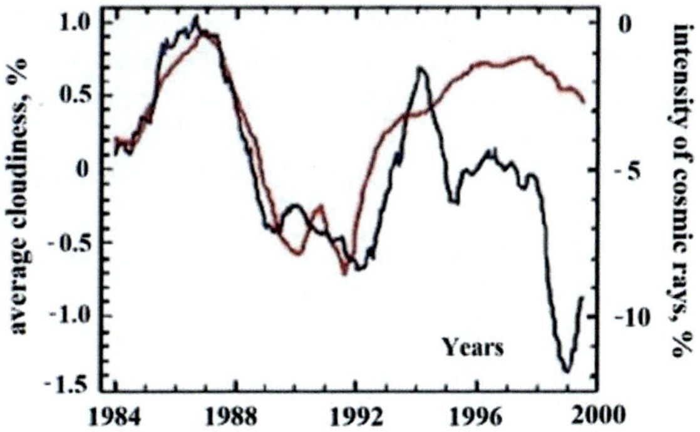

Fig. 10.9 Correlation between the anomalies of the cosmic-ray flux and the change in cloudiness in the lower atmosphere [66]. D—deviations from the average for the cosmic-ray flux, are given in black, deviations from the average cloudiness for the lower atmospheric layer are drawn in red

rays significantly influenced the cloud growth through formation of condensation nuclei, whereas then other channels appeared for water condensation nuclei in the atmosphere due to air pollution. This concept of the influence of cosmic rays on the climate was checked in experiments at the CERN accelerator [69], where the action of cosmic radiation on atmospheric air was simulated by the action of synchrotron radiation.

10.2 Anthropogenic Factors of Climate Change

10.2.1 Variations in Greenhouse Effect of Earth

Now, above 80% of the world energy results from combustion of fossil fuels. Forming carbon dioxide transfers in the atmosphere and its part remains there. As a result of transferring of carbon from the Earth's interior in the atmosphere, the concentration of atmospheric carbon dioxide increases and, since it is a greenhouse component, the radiative flux to the Earth's surface increases also. This concept is accepted as the only mechanism which lead to heating of the Earth. Moreover, this is a basis of the Paris agreement on climate of 2015 [70] and all the countries must support the fond of 100 billion dollars per year that has to use for development of noncarbon technologies for weak countries.

Indeed, this is a logically consistent concept. The above analysis allows to obtain numerical parameters of this concept. The numerical description operates often with the quantity equilibrium climate sensitivity (ECS) [71], that is the change in the global temperature if the concentration of atmospheric carbon dioxide molecules increases twice. Treating data of NASA of last 40 years for evolution the global temperature and the concentration of atmospheric CO_2 molecules, one can obtain that the value of ECS is given by formula (5.3.3) that is

$$\mathrm{ECS} = (2.1 \pm 0.4)^\circ\mathrm{C}$$

One can evaluate this value within the framework of the model of standard atmosphere [10] and using the HITRAN bank data for radiative parameters of CO_2 molecules located in atmospheric air. In this case, we have according to formula (7.3.16)

$$\mathrm{ECS} = (0.6 \pm 0.3)^\circ\mathrm{C}$$

We ignore here the data of climatological models [(formula (7.3.15)] which are the basis of the intergovernmental panel on climate change [72] data because of a wrong value for the change of the radiative flux, as it was analyzed in Sect. 7. Next, comparing the above svalues of the ECS, one can conclude that in spite of a large error in these values, CO_2 molecules are not the main greenhouse of the real atmosphere. This conclusion is authentic, and the contribution of CO_2 molecules in the change of the global temperature is estimated as $\sim 30\%$. Note that the contribution of CO_2 molecules to the radiative flux from the atmosphere to the Earth is 17% that is comparable with this value.

Of course, the first component as a candidate for this role is atmospheric water and we above formula (7.3.21) the possibility for this scenarium. As a result, approximately an increase in rate of

$$\frac{d\ln c(H_2O)}{dt} = 3\times 10^{-3}\mathrm{year}^{-1} \tag{10.2.1}$$

provides an additional heating of our planet. We below check this possibility.

It should be noted the troubles in determination of the increase rate for the concentration of atmospheric water because it is distributed in the atmosphere nonuniformly. In addition, according to Fig. 2.3, the amount of atmospheric water has season oscillations whose amplitude is of the order of 10% compared with the average amount. Nevertheless, Fig. 2.6 gives the results for determination this rate during a large time including several decades.

$$\frac{dc(H_2O)}{dt} = (7\pm 1)\times 10^{-3}\frac{\mathrm{g}}{\mathrm{kg}\cdot\mathrm{year}} \tag{10.2.2}$$

We now represent the water concentration for the standard atmosphere in these units g/kg (gram of atmospheric water per kg of atmospheric air). The total mass of atmospheric water is $M = 1.3\times 10^{19}$g [73–76] on average, and the total mass of atmospheric molecules is 5.1×10^{21} g, that is the concentration of atmospheric water in these units is 2.6 g/kg. Hence, Eq. (10.2.2) for the rate of an increase of the average concentration of water molecules takes the form

$$\frac{d\ln c(H_2O)}{dt} = (28\pm 4)\times 10^{-4}\mathrm{year}^{-1} \tag{10.2.3}$$

This coincides practically with Eq. (10.2.1) that convinces us in the possibility of this scenarium of the Earth's heating through the greenhouse effect where approximately 1/3 part of the heating rate is created by an increase of the concentration of atmospheric CO_2 molecules, and 2/3 of this rate is connected with an increase of the concentration of atmospheric water molecules. One can remember also that doubling of the concentration of CO_2 molecules will be through approximately 110 years at the contemporary growth rate. This leads to an increase of the global temperature by the value ECS = 2.1 °C according to formula (5.3.3).

10.2.2 Climate of Megapolis

The total power of industrial energy facilities on the Earth's surface is about 2×10^{13} W, which is about four orders of magnitude less than the power of solar radiation entering the Earth's atmosphere. Thus, human energy activity makes a small contribution to the energetics of the Earth and its atmosphere. However, energy installations are unevenly distributed over the Earth's surface, so that the greatest energy load falls on megacities, i.e., areas on the Earth's surface with a high population density, in reality, large cities. Therefore, it has long been clear that the climate of a megalopolis differs from that of neighboring regions on the Earth's surface.

As the demonstration of this statement, we show in Fig. 10.10 evolution of the September temperature of some Japanese towns during 100 years. As it follows from this figure, the indicated temperature of Tokyo increases for this time by 2.5 °C more than that for small cities. This proves that the temperature of megacities differs from that for neighboring areas. Note that according to the megalopolis definition we assume it and its environment have an identical altitude distribution of temperatures. In addition, a size of a megalopolis exceeds remarkably altitudes where atmospheric radiation to the Earth is formed. Therefore, a megalopolis size exceeds 10 km.

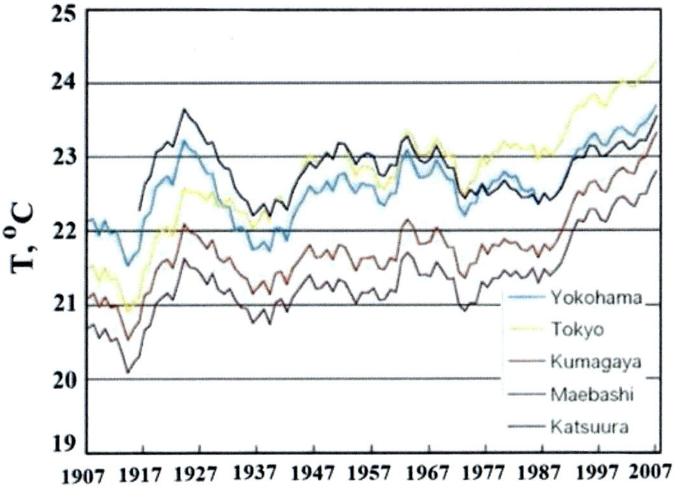

Fig. 10.10 September temperature of indicated Japanese cities averaged over the next 10 years [77, 78]

Let us consider a megalopolis as a heterogeneity on the Earth's surface with an increased population density. When making estimates related to the power output, for simplicity, we associate the energy released with the people who produce it. Then, the power of 2.3 kW accompanies each person and releases at a point of its residence. Evidently, this assumption is valid if the industry is concentrated in cities. Then, one can extract areas with the maximum density of population and evaluate on the basis of the above assumption which power refers to population of this area and which change of the global temperature corresponds to this area under these conditions.

It should be noted that the character of the development of civilization leads to accumulation of the population in small areas, mostly, in cities. Table 10.1 contains the population of the largest cities of the world, their area and the power release per unit area under the above conditions. The assumption used is that the power industry is located inside the city and the power of energetic facilities is proportional to the number of persons who are residents of this city. Possibly, the place of residence for this city is a rural area near it. Hence, it is necessary to take the value of the specific power of this table as a rough estimate. Nevertheless, from this, it follows that the city temperature may be differed from that of neighboring places by several degrees.

Evidently, the same situation takes place in some independent territories where an exchange by people with surrounding areas is restricted. Table 10.2 contains the population of self-sustained formation and their participation in power processes. As is seen, the specific power inside these territories is comparable with that of the largest cities under used conditions. One can expect heating of these territories compared with neighboring areas by several degree.

Let us consider from this standpoint the heating of the Tokyo area which is represented in Fig. 10.10 according to which its heating with respect to neighboring area is estimated as 2.5 °C. From Table 10.1, it follows that an additional specific power released in the Tokyo area under used assumptions is approximately 10 W/m^2.

Table 10.1 Population of the largest cities and their area for 2021 [79, 80]

City	Population, million persons	Urban area, 10^3 km^2	Specific power, W/m^2
Tokyo-Yokohama, Japan	37.3	8.23	10
Delhi, India	31.2	2.23	32
Shanghai, China	27.8	4.07	16
San Paulo, Brazil	22.2	3.11	16
Mexico City, Mexico	21.9	2.39	21
Dhaka, Bangladesh	21.7	0.46	108
Cairo, Egypt	21.3	2.01	24
Beijing, China	20.9	4.17	12
Mumbai, India	20.7	0.94	51
Osaka, Japan	19.1	3.01	15

Taking the climate sensitivity according to (5.3.5) as $0.5\ \text{K} \cdot \text{W/m}^2$, one can obtain that this corresponds to the surface heating by 5 °C. The above values coincide for the order of magnitude. In addition, the proportionality between an additional specific power released at a certain region and its heating are violated at large powers. Indeed, large cities usually are covered by a haze that changes the character of action of solar radiation on the heat regime of the Earth's surface. In addition, we assume that a size of a region between the city and surrounding areas where the temperature varies is small compared to a city size and in this manner ignores a heat exchange between the city and neighboring areas.

The effect of the Earth's heating may be remarkable for a total country. We include in Table 10.3 the parameters of countries with a high density population. The temperature change ΔT follows from the assumption that an average energy release proceeds in these countries with accounting for the average power of 2.3 kW per one habitant and taking into account the climate sensitivity value according to formula (5.3.5), where $S = 0.5\ \text{K} \cdot\ \text{W/m}^2$. As it follows from this, heating of countries with a high population density due to the human activity is below 1 °C.

Obviously, there is a limit for the number of people who can simultaneously live on our planet. Above, this value was determined by the level that is able to feed the population, and in the first half of the twentieth century, Southeast Asia approached this limit. However, in the middle of the twentieth century, the green revolution occurred associated with a sharp increase in crops, and the problem of feeding the world's population took on a different look. Moreover, progress in getting food more efficiently is expected to continue. Hence, we define the food limit for the

Table 10.2 Occupancy of populated self-sustaining territories [81]

Region	Area, 100 km^2	Density population, thousand/km^2	Specific power, W/m^2
Singapore	720	7.8	18
Hong Kong (China)	110	6.7	15
Macau (China)	30	21	48
Gibraltar (UK)	6.8	4.9	11
Monaco	2.0	19	44

Table 10.3 Occupancy of the most densely populated countries [81]

Country	Area, 100 km^2	Density population, thousand/km^2	ΔT, °C
Bangladesh	144	1.1	1.3
Taiwan	36	0.65	0.7
South Korea	100	0.52	0.6
Rwanda	26	0.46	0.5
Netherlands	42	0.42	0.5

world population by possibilities of the photosynthesis phenomenon, and this limit exceeds 100 billion persons who live simultaneously.

The proportionality between an additional extracted power due to human activity and the temperature change resulted from this activity violates under conditions of megacities because of changes in the atmosphere due to an extracted energy. Therefore, action of the human activity on the thermal state of the ground would be stronger if the extracted power is distributed over a large area. Megacities are growing, and eventually, they will be able to occupy the entire land territory. One can accept that in this limit the specific heat output power corresponds to the megacities with population of one million, that is approximately 6 W/m^2. One can estimate a typical time when this limit will be reached.

Of course, the above limit of the specific power will be reached at different times for different countries. For the sake of certainty, we will focus on China, where this limit will come earlier and estimate a time when this may happen. Since the area of China is 9.6×10^6 km^2, and its population currently exceeds 1.4 billion people [82], then the population density of China is approximately 150 people/km^2. This corresponds to a specific heat output of 0.4 W/m^2. The average population growth rate in China is 1.1% per year [83], and the average growth rate of energy consumption is 2.3% per year. The latter value follows from Fig. 5.8 as the average growth rate of energy consumption from 1965 to 2018. Hence, the growth rate of specific energy consumption for China is 3.4% per year, so that the doubling of the average specific power consumption for China occurs in 20 years, whereas at current growth rates in 110 years, this value will reach the value of 6 W/m^2, i.e., the characteristic specific capacity of a megalopolis.

Note that the time obtained practically coincides with the time of doubling the concentration of carbon dioxide in the atmosphere. However, in this case, the additional heating of the Earth due to the greenhouse effect, determined by the doubling of the concentration of carbon dioxide in the atmosphere, is an order of magnitude less than under the influence of anthropogenic energy sources. Thus, in the current state of civilization, energy related to human industrial activity can affect the thermal state of megacities. In about 100 years, the anthropogenic impact of energy will spread to countries.

Let us make one more estimation for a megapolis. If the temperature of a megapolis is large, one can decrease it by creating of an aerosol curtain that is a system of aerosols which hangs over the megapolis. In order to check the reliability of this action, we estimate below an amount of the aerosol mass under optimal conditions if an aerosol is of a radius of $r = 1\,\mu$m, and the aerosol absorbs solar radiation effectively, so that the absorption cross section of solar radiation by an aerosol is $\sigma_{abs} = 3 \times 10^{-8}$ cm^2. It is convenient to take as an aerosol a water droplet in which an absorbed material is dissolved.

Now, we consider the problem from another standpoint. Let us estimate for definiteness the aerosol amount which leads to the temperature decrease of 5 K that is comparable with temperature fluctuations. In accordance with the climate sensitivity $S = 0.5$ K m^2/W according to formula (5.3.5), this corresponds to a decrease of the solar radiation flux passed through the atmosphere as $\Delta J = 10$ W/m^2. Taking the solar radiative flux passed through the atmosphere as $J_s = 182\ \mathrm{W/m^2}$, one can find the optical thickness $u = \Delta J/J_s$ due to aerosol curtain which provides the above cooling as $u = 0.05$. This optical thickness corresponds to the droplet density $n = u/\sigma \sim 2 \times 10^6\ \mathrm{cm^{-2}}$. Because the mass of an individual droplet is $m = 4\pi r^3 \rho/3$, where the water mass density is $\rho = 1\mathrm{g/cm^3}$, this corresponds to the mass density of aerosols

$$\varrho = mn \sim 6\frac{\mu\mathrm{g}}{\mathrm{cm}^2} = 60\frac{\mathrm{kg}}{\mathrm{km}^2}$$

Let us return to a megapolis and take for definiteness its size $L = 50$ km, so that the curtain over it is $L^2\varrho \sim 200$ tonnes. This mass shifts under the wind action and must be restored. Taking a typical wind speed as $v = 10$ m/s, one can obtain that the restoring must proceed through a time of $\tau \sim L/v \sim 2$ h, so that the rate of aerosol replacing is

$$\frac{\mathrm{d}M}{\mathrm{d}t} = \varrho L v \sim 30\frac{\mathrm{kg}}{\mathrm{s}}$$

Note that this is a new technology that is an analogy to pour on streets at summer, as well as to aerial firefighting. The latter is close to the discussed technology according to its action, because the basis of both ones is an air tanker. In particular, the largest aerial firefighter on the basis of a Boeing 747 includes a supertanker contained 7.4×10^4 tones of water [84] that must be used during 20 min according to the above estimation. In contrast to aerial firefighting, in this case a close tanker contains a water solution rather pure water. A compressor allows one to inject this solution in air through nozzles for formation of the aerosols of a given size. Of course, creation of this technology requires to solve various problems. But the above estimations and the experience of aerial firefighting exhibit that this technology is available.

10.2.3 Greenhouse Effect from Renewable Energetics

Along with an energy which is used in the human activity, an energetic setup scatters an energy in an environment. Because of a small global power of the industrial energetics compared with the natural one, a strong action on environment in the course of the human activity may be resulted from a change of the rates of natural atmospheric processes. We now consider the action of hydroelectric plants on an environment through the change of the atmosphere composition that leads to the change in the greenhouse effect. Hydroelectric plants are considered as a type of the renewable energetics which do not act on the environment, because they transform the energy of moving water into the electric ones. In reality, a water reservoir accompanies any hydroelectric plant, and evaporation of water from this reservoir leads to an increase of the greenhouse effect which wall be considered below.

Change in the concentration $c(H_2O)$ of atmospheric H_2O molecules leads to the following change of the energy flux $J_\downarrow$ in accordance with formula (7.2.9)

$$\frac{dJ_\downarrow}{d\ln c(H_2O)} = 7\frac{W}{m^2}$$

Correspondingly, a new water reservoir causes an increase $\Delta J_\downarrow$ of the atmospheric radiative flux to the Earth's surface as

$$\Delta J_\downarrow = 7\frac{W}{m^2}\ln\frac{c'(H_2O)}{c(H_2O)}, \qquad (10.2.4)$$

where $c(H_2O)$ is the initial concentration of water in the near-surface atmospheric layer, and $c'(H_2O)$ is the final value of this concentration. Evidently, the air moisture over the water reservoir is $\eta'(H_2\mathbf{O}) = 100\%$. Let us determine the average atmosphere moisture η over land at the beginning.

One can find the average moisture η over the land on the basis of a simple model. We use that the average moisture over the Earth's surface is 80%. Taking into account that the average moisture of air located over the Earth's surface equals to 80%, and the moisture over oceans equals to 100%. Because 70% of the Earth' surface relates to oceans, and 30% corresponds to land, the average land moisture equals $c(H_2O) =$ 30%. From this, one can obtain that an average increase $\Delta J_\downarrow$ of the radiative flux to the Earth's surface as a result of the reservoir creation at the land is

$$\Delta J_\downarrow = 8\frac{W}{m^2} \qquad (10.2.5)$$

We now apply this formula to largest hydroelectric plants whose list is given in Table 10.4 [85, 86]. One can see that the change in the ratio of the power of an additional greenhouse phenomenon to the electric power of a plant is of the order of one on average. Their ratio depends on a locality, where the hydroelectric plant is constructed. If it is located in a gorge, the area of the water reservoir after the electric plant is relatively small. On contrary, a flat landscape is accompanied by a large area of the water reservoir after the electric plant. Hence, we consider separately some hydroelectric plants which are located in a flat land whose parameters are given in Table 10.5 [87–91].

It should be noted that the role of water located in a reservoir after an hydroelectric plant is not restricted by electric energetics. Its use for agriculture and shipping may be more important, especially, for a not large plants. Indeed, this water reservoir provides the stable navigation, as well as the reliable farming which does not depend on a weather. In addition, a flat landscape relates usually to a droughty land where the change of the radiative flux is larger than that due to formula (10.2.4). The above analysis gives a certain experience in the role of the greenhouse phenomenon in energetics.

One can compare the effect of action of hydroelectric plants due to injection of additional water in the atmosphere with a similar effect resulted from combustion

Table 10.4 Parameters of largest hydroelectric plants of the world and the power change ΔP_r of the change of the radiative flux from the atmosphere to the Earth due to the change of the air moisture over the water reservoir

Plant, country	Installed power	Area of reservoir	S/P,	$\Delta J_{\downarrow}(H_2O)$,
	P, GW	S, km^2	km^2/GW	GW
Three Gorges Dam (China)	22.5	1084	48	9.1
Itaipu Dam (Brazil, Paraguay)	14	1350	96	11
Xiluodu (China)	13.9	454	13	7.8
Guri (Venezuela)	10.2	4250	415	36
Tucuru (Brazil)	8.4	3014	360	25
Grand Coulee (USA)	6.8	324	48	2.7
Xiangjiaba (China)	6.4	95.6	15	0.8
Longtan Dam (China)	6.4	98.5	25	0.8
Sayano-Shushenskaya (Russia)	6.4	621	97	5.2
Krasnoyarsk (Russia)	6.4	2000	312	17

Table 10.5 Parameters of hydroelectric plants located in a flat districts and the change of the power ΔP_r due to the greenhouse phenomenon owing to the change of the air moisture resulted from formation of a water reservoir

Plant, country	Power P, GW	Area of reservoir S, km^2	S/P, km$^2/GW$	$S\Delta J_{\downarrow}(H_2O)$, GW
Saratov plant	1.4	1800	1.3	15
Volgograd plant	2.7	3100	1.2	26
Volga-Kama cascade	12	25,000	2.1	210
Aswan dam (Egypt)	2.1	5250	2.5	44

of carbon-contaning materials in air. For definiteness, we consider the following combustion process

$$CH_4 + 2O_2 \rightarrow CO_2 + 2H_2O + \Delta\varepsilon \tag{10.2.6}$$

If products of this reaction, as well as reactants, are found in the gaseous state, we have $\Delta\varepsilon = 8.4$ eV [92–94]. Assuming the CO_2 molecule which is the product of combustion of a CH_4 molecule, we below determine the energy of photons resulted from the greenhouse effect. If one CO_2 molecule is injected in the atmosphere, according to formula (7.2.4), one can obtain the power due to the greenhouse effect from one injected CO_2 molecule in the atmosphere

$$\frac{d\varepsilon}{dt} = \frac{2\,\mathrm{W}}{\mathrm{m}^2 \cdot n(CO_2)} = 4\,\mathrm{eV/year}, \tag{10.2.7}$$

where $n(CO_2) \approx 10^{26}\ m^{-2}$ is the number of CO_2 molecules per unit area for the standard atmosphere. Taking the residence time of CO_2 molecules in the atmosphere as $\tau \approx 4$ year [28], one can obtain the total energy of $\varepsilon \approx 16$ eV which is emitted by one CO_2 molecule during its location in the atmosphere. As is seen, this value is twice compared to the energy which is extracted as a result of combustion of the methane molecule. As is seen, the ratio between the energies which are consumed as a result of the greenhouse effects is of the order of the energy which is consumed as a result of the chemical reaction. This is compared with that for water which is injected in the atmosphere due to hydroelectric plants owing to a new equilibrium of atmospheric water with that located in water reservoirs.

Continuing this analysis, we now determine the energy which gives one water molecule evaporated from the Earth's surface and is located in the atmosphere the residence time that is approximately 9 days. By analogy with formula (10.2.7) and based on formula (7.2.9) instead of (7.2.4) for the radiative flux from the atmosphere due to additional water, one can obtain the radiative power due to the greenhouse effect involving one water molecule

$$\frac{d\varepsilon}{dt} = \frac{7\,W}{m^2 \cdot n(H_2O)} = 1 \times 10^{-26} W \tag{10.2.8}$$

Here, $n(H_2O) = 7 \times 10^{26}/m^2$ is the number of H_2O molecules per unit area for the standard atmosphere. From this, we obtain that one water molecule during its residence in the atmosphere creates infrared photons of the energy $\varepsilon \approx 2meV$ on average. Nevertheless, because of a large amount of water in the atmosphere, its presence is important for the greenhouse effect.

Another example of an action of renewable energetics on an environment is additional water evaporation during operations of a wind turbine which gives the electricity. The wind power is one of developing types of energy generation. At the end of 2020, an installed capacity of the world hydropower facilities was 1211 GW, while that for the wind power was 733 GW and for the solar power 714 GW [96]. A wind turbine transforms the wind energy in the electric one, and we consider below the spread contemporary wind turbine which is like the windmill that is represented in Fig. 10.11. This facility catches a horizontal wind that compels the propeller to rotate. Because the propeller rotates vertically, it directs partially the horizontal wind to the Earth that causes an increased evaporation of water.

We below show that similar to hydroelectric plant, action of the wind turbine creates an additional radiative flux from the atmosphere to the Earth's surface whose power is comparable with that of the wind turbine. Indeed, in the course of its working, a wind turbine under consideration converts horizontal winds into verical ones. A vertical wind enforces transport of wet air up and shifts an equilibrium between atmospheric and surface water. As a result, the rate of water evaporation increases, and the feedback causes a decrease of the surface temperature. Under optimal conditions for evaporation, the mechanical wind energy directed toward the surface is consumed on water evaporation.

Fig. 10.11 Character of operation of a wind turbine as a transformer of the wind energy in the electric one [95]

According to the construction of the wind turbine (see Fig. 10.11), only a part of a horizontal wind collided with a propeller moves down. In the same manner, only a part of the energy of this wind is transformed into the electric energy. One can assume for estimations, that these energies are comparable, that is the energy consumed on additional water evaporation is comparable with the electric energy created by the wind turbine. The energy for evaporation of one water molecule is 0.44 eV approximately. Subsequently evaporated molecule goes in neighboring regions and partakes in the greenhouse effect until it is washed out by rain. According to formula (6.2.15), the average energy of photons which are created by one water molecule during its residence in the atmosphere is 2.2 eV. Comparing these values, one can conclude roughly that the electric power of the wind turbine under consideration is comparable with the energy of an additional radiative flux from the atmosphere which occurs due to action of this wind turbine.

10.2.4 Climate Change and Human Activity

The power produced as a result of human activity, which is about 2×10^{13} W, is four orders of magnitude below the solar power that penetrates in the Earth's atmosphere. From this it follows that the human does not influence the energetic balance of the Earth directly. But this activity leads to water pollution in rivers and lakes, as well as to air pollution. This causes a change of the energy fluxes which determine the equilibrium between the Earth, its atmosphere and an environment. These changes act twofold. Pollutions change the chemical processes at the Earth's surface and atmosphere that influence the health of habitants. Variations of energy fluxes involving the atmosphere and Earth's surface change the thermal state of the Earth, that is change its climate. The latter is a topic of the subsequent analysis.

We represent the climate of future in the simplest form that reduces to use for its description a single parameter that is its temperature. Moreover, being guided by a long-term changes of the thermal state of the Earth, we characterize it by the global temperature, i.e., make an average over globe.

Figure 10.12 gives the rough dependence of the global temperature on time which is constructed on the basis of Fig. 5.18 where fluctuations of the global temperature are removal. In this case on the basis of data [97], we assume that before eighties variations of the global temperature do not exceed its fluctuations, whereas after them the global temperature grows linearly with the gradient (5.3.1). We also take into account Fig. 10.9 according to which the total pollution of the atmosphere took place in eighties, and after this external perturbation in the form of cosmic rays cannot influence the atmosphere state. As a matter, Fig. 5.18 demonstrates that before eighties an increase in the global temperature holds down to some factors, whereas after eighties, this resistance was overcame.

From another standpoint, one can suggest the mechanism of growth of the global temperature as a result of the greenhouse effect because of a change of the amount of

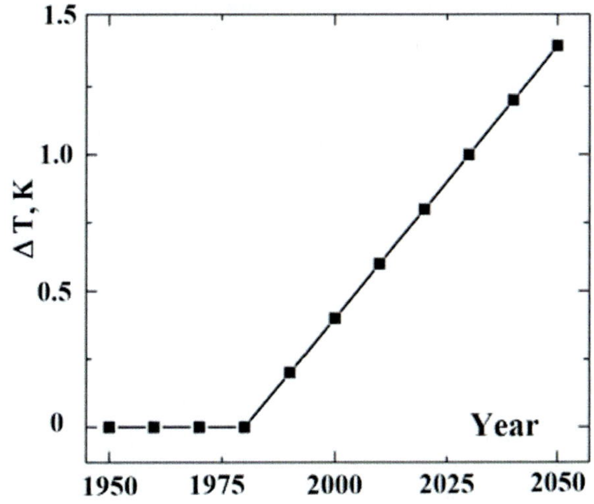

Fig. 10.12 Schematic character of evolution of the global temperature in our time

greenhouse components, i.e., CO_2 and H_2O molecules, as well as clouds consisting of water microdroplets mostly. Another mechanism of the change of the global temperature is associated with reflection and absorption of solar radiation in the atmosphere due to clouds and aerosols. Moving along this list, we note that we can determine reliable radiative fluxes to the Earth related to all the greenhouse components, as well as the change in the radiative flux due to carbon dioxide molecules. Determination of the change due to water molecules requires more precise data for real evolution of the water amount in the atmosphere. The accuracy of these data may be improved if this necessity will be understood.

Unfortunately, there are troubles in the analysis of clouds and aerosols in their interaction with solar radiation. The problem is that their distribution in the atmosphere is nonregular both in space and in time. The contribution of the global temperature change is estimated on average approximately in 10% of the change due to the greenhouse effect. In contrast to the greenhouse effect which leads to an increase of the global temperature as a result of an increase of the concentration of greenhouse components, an increase of the amount of clouds causes a decrease in the intensity of incident solar radiation as it passes through the atmosphere. In other words, the mechanisms of evolution of the global temperature in time are understood, as well as estimations for these mechanisms. Correspondingly, it is clear the way to improve the accuracy and reliability of the analysis of this problem.

Unfortunately, simplified and perverted information about reasons of the Earth's warming is spread widely in the interests of some financial groups. Such information leads to solutions which are not useful for the habitant. We mean the Paris agreement 2015 on climate [70]. The grounding of this agreement is that only increasing of the concentration of atmospheric carbon dioxide is the reason of the observed warming of the Earth. From the analysis of this book, one can conclude that this is a subsidiary reason of the warming. Next, the red line for the change of the global temperature is

2 °C, and then, the Earth's warming will become irreversible. But Fig. 5.19, Figs. 10.6 and 10.8 indicate that this change in past variations in the Earth's temperature was above 10 °C. Evidently, the basic goal of the Paris agreement 2015 was to create the fond of 100 billion dollars annually in order to develop noncarbon technologies. But up to now the information is absent in what manner was used 0.5 trillion dollars of this fond.

The previous experience of this kind with the use of a false concept in the interests of a certain group of people was contained in the Montreal Protocol [98], according to which the production of chlorine-containing freons was prohibited to preserve stratospheric ozone. Such freons were used in refrigeration units, and although they themselves have low chemical activity, when they enter the stratosphere, they can decompose under the influence of solar radiation to form chlorine atoms. Each chlorine atom in the stratosphere partakes in chain reactions with decomposition of a large number of ozone molecules. The simplest freon of this kind is the chemical compound CCl_4, which was widely used in chemical cleaning and was also a convenient filling in the neutrino counter, where the reaction of a neutrino with a chlorine atom leads to formation of an argon atom.

Ozone molecules are formed in excited air, and the ozone molecule is metastable, so that the conservation of ozone in oxygen is accompanied by the energy release. But collision of two ozone molecules does not lead to the formation of oxygen molecules, and ozone decomposition may proceed as a result of chain reactions. The balance of the ozone layer in the stratosphere was described by S.Chapman in 1930 [99]. He showed that ozone molecules are formed in the stratosphere as a result of attachment of oxygen atoms coming into the stratosphere from the ionosphere to oxygen molecules, and destruction of an ozone molecules results from its collision with an oxygen atom. This character of ozone decomposition is called the Chapman cycle or the oxygen cycle. In the early seventies, other cycles were discovered that lead to the conversion of ozone into oxygen, namely the nitrogen cycle, the hydrogen cycle and the chlorine cycle. For these investigations, P. Krutzen, M. Molina and F. Sh.Rowland received the Nobel Prize in Chemistry in 1995 [100].

In particular, a chlorine-containing type of freon molecules CCl_4 is chemically nonactive due to their closed structure, where four halogen atoms form a shell around a carbon atom. Therefore, these molecules are not danger for humans if they are located in the air. However, if these molecules are found in the stratosphere and higher, they can be destroyed by ultraviolet radiation, and then, chlorine atoms will participate in the destruction of ozone molecules. One can note that the total anthropogenic change in the ozone content in the stratosphere does not exceed several percent and this does not affect the operation of the ozone screen, which protects living organisms and plants from the action of solar ultraviolet radiation. The main anthropogenic destruction of stratospheric ozone occurs under the influence of nitrogen oxides formed at the stratopause as a result of aircraft flights. As for the chlorine-containing compounds that are formed on the Earth's surface, they are mainly washed out by rain, and chlorine atoms in the stratosphere are mainly formed when salts evaporate from the ocean surface.

Despite the reasonableness of the above arguments, the prohibition of chlorine-containing freons was adopted by the Montreal Protocol [98], which led to the ruin of the production of chlorine-containing freons and to the prosperity of the American chemical concern DuPont, which organized the production of bromine-containing freons and lobbied for the adoption of the Montreal Protocol. As a result, the value of this company has increased to 130 billion dollars [101]. This also led to an increase in the cost of energy equipment in refrigeration equipment. It should be noted that for a quarter of a century after the adoption of the Montreal Protocol, accurate measurements were not made on the evolution of the total mass of stratospheric ozone. After that, this problem ceased to be relevant. Apparently, the experience of the Montreal Protocol played a role in the adoption of the Paris climate agreement, the scale of which is significantly larger.

References

1. A.I. Shapiro, W. Schmutz, E. Rozanov et al., Astron. Astrophys. **529**, A67 (2011)
2. L.I. Dorman, Ann. Geophys. **30**, 9 (2012)
3. C.D. Camp, K.K. Tung, Geophys. Res. Lett. **34**, L14703 (2007)
4. P. Forster, V. Ramaswamy, P. Artaxo et al., *Changes in Atmospheric Constituents and in Radiative Forcing* (Cambr. Univ. Press, Cambridge, 2007)
5. T. Dudok de Wit, G. Kopp, C. Fröhlich, M. Schöll, Geophys. Res. Lett. (2017). https://doi.org/10.1002/2016GL071866
6. https://www.google.ru/search?q=Greg's+TSI+reconstruction+Dudok+de+Wit+2017+Solar+Irradiance
7. M. Milankovich, *Theorie Mathematique Des Phenomenes Thermiques Produits Par La Radiation Solaire* (Gauthier-Villars, Paris, 1920)
8. M. Milankovich, *Canon of Insolation and the Ice Age Problem* (Belgrade, 1941)
9. https://en.wikipedia.org/wiki/Solar-irradiance/media/File:Milankovitch
10. *U.S. Standard Atmosphere.* (Washington, U.S. Government Printing Office, 1976)
11. N.N. Semenov, Zs. Phys. **48**, 571 (1928)
12. Ya..B.. Zeldovich, D.A. Frank-Kamenestkii, Zhournal Fiz. Khim. **12**, 100 (1938)
13. Ya..B.. Zeldovich, D.A. Frank-Kamenestkii, Doklady AN SSSR **19**, 693 (1938)
14. S. Arrhenius, Zs. Phys. Chem. **4**, 226 (1889)
15. Ya.B. Zeldovich, *Theory of Combustion and Gas Detonation (Moscow-Leningrad)* (Izdat, AN SSSR, 1944). ([in Russian])
16. D.A. Frank-Kamenestkii, *Diffusion and Heat Transfer in Chemical Kinetics* (Plenum Press, New York, 1969)
17. Ya..B.. Zeldovich, G.I. Barenblatt, V.B. Librovich, G.M. Makhviladze, *The Mathematical Theory of Combustion and Explosions* (Consultants Bureau, New York, 1985)
18. B.M. Smirnov, *Fundamentals of Ionized Gases* (Wiley, Weinheim, 2012)
19. T.M. Cronin, *Paleoclimates?: Understanding Climate Change Past and Present* (Columbia Univ. Press, New York, 2010)
20. J.C. Zachos, M.Pagani, L.C.Sloan et al. Science *292*, 686(2001)
21. A. Lukeneder et al., Earth Planet. Sci. Lett. **296**, 103 (2010)
22. K. Stevens et al., Palaeogeography. Palaeoclimatology, Palaeoecology **417**, 164 (2015)
23. G. Faure, *Principles of Isotope Geology* (Cambr. Univer. Press, Cambridge, 1986)
24. A.P. Dickin, *Radiogenic Isotope Geology* (Cambr. Univer. Press, Cambridge, 1995)
25. S. Ruben, M.D. Kamen, W.Z. Hassid, Science **90**, 510 (1939)
26. S. Ruben, M.D. Kamen, Phys. Rev. **57**, 549 (1940)

27. S. Ruben, W.Z. Hassid, M.D. Kamen, J. Am. Chem. Soc. **62**, 3443 (1940)
28. M. Grosjean, J. Goiot, Z. Yu, Global Planet Change **152**, 19 (2017)
29. V.A. Zubakov, I.I. Borzenkova, *Global Paleoclimate of the Late Cenozoic* (Elsevier, Amsterdam, 1990)
30. M. Benton, D.A.T. Harper, *Introduction to Paleobiology and the Fossil Record* (Wiley-Blackwell, Chichester, 2009)
31. https://en.wikipedia.org/wiki/Paleontology
32. https://en.wikipedia.org/wiki/Paleoclimatology
33. M.L. Bender, *Paleoclimate* (Princeton, Princ. Univer. Press, 2013)
34. G. Faure, D. Mensing, *Isotopes-Principles and Applications* (Wiley, Hoboken N.J., 2005)
35. https://en.wikipedia.org/wiki/Geochronology
36. E.P. Borisenkov. *Climate and Its Changes* (Moscow, Znanie, physical series, N6, 1978; in Russian)
37. https://en.wikipedia.org/wiki/Eocene
38. S.L. Wing, D.R. Greenwood, Phil. Trans. Roy. Soc. **341B**, 243 (1993)
39. P. Wilf, S.L. Wing, D.R. Greenwood, C.L. Greenwood, Geology **26**, 203 (1998)
40. A.F. Diefendorf, K.E. Mueller, S.L. Wing et al., Proc. Nat. Acad. Sci. **107**, 5738 (2010)
41. D.J. Peppe, D.L. Royer, B. Cariglino et al., New Phytologist **190**, 724 (2011)
42. A. Sluijs et al., Nature **441**, 610 (2006)
43. K. Moran et al., Nature **441**, 601 (2006)
44. C. Jaramilo et al., Science **330**, 957 (2010)
45. F.A. McInerney, S.L. Wing, Ann. Rev. Earth Planet. Sci. **39**, 489 (2011)
46. E.D. Currano, P. Wilf, S.L. Wing et al., Proc. Nat. Acad. Sci. **105**, 1960 (2008)
47. D. Martyn, *Climates of the World* (Elsevier, Amsterdam, 1992)
48. W.F. Ruddiman, *Earth's Climate: Past and Future* (Freeman, New York, 2000)
49. H.G. Jones, *Plants and Microclimate: A Quantitative Approach to Environmental Plant Physiology* (Camb. Univ. Press, Cambridge, 1992)
50. D. Lawrence, K. Vandecar, Nature Climate Change **5**, 27 (2015)
51. P. Schulte, L. Alegret, I. Arenillas et al., Science **327**, 2014 (2010)
52. https://en.wikipedia.org/wiki/Chicxulub-crater
53. P.R. Renne, A.L. Deino, F.J. Hilgen et al., Science **339**, 684 (2013)
54. H. Palike. Science **339**, 655(2013) - chronology
55. S. Ohno et al., Geoscience **7**, 279 (2014)
56. http://edition.cnn.com/2016/03/04/world/scientists-drill-impact-crater-irpt
57. http://www.bbc.com/earth/story/20160415-what-really-happened-when-the-dino-killer-asteroid-struck
58. A.N. Yoder, Science **339**, 6565 (2013)
59. A. Grinsted, The Cryosphere **7**, 141 (2013)
60. http://climate.nasa.gov
61. A. Zecca, L. Chiari, Comets and climate. J. Atmos. Sol. Terr. Phys. **71**, 1766 (2009)
62. E.A. Kasatkina, O.I. Shumilov, M. Krapiec, On periodicities term climatic variations near 68° N, 30° E. Adv. Geosci. **13**, 25 (2007)
63. M.G. Ogurtsov, O.M. Raspopov, Possible impact of interplanetary and interstellar dust fluxes on the Earth's climate. Geomagn. Aeronomy **51**, 275 (2011)
64. M.B. Enghoff, J.O.P. Pedersen, U.I. Uggerhøj et al., Geophys. Res. Lett. **38**, L09805 (2011)
65. J. Kirkby, J. Curtis, J. Almeida et al., Nature Lett. **476**, 429 (2011)
66. H. Svensmark, E. Friis-Christensen, J. Atmos. Terr. Phys. **59**, 1225 (1997)
67. H. Svensmark, Proc. Roy. Soc. **463A**, 385 (2007)
68. H. Svensmark, T. Bondo, J. Svensmark. Geophys. Rev. Lett. **36**, L151001 (2009)
69. H. Svensmark, M.B. Englo, J.O.P. Pedersen, Phys. Lett. **377A**, 2343 (2013)
70. http://unfccc.int/resource/docs/2015/cop21
71. https://en.wikipedia.org/wiki/Climate-sensitivity
72. Intergovernmental panel on climate change. Nature **501**, 297;298 (2013) http://www.ipcc.ch/pdf/assessment?report/ar5/wg1/WGIAR5-SPM-brochure-en.pdf

73. I.A. Shiklomanov, in *Water in Crisis : A Guide to the World's Fresh Water Resources*. ed by P.H. Gleick (Oxford, Oxford Univ. Press, 1993), pp. 13–24
74. I.A. Shiklomanov, J.C. Rodda (eds.), *World Water Resources at the Beginning of the Twenty-First Century* (Cambridge Univ. Press, Cambridge, 2003)
75. R.W. Healy, T.C. Winter, J.W. Labaugh, O.L. Franke. *Water Budgets: Foundations for Effective Water-Resources and Environmental Management* (Reston, Virginia, U.S.Geological Survey Circular 1308, 2007)
76. https://en.wikipedia.org/wiki/Atmosphere-of-Earth
77. https://en.wikipedia.org/wiki/Urban-heat-island
78. https://www.data.jma.go.jp
79. https://en.wikipedia.org/wiki/List-of-largest-cities
80. https://www.archdaily.com/906605/the-20-largest-cities-in-the-world-of-2021
81. https://en.wikipedia.org/wiki/Population-density
82. http://www.worldometers.info/world-population/china-population/
83. https://www.google.ru/newwindow=1q=population+rate
84. https://en.wikipedia.org/wiki/Aerial-firefighting
85. https://www.power-technology.com/features/feature-the-10-biggest-hydroelectric-power-plants-in-the-world
86. https://en.wikipedia.org/wiki/List-of-largest-hydroelectric-power-stations
87. http://referatwork.ru/category/ekologiya/view/538545-vodohranilischa-volzhsko-kamskogo-basseyna
88. https://en.wikipedia.org/wiki/Volga-Hydroelectric-Station
89. https://en.wikipedia.org/wiki/Saratov-Hydroelectric-Station
90. https://en.wikipedia.org/wiki/Saratov-Reservoir
91. https://en.wikipedia.org/wiki/Aswan-Dam
92. L.V. Gurvich, I.V. Veyts, C.B. Alcock, *Thermodynamic Properties of Individual Substances* (Hemisphere Publishing, New York, 1989)
93. M.W.Chase, NIST-JANAF themochemical tables, Fourth Edition. J. Phys. Chem. Ref. Data. Monograph **9**, 1 (1998)
94. J.A. Manion, J. Phys. Chem. Ref. Data. **1**, 123 (2002)
95. https://en.wikipedia.org/wiki/Wind-turbine
96. https://en.wikipedia.org/wiki/Renewable-energy
97. J. Hansen, M. Sato, R. Ruedy et al. http://www.columbia.edu/~jeh1/mailing/2016/20160120-Temperature2015
98. https://en.wikipedia.org/wiki/Montreal-Protocol
99. S. Chapman, Mem. Royal. Meteorol. Soc. **3**, 103 (1930)
100. Nobel Lectures in Chemistry 1991–1995. Ed. Bo. G. Malmström (Singapore, World Scientific, 1997)
101. https://en.wikipedia.org/wiki/DuPoint

Chapter 11
Conclusion

Let us summarize the results of the theoretical analysis of this book devoted to the physical processes that are responsible for the thermal state of our planet. Our task was to compose the physical picture of processes responsible for the evolution of the Earth and its atmosphere. In our analysis, we relied on physical laws, information obtained from monitoring the atmosphere and measuring its parameters, as well as on contemporary information. The physical approach, aimed at identifying the dominant elements of the problem and neglecting other elements, ensures the simplicity, transparency and reliability of the conclusions obtained. Together with numerical parameters or estimates of the properties and processes under consideration, this allows for a deeper understanding of some aspects of atmospheric physics compared to the standard climatological approach, which seeks to take into account all the factors affecting the problem under consideration. Briefly presented in the proposed book, the physical picture of the global energy of the atmosphere is reduced to the following.

The main components that determine the energetics of the atmosphere are water and carbon dioxide. The amount of carbon dioxide in the atmosphere is associated with the equilibrium between the carbon of the atmosphere and the biosphere. A typical time for establishing this equilibrium is several years, while a typical time for establishing the equilibrium between the biosphere and underground carbon is estimated as thousands of years. Note the important role of the photosynthesis for the equilibrium of carbon between the atmosphere and the biosphere. As a result of this is a relatively small concentration of carbon dioxide in the atmosphere (about 0.04%), unlike other planets of the solar system. An increase in the concentration of carbon dioxide in the atmosphere is associated with human economic activity. This influence is primarily associated with an extraction and burning of fossil fuels—coal, oil and methane. As a result, carbon is taken from the Earth's interior and is included in the balance between the atmosphere and biosphere.

The weakness of the anthropogenic factor in an increase in the concentration of atmospheric carbon dioxide is manifested in the fact that the rate of its release into the

B. M. Smirnov, *Global Energetics of the Atmosphere*, Springer Atmospheric Sciences, https://doi.org/10.1007/978-3-030-90008-3_11

atmosphere as a result of burning fossil fuels currently amounts to 5% of the transition rate for carbon dioxide from the biosphere to the atmosphere as a result of natural processes. However, over several centuries of the industrial epoch, the concentration of carbon dioxide in the atmosphere has increased by almost 50%. A small contribution to this increase (about $(10 - 20)\%$) is made by human economic activity, mainly related to deforestation. This activity shifts the carbon balance between the atmosphere and biosphere. Thus, an increase in the concentration of carbon dioxide in the atmosphere is anthropogenic in nature.

An increase in the concentration of atmospheric carbon dioxide is also governed by a change in the greenhouse effect of the atmosphere, which leads to the warming of the Earth and an increase in the global temperature. A current understanding of this problem, data on changes in the concentration of carbon dioxide and data on radiative transitions of carbon dioxide molecules from the HITRAN bank, allow us to determine reliably changes in radiation fluxes, both for a given frequency and integrated over the spectrum. Hence, from this it follows the rate of an increase in the global temperature. Comparing this value with the observed rate of the global temperature growth, we come to the conclusion that the processes considered with participation of carbon dioxide make a significant contribution to change in the global temperature, providing a smaller part of the rate of warming of our planet (about one third). This is somewhat different from the popular opinion that the observed increase in the global temperature is entirely determined by carbon dioxide. Moreover, we have shown that this opinion is based on erroneous climatological calculations, where for simplicity the Kirchhoff law is ignored, which in this case is of fundamental importance.

Another important component of the atmosphere is atmospheric water, both in the form of free water molecules and in the condensed phase, mainly in the form of micron-sized droplets. For the equilibrium of water between the Earth's surface and the atmosphere, we keep to the old concept of water circulation in the atmosphere, according to which water evaporates from the Earth's surface in the form of molecules, and returns in the form of precipitation, i.e., in the form of rain and snow. Moreover, considering the exponential dependence of the atmospheric water density on the altitude, one can obtain the relationship between the water fluxes to the Earth's surface for microdroplets and molecules, as well as the parameters for the altitude dependence for the densities of water and air molecules. The flux of water molecules to the Earth's surface as a result of precipitation is about four times greater than this flux in the form of free molecules.

Water is the main greenhouse component, so that the contribution of water molecules to the flux of infrared radiation generated by the atmosphere and absorbed by the Earth's surface is about 60%, and the contribution of clouds consisting of micron-sized water droplets is about 20%. Carbon dioxide creates about 20% of the flux of infrared radiation from the atmosphere to the Earth's surface. One can note that the distribution of molecules of different species over altitudes above the Earth's surface, as well as their densities near the Earth's surface, are known well enough, which allows, using HITRAN bank data for the parameters of radiative transitions of molecules, to calculate the radiation fluxes from each molecular component with an accuracy of several percent. In the case of clouds, there is no reliable information of this kind, so the contribution of clouds to the radiation of the atmosphere and the atmosphere is determined by the residual principle, subtracting from the total flux of infrared radiation to the Earth's surface, which follows from the energy balance of the Earth and its atmosphere, the calculated radiation fluxes created by molecules. We add to this that the greenhouse trace gases, which include methane, nitrogen dioxide and ozone molecules, give the total contribution to the infrared radiative flux from the atmosphere of about 1%.

Changes in the radiation fluxes from the atmosphere to the Earth's surface for the real atmosphere, which occur as a result of changes in the concentration of greenhouse components, are of interest. If we ignore the change in the density of aerosols that reflect and absorb solar radiation, then the increase in the flux of infrared radiation from the atmosphere to the Earth's surface is determined by about a third by carbon dioxide molecules and by two-thirds by water molecules. If we take into account the continuous increase in the mass of aerosols in the atmosphere, the role of atmospheric water in changing the global temperature increases. Note that the contribution of carbon dioxide to the change in the global temperature is reliably calculated, while the contribution of atmospheric water is determined by the residual principle, when the part associated with carbon dioxide is subtracted from the total rate of global temperature change. This corresponds to the rate of increase in the water amount in the atmosphere $0.3\%/yr$. If we take into account that the global temperature changes with an increase in the mass of atmospheric water, then the increase in air humidity is $0.2\%/yr$. If we take the average humidity of atmospheric air to be 80%, then the current changes in humidity can continue in this manner for 100 years.

Thus, at the contemporary rates of atmospheric processes, the evolution of the atmosphere in the existing regime can last about 100 years. This applies to the time of doubling the concentration of carbon dioxide in the atmosphere, the rate of growth of the mass of atmospheric water and for other processes. The obtained conclusions relate to the standard atmosphere, i.e., to the atmosphere with averaged parameters. This model is based on fast horizontal mixing of atmospheric air. Let us consider another limiting case, when the horizontal mixing of atmospheric air is weak. In this case, the horizontal distribution of the components becomes inhomogeneous, and we will demonstrate this by an example when a greenhouse instability occurs over a certain area of the Earth, in which the radiation of one water molecule during its residence in the atmospheric air causes the evaporation of at least one more water molecule. With the development of greenhouse instability, there is a transition to a new thermal regime with an increased temperature of the atmosphere in this area, as it happened in the Eocene epoch (see Fig.10.8). Then, the subsequent slow process will lead to the flux of atmospheric water into neighboring areas with an increase in the temperature there. Thus, the greenhouse instability developed over a certain area of the Earth's surface leads to an increase in the temperature of the atmosphere by several degrees both over this area and over neighboring ones.

We also give other values which accompany the physical picture of atmosphere. These values are combined with data resulted from measurements and monitoring of the atmosphere which followed from the analysis of the greenhouse effect. In contrast to climatological models, laws of physics are included in the analysis. Along with the contribution of the greenhouse components to the radiative flux from the atmosphere, the above analysis gives changes of this flux and the global temperature as a result of change for the concentration of greenhouse components. The total change of the global temperature ΔT as a result of doubling of the concentration of CO_2 molecules in the real atmosphere, as well as the contribution of CO_2 molecules $\Delta T(CO_2)$, the contribution of H_2O molecules $\Delta T(H_2O)$, and aerosols $\Delta T(aer)$ are equal correspondingly

$$\Delta T = (2.1 \pm 0.4)\,\mathrm{K},\ \Delta T(\mathrm{CO_2}) = (0.6 \pm 0.3)\,\mathrm{K},\ \Delta T(\mathrm{H_2O}) = (1.7 \pm 0.4)\,\mathrm{K},\ \Delta T(aer) \approx -0.2K \quad (11.0.1)$$

In this case, we account for that in the real atmosphere a change in the concentration of CO_2 molecules is accompanied by changes of concentrations of other components. One can note a more weak influence of an increase in concentration of greenhouse components. This is reflected in the relationships for the radiative flux $J_\downarrow(\mathrm{CO_2})$ from the atmosphere to the Earth due to CO_2 molecules and the same radiative flux $J_\downarrow(\mathrm{H_2O})$ due to H_2O molecules

$$\frac{\partial \ln J_\downarrow(\mathrm{CO_2})}{\partial \ln c(\mathrm{CO_2})} \approx 0.03, \quad \frac{\partial \ln J_\downarrow(\mathrm{H_2O})}{\partial \ln c(\mathrm{H_2O})} \approx 0.03, \quad (11.0.2)$$

where $c(\mathrm{CO_2})$, $c(\mathrm{H_2O})$ are the concentrations of CO_2 and H_2O molecules correspondingly. From the formulas, it follows that the contributions of added molecules is lower than those located in the atmosphere. These formulas exhibit also in what manner the efficiency of emission of greenhouse molecules falls as their concentration increases.

One more importance of the physical picture of global atmospheric processes is the possibility to connect various aspects of this picture. In particular, above we estimate the average density of atmospheric condensed water which is located in the atmosphere in the form of water microdroplets. These microdroplets determine electric processes in the atmosphere, so that falling of charged microdroplets leads to electric charging of the Earth. The average density of condensed water in atmosphere is estimated as 10 mg/cm^2 and follows from passing of infrared radiation of the Earth through the atmosphere and reflection of solar radiation by the atmosphere. It is clear that a direct determination of this value is problematic. For comparison, the content of water in the atmosphere is 2.5 g/cm^2; i.e., the mass of condensed water in the atmosphere is estimated as 0.4% of that of the atmospheric water vapor in the form of free molecules.

The understanding of global atmospheric processes allows one to create a new technology to act on a local climate by analogy with a change of the temperature inside buildings on the basis of air-conditioners. According to the above estimations, supertankers for aerial firefighting may be transformed for formation of an aerosol curtain over a megapolis for a temporary decrease of the temperature.

Thus, this book represents the physical picture of powerful global atmospheric phenomena which is based on the monitoring of some atmospheric parameters along with the theory of transport of infrared radiation in the atmosphere. Understanding the nature of global processes allows one to analyze the character of climate change, i.e., to a change of the global temperature. It is shown that a long-term change of the global temperature results from a strengthening of the greenhouse effect due to an increase of the concentration of atmospheric water and carbon dioxide molecules. This follows mostly from the human activity. On the basis of this understanding, the technology may be created to decrease the temperature of a megapolis.

Index

B. M. Smirnov, *Global Energetics of the Atmosphere*, Springer Atmospheric Sciences, https://doi.org/10.1007/978-3-030-90008-3